D0618463

THE
BOOK
OF
AGES

by the same author

THE BIOLOGY OF ART
MEN AND SNAKES, *with Ramona Morris*
MEN AND APES, *with Ramona Morris*
MEN AND PANDAS, *with Ramona Morris*
THE MAMMALS: A GUIDE TO THE LIVING
SPECIES
PRIMATE ETHOLOGY, *editor*
THE NAKED APE
THE HUMAN ZOO
PATTERNS OF REPRODUCTIVE BEHAVIOUR
INTIMATE BEHAVIOUR
MANWATCHING
GESTURES, *with Peter Collett, Peter Marsh and Marie
O'Shaughnessy*
ANIMAL DAYS
THE SOCCER TRIBE

THE
BOOK
OF
AGES

DESMOND MORRIS

THE VIKING PRESS   NEW YORK

Copyright © 1983 by Desmond Morris
All rights reserved
Published in 1984 by The Viking Press
40 West 23rd Street, New York, N.Y. 10010

LIBRARY OF CONGRESS CATALOGING IN PUBLICATION DATA

Morris, Desmond.
   The book of ages.
   Includes index.
   1. Life cycle, Human—Chronology.   2. Human growth—
Chronology.   3. Maturation (Psychology)—Chronology.
4. Life span, Productive—Chronology.   5. Biography—
Anecdotes, facetiae, satire, etc.   I. Title.
HQ799.95.M67   1984   305.2   83-47907
ISBN 0-670-10948-7

Printed in the United States of America
Set in Ehrhardt

TO MY MOTHER
who at 85
finally admits to being
'over 39'

Age Records

Great care has been taken to achieve accuracy in ages wherever possible. Unfortunately, famous individuals from bygone centuries often have no exact birth dates, and in such cases precision is slightly blunted. But with later figures this can be avoided. Sadly it has not always been done in the past, even in reputable encyclopaedias and reference works, and double and treble checking has often been necessary.

One of the most common errors has stemmed from the habit of arriving at an age simply by subtracting the year of birth from the year of death: born 1900, died 1960, gives a life-span of 60, for example. But this is a faulty system. If the man was born in December 1900 and died in July 1960, he was only 59 when he died and should have been recorded as such. In the past, this degree of precision has often been ignored.

Another fault stems from the use of the phrase 'she was in her 34th year'. Most people, without thinking too hard about it, imagine that this means the woman in question was 34, but she was not. She was 33. This is because when she was 0 she was in her first year, when she was 1 she was in her second year, and so on.

Acknowledgments

My thanks are due to the following: to Peter Medawar for first arousing my interest in the subject when I was a zoology student at Birmingham University; to my wife, Ramona, for a tremendous amount of research assistance; to my son, Jason, for his help in analysing many tables of figures; to Dr W.M.S. Russell for supplying valuable references and documents; to Valerie Buckingham for her painstaking copy-editing; to Ian Craig for his skilful approach to the book's design; and, as always, to my publisher, Tom Maschler, for his enthusiasm and unflagging encouragement.

Introduction

How old are you?
Are you young for your age?
How do you compare with other people?
Is your brain working better than your body?
At what age do the famous achieve success?
At what age do pressures strike down the famous?
What kind of people manage to slow down the ageing process?

These questions – about ourselves and our friends – often flit through our minds, for the one thing we all share is a life-span. The one subject we all have to face is our mortality, the fact that life is just a little time to spare between the light of birth and the dark of dying.

Age is such a central topic in human thought that it comes as something of a surprise to discover that, until now, nobody has ever set out in a book the events of life on a yearly basis. A year-by-year book showing who did what at each age, from zero to a hundred, has an obvious fascination, both at the level of satisfying idle curiosity and at the deeper level of detecting trends and variations in the ageing process. It also, incidentally, provides a useful reference work for quickly checking the ages of the famous at their peaks of creativity and at the moments of death.

At a personal level we all view ageing as something exciting when young – the eager anticipation of growing up – and as something depressing when old – the horror of physical decline. With typical human perversity, we nearly always want to be an age we are not. One lesson to be learnt from this book is that every age has something special to offer, if we accept it for what it is. If we dream of being older and wiser and more experienced, or if we crave eternal youth and struggle against nature to regain it, we will be making serious errors. We will be failing to live in the present, *our* present, and to use it to the full.

Longing for the future and nostalgia for the past are the two great enemies of a rewarding life. It is clear from reading the age-pages which follow, that the most successful members of society have broken all the traditional age-rules. They have done things when they wanted to do them and have ignored any imagined limitations of time. They have succeeded early or triumphed late, lived short busy lives or long happy ones. The lesson they teach us is to eat life up greedily, never to procrastinate and never to think that it is too late.

Ageing, like pain, is a protection device for which we are less than grateful. Just as pain warns the individual about dangers to the body, so ageing protects the species from the hazards of inflexibility. Looked at from the point of view of our genes, our bodies are merely temporary containers, disposable housing units that must be shed from time to time. As generations pass, our immortal genes are able to tumble and jumble themselves together in different combinations. This means that they are always ready to adapt to shifting demands of the environment, always ready to produce new combinations, some of which will do better than others when conditions change.

The only way the genes seem to be able to keep up the necessary level of flexibility, permitting a slow evolutionary process to occur, is by putting a time limit on efficient cell replacement. If our body cells, which keep on replacing themselves as we grow older, were super-efficient at this rebuilding of our bodies, we would never grow old. And there is, in theory, no reason whatsoever why we should not go on living for ever (barring fatal accidents).

There is no great mystery about mortality and immortality. It is simply that being mortal is more efficient, biologically, than being immortal. So our cells have a built-in obsolescence; a deliberate decline in their efficiency at renewing themselves. In old age we therefore start to shrink and become more fragile, more prey to ailments of many kinds. It is not, for us, the best way to go. Much better if the genes had devised some way in which we could stay physically young until the very end and then, perhaps, explode in a split second. But gradual ageing is what we have, and the best way of handling it is to make the most of each stage as we pass through it.

Up to the present, nobody has found a way of preventing the decline in quality of cell replacement, although many have tried and some have even made wild claims and opened expensive clinics. One day someone will no doubt find a solution, the long-sought elixir of perpetual youth. There is nothing mystical about such a proposition – it is merely a matter of scientific technique. Eventually it will almost certainly be possible for particular individuals to have themselves treated against ageing, like Dorian Gray but without an ageing portrait in the attic. It will, of course, create an overpopulation nightmare. Today's problems will seem trivial by comparison. Breeding will become a major crime. Once again we will have set ourselves a fascinating challenge and the ingenuity of our species will be taxed anew. In the meantime, it remains of great interest to study the way we age and the things we do – or do not do – in each of our advancing years on this planet.

0

Zero is the year of the baby – the pre-walking, pre-talking phase of life. Zero is the only age that is never spoken of in years. No one ever refers to a child as being zero years old. At this stage ages are given in weeks or months.

Growth **Weight** at birth for the average baby is between six and eight pounds. By 5 months this weight has doubled and by the end of the first year of life the typical baby weighs three times its birth-weight. To give some idea of the growth that lies ahead, the weight at birth is about 5 per cent of the adult weight or, to put it another way, twenty new-born babies weigh as much as one average adult.

Height at birth is usually between 19 and 22 inches and by the end of the first year this has increased by about 10 to 12 inches. A new-born baby's height is about one-third of that of the typical adult.

The brain more than doubles its weight in the first year. At birth it is 25 per cent of its adult weight; at 6 months nearly 50 per cent and at the end of the first year 60 per cent.

The head is a quarter of the total body length at birth, but becomes *relatively* smaller with advancing age, being only one-eighth of the total length in the typical adult.

The skull bones come together during the first year of life, closing off the soft spot on the top of the baby's head called the fontanelle.

The teeth usually start to erupt from the teeth-buds towards the end of the first year.

The heart of the new-born baby weighs less than an ounce. By the end of the year it has grown to weigh 1·6 ozs. The pulse rate, like the breathing, is very fast at birth, an average of 140 heartbeats per minute, falling to 115 at the end of the first year.

The Senses Shortly after it is born the baby can focus its eyes. The best focus point is about eight inches away. It can follow a moving object with its eyes, a skill that improves rapidly. It can discriminate some colours at 2 weeks, but whether it can do this at birth is still not certain.

The baby can hear well, especially in the pitch and loudness range of the human voice. It is particularly soothed by the rhythmic sound of the adult human heartbeat – a sound to which it was exposed during its days in the womb and which now spells security and comfort. It can locate the direction from which sounds come by the age of 6 months and probably much earlier.

There are also strong senses of taste, touch and smell. The baby can tell the difference between salt and sweet, preferring the latter, and can also distinguish between sour and bitter.

Reflexes At birth the baby has a number of characteristic reflex reactions. The startle reflex is a reaction to a very loud noise or a sudden loss of support.

The baby flings out its arms with its fingers spread and then closes its arms as if in an embrace. It is a remnant of our monkey days, when babies clung to the mother's fur in fear or panic. In humans it vanishes after about three months. The grasp reflex is a strong gripping action seen in both the hands and the feet. If the foot is stroked behind the toes, they curl in and try to grasp the stroker. Even at birth the fingers are strong enough to enable the baby to be pulled up into a sitting posture while clinging to the adult's hands. The stepping reflex is shown when the baby is held up so that the feet just touch the ground. In this position it makes walking movements with its legs, a reflex action that disappears by the end of the first month. The rooting reflex, a nipple-searching action, is shown when the baby's cheek is lightly touched. It responds by turning its head towards the touched side and 'rooting' around with its mouth. The swimming reflex, seen when babies are immersed in water, consists of an efficient control of respiration, the breath being held when the face is under the water. The babies relax and move placidly, with their eyes wide open beneath the surface, gazing about them. There is no struggling and no fear, and this response has been used in support of the idea that human beings had an aquatic phase in their ancient past. It is a response that vanishes after about the eighth month, to be replaced by disorganized or struggling activity. A baby nearing the age of 1, placed in the water for the first time without support, will panic and thrash about.

Sleeping New-born babies sleep for about two-thirds of each 24-hour period, not in a single bout, but in short bursts interspersed with demands for feeding. As one tired parent remarked: 'People who say "I slept like a baby" usually don't have one.' By the age of 6 months the total sleeping time has shrunk to about fourteen hours and a strong preference for night sleeping has developed, although there are still frequent daytime naps.

Vocalizing CRYING is the first sound made by the human baby, and it is the only noise to be heard for some time. After about four weeks the baby can be seen to mimic its mother's spoken mouth movements, forming its own mouth into different shapes. This is followed very soon by the first real smiling and then, at about 2 months, BABBLING becomes a common sound in the nursery, as if the baby is trying out its vocal apparatus. Around the third month LAUGHTER can be heard for the first time and this heralds the phase in which 'mother' becomes distinguished from other adults. Smiling now becomes specific; strangers upset the baby, making it cry when only a few weeks earlier they made it grin. By the sixth month LALLATION starts, in which the baby utters repeated sounds such as ma-ma-ma-ma, or ba-ba-ba-ba. This leads into the stage of vocal mimicry at about 10 months, when the baby tries to copy sounds made to it by the parents. But it is still not able to employ words in a useful way as signals. Even by the end of the first year there will only be one or two real words used as specific signals by the baby.

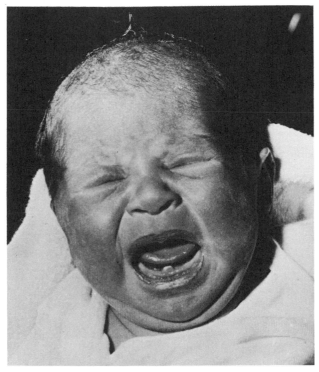

A rare occurrence – a baby born with a tooth

Body Movements

Body co-ordination improves at a steady rate during the first year. For example, the new-born turns its whole body towards a stimulus, but after a few months it is turning only its head. At first, a baby swats at objects it wants to grasp, using gross arm movements, but within a few months it is co-ordinating its reaching and grasping actions much more efficiently. By 4 months it is sitting up with support from its parent. By 5 months it is grasping objects firmly. By 7 months it can sit up alone. By 8 months it stands up with parental support. By 10 months it can creep along on the ground. By 11 months it can walk when led. And by the end of the first year it may achieve the momentous climax when, for a few seconds, it stands up alone and unaided. There are slight variations in the speed with which these stages are reached, but these are the average figures.

Learning

During this first year of life the baby is learning what it is like to be alive. All it wants from its parents at this stage is total security, comfort and love. It needs a great deal of gentle body contact – cuddling, rocking, stroking, cooing and feeding. It is too young to understand strict discipline or rigid regimes. Yet certain infamous experts have in the past insisted on parents severely rationing their love, restricting their attentions and limiting the feeding times. Many parents listened to this advice and damaged their babies in the process, producing children who became unresponsive to parental demands in later years and who suffered from intense feelings of insecurity all their adult lives.

Life Expectancy

A baby born alive in Britain today can expect to live for about 69 years if he is a boy, or about 75 years if she is a girl. These are great improvements on the figures for the turn of the century, when, in 1901, a baby boy had an expectation of 48 years and a baby girl 52 years.

Despite enormous improvements in hygiene during recent years, between 1 and 2 per cent of live births still die within their first year of life. The worst figure for an advanced country is Austria (2·08 per cent) and the best is Sweden (0·83 per cent). The UK figure is 1·77 per cent and the USA figure is 1·61 per cent.

Special Events at 0

Among the famous, even the most precocious can achieve nothing on their own behalf at this tender age, but fame is sometimes thrust upon certain unsuspecting babies. For example:

The Buddha was 5 days old when, for his name-giving ceremony, 108 Brahmins were invited, among whom were eight specialists at interpreting bodily marks. Of these eight, seven predicted two possibilities: if the child remained at home, he would become a universal monarch, but if he left home, he would become a buddha. The eighth specialist predicted that the baby would definitely become a buddha. The baby was the son of Prince Suddhodana from a province of Northern India near Nepal, and his father raised him in great luxury and comfort in an attempt to bias him in favour of the worldly life, so that he would not leave home, but he failed. His child became a wanderer seeking truth and went on to become the spiritual leader of Buddhists throughout the world, now numbering hundreds of millions.

Edward VII, the British King, was created Prince of Wales by his mother, Queen Victoria, when he was only 1 month old, but he had to wait a further fifty-nine years before he came to the throne. As a young man he displeased his mother, who thought him irresponsible. She refused to give him any official duties and he wasted his life away as a playboy.

James, the Old Pretender, son of James II, King of England, was never able to assume the throne himself because it was claimed that he had been slipped into the Queen's bedchamber in a warming-pan as a new-born infant, to give the royal couple a male heir. Because of this, later queens always had witnesses present at royal births to vouch for the validity of the royal babies.

One is the year of the toddler – the year of bipedalism, of learning to walk upright. It is also the year of telegraphic speech – the start of the great journey towards articulate verbal communication. And it is the year of family attachment when the young human learns the kin group to which it belongs.

Growth
Weight increases now at a much slower rate than in the first year of life. Whereas it trebled in the first twelve months, it only increases by about 25 per cent in this second twelve months. For example, average weights for male toddlers in America rise from 22·2 lbs at age 1, to 27·7 lbs at the close of the year. For female toddlers the figures rise from 21·5 lbs to 27·1 lbs.

Height increases from just over 30 inches to a point at the end of the year where the toddler is almost exactly half its adult height. To be more precise, male toddlers at age 1 are 42 per cent of their adult height and by the end of the year are 49 per cent of it. Female toddlers go from 45 per cent to 52 per cent.

Sleeping
The total sleeping time required by the 1-year-old has shrunk to about 13 hours out of 24. Daytime naps are greatly reduced in number, usually only a morning and afternoon nap in the early part of the year and then only a single afternoon nap.

Vocalizing
This is the year in which the miraculous process of acquiring speech begins in earnest. Understanding words spoken by parents comes first, then the early attempts to utter them. The child seems to be possessed by a powerful urge to experiment with verbal messages and the intensity of this motivation is clearly an inborn quality of our species, setting us apart from the other monkeys and apes. The 1-year-old human begins the year with only one or two real words in its vocabulary, but this number rises to nearly 300 by the end of the year and a major communication threshold has been passed. A recent survey gave the following average figures:

1 year 0 months	3 words
1 year 3 months	19 words
1 year 6 months	22 words
1 year 9 months	118 words
2 years 0 months	272 words

It is obvious from these figures that it is during the second half of year one that infants delight in word-interaction with their parents. The toddlers are hungry for words at this point and are struggling hard to understand their meanings. This is the stage when telegraphic sentences are being put together, and the two-word sentence of a 1-year-old is very different from the two-word sentence of an adult. When the toddler says 'Mummy book' it is making a long statement, reduced to a level of verbal simplicity and relying heavily on context. It may mean, 'Mother, I want to show you this book I am holding', or, 'Will you please hand me that book over there', depending on the circumstances. Parent and child develop a partnership: the parent tries to tune in to these simple statements while the child tries to improve them and make them more explicit.

Body Movements
The year starts out with the momentous discovery of how to stand up on two feet and then, by about 15 months, the first unaided walking occurs. For parents this is a threshold as important as Neil Armstrong setting foot on the moon. For the infant, the first small steps create a problem: how to get down on the ground again. The undignified solution at this stage is to collapse backwards with a bump into a sitting position. But walking has begun and can now be improved, with the legs being brought gradually closer together as they make forward steps, and with the irregularity of the walking actions disappearing to be replaced by a smooth, efficient walking rhythm. There is also a hurried walk which looks like running but is not.

The precision grip between thumb and forefinger is improving rapidly and the 1-year-old discovers a new range of activities that this manipulation ability opens up, including picking up small toys, turning the pages of a picture-book, holding a crayon to make simple scribble patterns. Food can be carried to the mouth on a spoon, and drinking from a cup held in

both hands can be mastered. Such developments are typical of the 18-month-old infant, and he or she begins to show a marked preference for one hand over the other when carrying out these actions.

Increased mobility makes this a dangerous period of life, with the toddler keen to investigate potentially damaging objects and able now to reach them.

Learning
Learning is impressively fast and it is crucial that the infant be allowed a rich variety of experimentations if its curiosity is to develop prop-

erly. If the 1-year-old child hears the phrase 'Don't do that!' too often it may obey to the extent of becoming an ineffectual adult.

Social learning involves the multiple attachments to the members of the nuclear family – brothers, sisters, grandparents, uncles and aunts – in addition to the attachments to the parents that developed during year nought. This attachment to close kin occurs at around 18 months in the average infant. It gives rise to an increase in anxieties about strangers and fears of separation, as the other side of the coin.

Accession at 1
James V became King of Scotland when he was 1 year and 5 months old on September 9, 1513, when his father was killed in battle. As a child he was the pawn of a political power struggle. As King he produced a daughter who was to become Mary, Queen of Scots.

James VI became King of Scotland at an even earlier age than his grandfather, James V, when he was only 1 year and 1 month old. This happened on July 24, 1567, following the abdication of his mother, Mary, Queen of Scots. A year later she left the country and never saw her son again. When Elizabeth I of England died in 1603, James VI went south to become James I of England.

Misfortune at 1
Helen Keller, the American author and lecturer, suffered what seemed an insurmountable misfortune at the age of 1 year and 7 months when, as a result of an attack of scarlet fever, she was rendered both blind and deaf. But she refused to allow her handicaps to inactivate her. She doggedly learned to read Braille, to write and to speak. She even obtained a college degree and went on to become a public figure who was a symbol of human determination and an inspiration to the handicapped everywhere.

Jean Baptiste Racine, the French poet and master of French tragedy, suffered his own personal tragedy at the age of 1, when his mother died, followed before long by his father. The orphan was brought up by his grandmother and, either *despite* his early traumas or *because* of them, went on to become the first French playwright born without rank or money to rise to the level of a courtier, a noble, and a man of means and influence. It is hard to predict the impact of the early loss of parents, but it is remarkable how many highly successful men and women were orphaned at a tender age.

Life-span of 1
Charles Lindbergh Jnr, the infant son of the famous American aviator, Colonel Charles Lindbergh, who flew solo from New York to Paris in May 1927, was kidnapped from his cot in 1932 when 1 year and 8 months old. A ransom of $50,000 was demanded and was paid by the distraught parents. After the money had been handed over, the baby's body was accidentally found in a wood, decomposed and with a fractured skull. Some of the money was traced to a man called Bruno Hauptmann who was convicted of the kidnapping and electrocuted in 1936, vigorously protesting his innocence.

The Wichita Eagle

VOLUME XCIII   PRICE: In the City, 3c; Outside City, 5c   WICHITA, KANSAS, WEDNESDAY MORNING, MARCH 2, 1932   FOURTEEN PAGES   NUMBER 74

LINDBERGH BABY IS KIDNAPED

JAPS ROUT CHINESE ARMY AT KIANGWAN IN BLOODY BATTLE

Capture Tachang, Four Miles Westward, with Claimed Losses of Only 60 Men While 1,800 Defenders Are Mowed Down by Newly Arrived Troops

BOMBS SEVER SHANGHAI-NANKING RAILWAY

Natives Attempt to Prevent Landing of 12,000 Reinforcement at Liuho; Entire War Area Rocked by Artillery Duel; Chapei Blazing Inferno

SHANGHAI, Mar. 2.—(Wednesday)—(AP)—The Chinese retreat became general at 1:30 p. m. today when soldiers holding the front lines of Chapei began withdrawing toward the vicinity of Liuying road, the official Japanese army report said.

SHANGHAI, Mar. 2.—(Wednesday)—(AP)—Japanese troops said they had the Chinese army in full retreat northward and westward of the Chapei-Kiangwan battle line today. The Chinese force was completely demoralized, the enemy said, with the Japanese in pursuit. Japanese headquarters also reported their forces had occupied Miaochungchen, northwest of Kiangwan.

BY MORRIS J. HARRIS
(COPYRIGHT, 1932, BY THE ASSOCIATED PRESS)
SHANGHAI, Mar. 2.—(Wednesday)—Fresh Japanese storm troops beat China's army into rout on the

Sino-Jap War News in Brief

(By the Associated Press)
JAPANESE drove the Chinese army at Shanghai into retreat today and were reported to have occupied Tachang, four miles west of Shanghai.

Six Japanese airplanes at about the same time bombed and said they cut the Shanghai-Nanking railway near Kunshan, 35 miles west of Shanghai, in pursuance of a threat to destroy it if the Chinese continued to bring in reinforcements.

On the north thousands of fresh Japanese reinforcements, just off ships in the Yangtze river, were attempting to advance southward for a great encircling movement.

Over Chapei, at the Shanghai end of the battlefront, artillery of both sides pounded each other's positions.

Weakened by a battering assault which lasted from dawn till dusk yesterday, the Chinese fell back to the south, fighting for every foot of ground.

Japanese headquarters announced that during yesterday's action the whole line between Kiangwan and Miaochungchen, a distance of about two miles, had

(Continued on Page 14)

GIVE AWAY HALSTEAD HOSPITAL

THEIR BABY STOLEN FROM CRADLE

COL. AND MRS. CHARLES A. LINDBERGH, JR., whose baby, Charles A., Jr., 19 months old, was whisked out of his cradle by kidnapers last night. No clue as to the persons responsible or their motive has been found. Police of New York and New Jersey are scouring the highways, searching all cars, in a frantic effort to apprehend the abductors before they can leave the vicinity.

THE EDITOR SPEAKING:
Next Friday London will send $150.-

INFANT SON OF FAMED FLYING PAIR SNATCHED FROM CRADLE IN HOME

Police of Two States Scour Highways and Guard Tunnel in Effort to Apprehend Abductors Before They Escape from Hopewell, N. J., Area

NEW YORK COPS COMB UNDERWORLD FOR CLUE

Mystery Surrounds Disappearance of Child from Nursery; Gone from Bed When Someone Glances into Room Several Hours After Tot Retires

BY SAMUEL BLACKMAN
(COPYRIGHT, 1932, BY THE ASSOCIATED PRESS)
HOPEWELL, N. J., Mar. 2.—(Wednesday)—(AP) —Charles Augustus Lindbergh, Jr., 20-month-old son of the flying colonel, was kidnaped last night from his nursery in the Lindbergh country home near here.

Police said he apparently was spirited away in an automobile, which they have not yet identified; an automobile which contained two men, stopped at least two persons prior to the kidnaping, and its occupants asked directions to the isolated Lindbergh home.

Within an hour after Colonel Lindbergh himself telephoned the first alarm, police squad cars blockaded every Jersey road for miles. They had orders to stop

DEMOS TO

Two is the year of make-believe play. It is also the year of the temper tantrum – the age when disputes erupt in the nursery, ownership of toys becomes a matter of fierce competition and the human ego begins to assert itself with both vocal and physical vigour. The word 'I' suddenly assumes a major role in the vocabulary. There are long periods of energetic activity in which interest in one subject is short-lived and there are constant switches from toy to toy, game to game, or action to action. It is as if, for the 2-year-old, there are simply too many exciting things to do in the world and too little time in which to do them. Frustration, from whatever source, when this great urge to explore and investigate and test the environment is blocked, is expressed noisily and forcefully. In other words, 2 is the age at which the true colours of mankind, as a species, begin to show.

Growth

The average weight increases from about 27 lbs when the toddler reaches 2, to around 32 lbs by the end of this year. The average height increases from roughly 34 inches to 37 inches: the 2-year-old is approximately half-way to the adult height.

In the middle of this year the toddler's brain has reached 75 per cent of its adult weight.

Sleeping

There is little change in the sleeping pattern from that of the previous year. At $2\frac{1}{2}$, children have an average sleep-time per day of 12·9 hours. There is enormous individual variation, however, ranging from as little as 8 hours to as much as 17.

Vocalizing

The vocabulary continues to expand at a dramatic pace. By the time it has reached its second birthday the average child is using 272 words. This becomes 446 words by $2\frac{1}{2}$ and 896 by the end of the year. There is a great deal of chattering now as this wonderful new ability is exercised to the full.

Body Movements

Two is the age when walking develops into running. Other forms of locomotion such as skipping, climbing and jumping are explored and enjoyed. Jumping is still rather primitive – the take-off is two-footed.

The 2-year-old also discovers that it can walk backwards, balance on tiptoe and pick things up from the floor without over-balancing. The toddler at this age is capable of walking up and down stairs on its own, but still prefers to hold on to a rail when descending and usually puts two feet on each step before moving to the next one.

There is a dramatic improvement in avoiding obstacles when walking or running. Doors cease to be the safety barriers they once were, following the discovery of how to turn door handles.

The 2-year-old tries to catch and throw a ball but is usually inefficient at it. Footballs can be kicked at this age, however, and large toys pushed and pulled. Building-bricks can now be built up into towers of six or seven cubes, and hammering, banging and drumming actions are popular.

Learning

This is the age of playing on the floor with parents and toys, of 'pretend' playing and story-telling, the development of fantasy and of concept. It is a time for classifying the things in the world around – for discovering that not all four-legged creatures are dogs and of telling the difference between dogs, cats, cows and horses. There is an insatiable appetite for information about the world and heavy demands are put upon parental teaching.

An area of ignorance at this phase: at $2\frac{1}{2}$ years most children are not positive about their sex.

Princess Anne, 1952

Social Relations

Aggression towards one another first appears in children now. A 2-year-old, when someone grabs its favourite toy, may beat the other child over the head with a fist or a stick without a moment's hesitation. The primitive over-arm blow of our species makes its début at this tender age.

This is also the stage of life, however, when altruistic acts first appear, usually in the form of offering to share a toy or some food. In other words, competition and co-operation begin to show themselves as social patterns at about the same time. But these are early days for social interaction and neither pattern is prolonged. They are fleeting moments rather than organized sequences. At 2, children love to watch each other's activities and to play alongside one another, but complex, co-operative play with other children is still rare.

Relations with adults at this point depend very much on the adults' *style* of behaving. As the song says: it's not what they do, it's the way that they do it. For example, at the age of 2, Kibbutz-raised children are as attached to their real parents as children in other cultures living in their own family homes. The reason is that the *quality* of the brief contact with the real parents is more tender and loving than the extended contact with the busy and efficient Kibbutz nursemaids.

Early Start at 2 Princess Anne began riding
lessons at the age of $2\frac{1}{2}$, at first taught by her mother, Queen Elizabeth II. She was strongly competitive with her older brother Charles and this expressed itself in her single-minded determination to succeed at riding. Her remarkably early start in this direction appears to have stood her in good stead, for, nearly twenty years later, she won the 1971 European Championships at Burghley and was voted Sportswoman of the Year.

Joseph Grimaldi, regarded as the greatest of all clowns, also made an early start, his first role being as a baby clown with his family in 1781 at the age of 2. In later life, his white-faced character 'Joey' strongly influenced the whole development of the circus clown.

Misfortune at 2 Queen Elizabeth I, when
she was the 2-year-old Princess Elizabeth, suffered the horrific experience of having her mother, Anne Boleyn, beheaded on the orders of her father, Henry VIII. This may account for the fact that, as a young child, she insisted 'I will never marry', adhered to this as an adult, and died the 'Virgin Queen', despite great efforts by those around her to persuade her to take a husband. There was apparently no physical reason why she should not have married and had children, and she enjoyed several romantic friendships with attractive men, but her trauma at the age of 2 appears to have left a lasting mark on her personality.

Samuel Johnson, the great English lexicographer and author, was not a healthy child. His eyes were weak and at the age of 2 he was suffering from a tubercular infection in the glands of the neck commonly known as 'The King's Evil'. It was believed that the disease could be cured by the monarch's touch, so Mrs Johnson travelled to London with Samuel in March 1712 where the $2\frac{1}{2}$-year-old boy was duly touched by Queen Anne. Later in life he retained 'a sort of solemn recollection of a lady in diamonds and a long black hood'. Whether he received any physical benefit from the royal touch is not clear, but the gold amulet which the Queen hung around his neck remained there until his death.

Edgar Allan Poe, the master of the macabre tale and the creator of the detective fiction genre, experienced the misfortune of his mother's death when he was 2. He was brought up by his god-parents who educated him well, but he suffered from personality conflicts for the rest of his life. Although gentle and devoted to those he loved, he was harshly critical of others, suffered from feverish dreams, was obsessed with the morbid and deeply involved with the occult and the satanic. He was also self-centred, unprincipled, an alcoholic and a compulsive gambler whose losses as a student were so heavy that his infuriated guardian removed him from college. How many of these qualities were due to his being orphaned at 2 it is hard to say, but it seems highly probable that his early misfortune played a major part in forming his character – and also, incidentally, gave the world gothic horror-tales of lasting brilliance.

Edgar Allan Poe

Three is the year of dexterity – the year when the child struggles fiercely against the clumsiness of infancy. It is the age of undoing buttons, washing one's own hands and walking on tiptoe. In general it is an age of investigation and finding out how things work, even if this often means destructively taking things to pieces.

Growth The average weight increases from about 32 lbs to 36 lbs during the year. The average height increases from 37 inches to 40 inches. This makes the child just over 55 per cent of the adult height (54 per cent in boys, 57 per cent in girls). It is interesting that Leonardo da Vinci, writing in the fifteenth century, claimed that 'Every man at three years old is half his height', which would make the average adult of that period something like 6 feet 2 inches. Obviously this was not the case and the explanation must be that in those days infant growth was slightly stunted compared with the present time. This is not surprising in view of our greater understanding of the nutritional needs of young children. Today it is the 2-year-old child who is half-way to adult height.

Elizabeth Taylor with her mother

Vocalizing The typical 3-year-old knows about 1,000 words. Accurate studies reveal a climb from 896 words at the start of the year to a level of 1,222 at its end. Chattering increases to the point where the child can often be heard talking vigorously to itself while playing alone.

Body Movements The main advance with the 3-year-old is in manual control. Simple jigsaw puzzles can be mastered and keys can be turned in addition to door handles. Most children can now use a knife and fork with some success, but this is not yet the age of 'table manners', regardless of how hard parents try to enforce these rules of conduct.

With gross body movements, the greatest joy at 3 is riding a tricycle successfully. Walking becomes more erect and delicately controlled, as does stair climbing. The 3-year-old can also balance for some seconds while standing on one foot.

A major triumph for the 3-year-old is the production of the first pictorial images while drawing and painting. Out of the earlier scribbling, specific representations emerge, the most common one being the 'cephalopod' – a face with arms and legs sprouting from it. This appears in the drawings of children of all cultures studied so far and is a basic stage in the development of human art.

Learning This is an age for learning about the physical properties of objects – for playing with mud, clay, finger-paints, Plasticine and dough. Water is no longer splashed, it is now poured, trickled and sprayed – its properties are under analysis. Mechanical objects are also analysed, often to destruction. This creates a problem for the parents. Too much control stifles the natural growth of the child's scientific curiosity; too little control can be potentially damaging both for child and property. The parent must walk a tightrope and soon discovers how easy it is to deflate the 3-year-old and produce a sulk or an outburst of tears.

Erik Erikson has summed up the situation: 'If parents reward the child's successful actions at this stage and do not shame his or her failures (say in bowel and bladder control) the child's sense of autonomy will outweigh the sense of shame and doubt.'

Three is the year when the average child knows its own age, and two out of three children know their own sex.

Many cultures start to send their infants off for nursery schooling at 3. There are however some rather striking national differences. Here are figures showing the percentage of 3-year-olds in nursery schools in eight different countries: BELGIUM 90 per cent, FRANCE 61 per cent, JAPAN 17 per cent, USA 13 per cent, SPAIN 12 per cent, UK 6 per cent, CANADA 5 per cent, IRELAND 0 per cent.

Social Relations Three is the age of jealousy and rivalry, when the infant feels strongly possessive about its parents. It is still not fully at ease when playing co-operatively with other children. As at 2 years of age, it still prefers to play alongside them rather than with them, although an interest in more social play does surface from time to time, before it becomes submerged in the battle to conquer the inanimate world.

One social embarrassment at this age is connected with the development of phobias. Three is the peak age for irrational fears – fears of certain animals, of thunderstorms, of night-time terrors and of getting lost. The parent who is not ready for this eruption of groundless fears may feel ashamed of the child who, when younger, seemed delightfully fearless. Lack of parental understanding in such cases, due to feeling embarrassed in the presence of other adults, only increases the acuteness of the problem.

Another confusion for the parent stems from the 3-year-old's pendulum swings between being baby-ish and very grown up and the parent sometimes gets out of phase. As the child struggles to mature, this is an alternation that will crop up again and again in the years ahead and it is a wise parent who learns to adjust to the swings of the 'ageing pendulum' at an early stage.

Early Start at 3
Among the famous, 3 is the earliest age at which precocious genius surfaces.

Thomas Macaulay, the British historian, was an infant prodigy who at 3 read incessantly and developed a photographic memory. Throughout his life he read voraciously and forgot nothing. As an adult he was known for his 'torrents of talk' which often proved exhausting for his companions.

John Stuart Mill, the British philosopher, was another early verbalizer. At 3 he was already learning Greek under the tuition of his eager father. Under this parental pressure the child made extraordinary strides in learning, but in so doing also sacrificed many of the normal pleasures of childhood.

Wolfgang Amadeus Mozart

Wolfgang Amadeus Mozart, the Austrian composer, was the most precocious genius in the history of music. He was already playing the harpsichord and could memorize musical passages simply by listening to them once.

Elizabeth Taylor, the London-born actress later to become a world famous star, gave her first Royal Command performance at 3, when she danced (with her ballet class) before King George V and Queen Mary in 1935.

Peter Falk as Columbo

Misfortune at 3
Peter Falk, the American actor, was not precocious at 3 but instead suffered an accident that would have proved a major setback to most members of the acting profession. He developed a tumour in his right eye and, tragically for a 3-year-old, lost the eye. This did not stop him, however, from going on to become an internationally popular television star as the shabby and modest, but deviously clever, detective Columbo, in the series of that name.

Benito Juarez, President of Mexico and 'Father of the Mexican Nation', suffered the loss of both his parents when he was 3. Reared by an uncle, he received no formal education until he was 12. Despite this he went on to obtain a degree in law and eventually to lead his country's struggle to create a democratic federal republic.

Four is the year for asking interminable questions. It is an adventurous age, sometimes boastful and bossy, as the child begins to assert itself as an individual. It is a time of great energy and enormous pride is taken in any achievements, especially if these are admired by an adult. It is also an age of increased concentration and patience, when there seem to be endless fascinating problems to solve and the world is full of exciting mysteries.

Christopher Morley wrote a short poem about this age, called 'To a Child': The greatest poem ever known/Is one all poets have outgrown:/The poetry innate, untold,/Of being only four years old.

Growth
The average weight increases from about 36 lbs (on reaching 4) to 40 lbs (by the end of the year). The average height increases from 40 inches to 43 inches. This makes the child about 60 per cent of the adult height (58 per cent in boys, 62 per cent in girls) at the start of the year.

Vocalizing
The typical 4-year-old knows about 1,500 words. Average figures given in an accurate study are: 4 years 0 months: 1,540 words. 4 years 6 months: 1,870 words. End of the year: 2,072 words.

This is the year when language mastery meets the point where such words as 'on', 'under', 'inside' and 'in front' can be fully understood.

Body Movements
The child can now run fast and can turn and swerve with skill. Stairs are now climbed in adult fashion. Hopping, skipping and jumping become easy tasks and the 4-year-old can even run on tiptoe. Faces can be washed, hands dried without help, hair combed and teeth brushed. The child can also dress and undress itself, with a little help at difficult moments.

A major advance for the 4-year-old is the development of much stronger legs. Out of doors this makes ladders and trees enticing and although this new climbing ability is exciting, it can also lead to dangers if parents are unwary. In the water, the strong legs mean that efficient swimming can take place and 4-year-olds who are given the opportunity can usually swim unaided.

With pencil and paper, an improvement in the representation of the human figure means that the old 'cephalopod' shapes – faces with arms and legs sprouting from them – now become heads with proper bodies beneath them.

Kicking and throwing a ball improves, but catching a ball still poses problems and is usually dealt with by clasping the arriving ball to the chest.

Learning
This is the age for curiosity about the basic mysteries of life – God, birth, sex and death – and the 4-year-old is often startlingly direct in its questioning. It knows exactly how to express the questions, but it is not so easy for parents to find answers in terms that the 4-year-old can understand. In particular the child becomes fascinated by its own body and its gender. A boy may demand to know whether it can have a baby and a girl may enquire whether she will grow a penis. Erik Erikson's advice: 'If parents accept the child's sense of curiosity [about sexual and other matters] and do not belittle its need to know and question, the child's sense of initiative will outweigh the sense of guilt.'

Nursery school attendance figures still vary considerably from country to country at this age: BELGIUM 95 per cent, FRANCE 87 per cent, HOLLAND 84 per cent, JAPAN 61 per cent, IRELAND 60 per cent, SPAIN 43 per cent, UK 35 per cent, USA 28 per cent, CANADA 20 per cent.

Social Relations
Four-year-olds begin to play together a little but real social play is still some way off. When interacting with one another, 4-year-olds all want their own way and this often leads to serious quarrelling, especially over turn-taking, or toy-sharing.

Louis XIV

Accession at 4
Louis XIV, the 'Sun King' of France, succeeded his father at the age of 4 years and 8 months. According to the laws of his kingdom, this made the 4-year-old infant not only the master of 19,000,000 people, but also the owner of their bodies

and property. However, although officially saluted as a 'visible divinity', he was in reality a neglected child, given into the care of servants, and he narrowly escaped drowning in a pond because no one was watching him.

Early Obsession at 4

Paul Klee, the Swiss genius of modern art, was already drawing so vividly at the age of 4 that he was forced to take refuge near his mother when the devils he was scrawling on scraps of paper seemed to him to come to life. Sitting in his uncle's café, he would day-dream over the sinuous patterns in the marble table-tops which suggested to him fantastic designs.

Jean-Paul Sartre, the French author and existentialist philosopher, retreated into fantasy at the age of 4, when he discovered books. His father had died when he was 1 and his mother over-protected him, probably because he was such an unattractive child and she feared what others might do to him. He was small, timid, ugly and wall-eyed and had virtually no childhood friends, growing up in his grandparents' Parisian home. But with the aid of books he found he could escape from the world and this early obsession clearly set the direction of his whole life.

The young Jean-Paul Sartre with his family

Early Start at 4

Gustav Mahler, the Bohemian composer, revealed his musical talent early. He began composing at 4, having become fascinated by the military music at a barracks in his village, and by the folk music sung by the working people. He reproduced both of these on the accordion and the piano and it was this that led him to start inventing his own music at this early age. Although intensely creative, he was a tormented child. As Jews, his family suffered racial tensions; his fiercely self-made father beat his delicate, cultured mother; and among his eleven brothers and sisters there was much illness and several deaths. The impact of these happenings on the young child may partly explain the intense emotionalism of much of his adult work and its sudden and dramatic changes of mood.

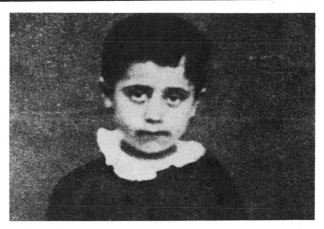
Gustav Mahler

Misfortune at 4

Adam Smith, the Scottish economist and author of *The Wealth of Nations* – the pioneer work on political economics – whose father had died before he was born, was carried off by gypsies when he was 4. Fortunately for the history of his subject, a pursuit was mounted and the boy was abandoned by his captors and rescued.

Malcolm X

Malcolm X, the American black militant leader, was 4 years old when he saw his house burned down by the Ku Klux Klan and heard his father, a Baptist minister, preach the back-to-Africa gospel. He grew up to become an extremist who advocated violence for self-protection.

5

Five is a friendly year. It is also a more serious, competent year. It is the age when special efforts are made to be independent, but when there remains a powerful need for repeated contact with the protective parent. The 5-year-old likes to help its parents and will happily run simple errands for them. It innocently tells tales about its child companions, without realizing that it is being 'disloyal' to them.

According to some schools of thought, the human personality is already fully formed at 5. A Hindu proverb states: 'What the mind is at the age of five it will be at the age of twenty-five.' Although this may apply to the individual's personality, others see 5 as the last year of childhood purity, after which behaviour deteriorates into adult sophistication. Clifton Fadiman: 'All children talk with integrity up to the age of five, when they fall victim to the influences of the adult world and mass entertainment. It is then that they begin, all unconsciously, to become plausible actors. The product of this process is known as maturity, or you and me.'

Growth

The average weight increases from about 40 lbs, when 5 is reached, to 44 lbs as the year closes. The average height increases from 43 inches to just over 45 inches. This makes the average child about 64 per cent of its adult height (62 per cent for boys, 66 per cent for girls) at the start of the year.

The brain of the child, which weighed only 25 per cent of its adult weight when it was born, now weighs 90 per cent.

Vocalizing

The vocabulary at 5 has an average of 2,072 words, rising to 2,289 words at $5\frac{1}{2}$ and then to 2,562 as the year ends.

This is the age when the child loves listening to words and will sit in rapt attention for story-tellers.

Several new verbal skills emerge: counting up to ten; naming coins; naming several colours; and the beginnings of telling the time.

Body Movements

This is the year for super-confident, super-fast tricycle riding and also for the very first bicycle rides.

Manual dexterity improves to the point where scissors can be used to cut out paper shapes, and knife, fork and spoon can be used with ease, allowing for rapid wolfing of food. Table manners can be obeyed at 5, but are often ignored.

Swinging, sliding and skipping become commonplace and children begin to dance in time to music.

Despite these advances, bed-wetting still occurs in 15 per cent of boys and 10 per cent of girls.

Learning

The 5-year-old learns a great deal about human behaviour from fantasy play in which it acts out the lifestyle of imaginary figures – both boys and girls delight in dressing up. At school, there is the problem of adjusting to being just a member of a group, now sharing a 'parent' – the teacher.

Five is the compulsory starting age for formal education in the UK, France and Belgium. Other countries tend to favour 6 as a more suitable starting point. Percentages for children in school in other countries are: HOLLAND 96 per cent, IRELAND 90 per cent, CANADA 85 per cent, USA 80 per cent, JAPAN 80 per cent, SPAIN 49 per cent.

Social Relations

Five is the year when play between children takes an important step forward. Although the 5-year-old is still a poor *group* member, *pairs* of children at this age often indulge in real co-operative play for the first time.

Like the earlier years of infancy, 5 remains a non-sexist year, with boys and girls associating with one another freely. There is as yet no sign whatever of male gangs and female gangs splitting off from one another.

Life Expectancy

If you have reached the age of 5 and you are a boy, then you can expect to live for another 65 years; if a girl, another 71 years. This is a great improvement on the turn of the century when, in 1901, a 5-year-old boy could only expect, on average, another 55 years, and a girl 58 years.

In the animal world, 5 years is the maximum lifespan you can expect if you are a rat or a mouse. Surprisingly, the record for a pet mouse at present outstrips that for the much larger rat, although this may simply reflect the mouse's greater popularity as a pet animal: the record mouse is one called Hercules who died in Surrey in 1976 at the age of 5 years and 11 months; the record rat is one that died in the USA in 1924 aged 5 years and 8 months.

Accession at 5

The 14th Dalai Lama was enthroned at the age of 5, as the spiritual and temporal leader of Tibet, in 1940, but was later forced to flee the country with most of his court, following invasion by Chinese communists. He is still waiting to return to his homeland.

The Dalai Lama

Lillian Gish in *Birth of a Nation*

Early Obsession at 5

Cardinal de Richelieu, Chief Minister to Louis XIII of France, and the chief architect of France's greatness in the seventeenth century, became inspired with a great passion at the age of 5. This came about as a result of the death of his father, an event which revealed to the small boy the ruinous state of his family's finances. The family estates had been brought to ruin by inflation, extravagance and mismanagement and the boy became determined to restore the honour of his house. This reaction, so amazingly adult for a 5-year-old, was eventually expanded to become a personal crusade to restore the honour of the whole of France, which he was to achieve as a powerful and influential adult.

Early Start at 5

Five is a great year for precocious 'early starters' of many kinds, but perhaps the most extraordinary example is the little 5-year-old girl who was delivered of a baby by Caesarian section. This abnormal case is the youngest known instance of a human being giving birth, and is a rare example of unusually accelerated hormonal development.

Other interesting but less bizarre instances of 5-year-old precocity include the following:

Busby Berkeley, the famous Hollywood choreographer, who revolutionized the film musical in the inter-war period, was already performing professionally on the stage at the age of 5.

Blondin, the French acrobat and tightrope walker, made his first public appearance as the 'Little Wonder', after six months' training at the École de Gymnase in Lyons. From these beginnings in 1829 he went on to become the world's most famous highwire performer, crossing Niagara Falls on a number of occasions.

Lillian Gish, the American actress who achieved lasting fame for her performance in D.W. Griffith's *Birth of a Nation*, made her stage début at the age of 5.

Ivan the Terrible, formally Grand Prince of Russia since the age of 3, began to take part in court affairs at 5, and was the centre of a furious power-struggle among the prominent members of the nobility.

Wolfgang Amadeus Mozart, the Austrian composer, was already beginning to compose minuets and other pieces. Those which have survived reveal his immediate grasp of musical forms.

Sergei Prokofiev, the Russian composer, was also a highly gifted child and, at the same age as Mozart, was composing his own little pieces, the first of which was called 'Indian Gallop'. It was carefully written down for him by his mother, who was then his piano teacher.

Vesta Tilley, the English music-hall star, was already performing on the stage in Nottingham, dressed as a tenor, complete with moustache and singing 'The Anchor's Weighed'.

Natalie Wood, the American film actress, was the daughter of a ballet dancer and was taking dancing lessons even before she could walk properly. At 5 she had her first film part, in *Happy Land*, and so impressed the director that he sought her out a few years later for a featured role in an Orson Welles film. From there she went on to become one of Hollywood's top child stars.

The young Natalie Wood

6

Six is the year when reading becomes a pleasure. It is an imaginative age, an age for thinking things out in one's head before performing an action. It is a time for internalizing one's thoughts – a major threshold for the human being to pass. This leads to an increased interest in school subjects and a decreased interest in toys. Books, stories, television and films suddenly become more absorbing because of the child's expanding imagination. As so much more is going on inside the head of the 6-year-old, he or she often seems rather preoccupied, compared with younger children; also more apprehensive and sometimes hesitant. The 6-year-old often dawdles where the 5-year-old would have rushed.

Despite these changes, the 6-year-old child is still a delightful, if rather demanding, companion.

Growth

The average weight increases during the year from about 44 lbs to 49 lbs (48 lbs in girls, 50 lbs in boys). The average height increases from 45 inches to 48 inches. This makes the average child about 68 per cent of its adult height when it reaches 6 (just over 65 per cent for boys and just over 70 per cent for girls).

Six has been called the ugly duckling stage of childhood because it is the year when the baby teeth start falling out and a few of the permanent teeth start breaking through, giving the face a tooth-gap grin. There are some compensations, however – the body loses its knock-knees and its protruding abdomen, for example.

Vocalizing

The 6-year-old enjoys carrying on long conversations and has an excellent vocabulary with which to do this. A detailed study of the vocabulary of a large number of children gave an average figure of 2,562 words.

Body Movements

Physically, this is a very active age. Most of the basic muscle skills have been acquired now, and 6 is the year for practising them and combining them. One action still found difficult, though, is the accurate catching of a small ball. Throwing a ball is done with force and with some precision, but estimating the ball's trajectory successfully still eludes most 6-year-olds.

Learning

Six and a half is said by some to be the best age at which to start the formal teaching of reading in school. Other authorities (particularly in France) feel that such teaching should begin earlier, but the brain does not seem to be ready for it until now, despite claims to the contrary.

One important point about early reading: even at 6, the eyes have not yet reached their final shape and size, and many children are still far-sighted, a condition that corrects itself naturally during the next few years. The 6-year-old should therefore be given book print that is larger than that used by older children.

With numbers, the 6-year-old should be able to count up to about thirty.

This is the year when the majority of advanced countries make it compulsory for children to start formal education. This is so, for example, in Australia, Austria, Canada, France, Ireland, Italy, Holland, Japan, Spain, Switzerland and West Germany.

Social Relations

The 6-year-old still needs adult supervision. Play in groups continues to cause trouble, with individuals refusing to become proper 'group members' at this stage. There is still no group loyalty or group cohesion. However, a close friendship with one other child may develop.

Accession at 6

K'ang Hsi, one of the supreme monarchs of China, who laid the foundation for a long period of dynastic stability, became Emperor in 1661 when he was 6 years and 9 months old. His young father had died suddenly from smallpox and the small boy was immediately raised to the Imperial Throne, where he continued to rule for sixty-one years, with three Empresses who, between them, gave him thirty-five sons of his own.

Early Start at 6

Although a human baby has been born to a 5-year-old girl by Caesarian section, 6 is the youngest age at which a mother has delivered a baby normally. In April 1939 a Peruvian woman living in the foothills of the Andes brought her young (70 lbs) daughter to a local hospital saying that she was ill, possessed of evil spirits in her belly. Examinations revealed that she was eight months pregnant and she was later delivered of a healthy 6½ lbs baby boy. According to her birth certificate, the mother, Lina, was 4 years and 8 months old, but this was not accepted by Dr Geraldo Lozado who attended her. A study of her bone development and her teeth, using X-rays, established that she was in reality 6 years old. The rare cases of this type of sexually precocious development are usually due to an abnormal speeding up of hormonal maturation.

Other early starts at 6 include:

Ferdinand Foch, Marshal of France and the Allied Commander-in-Chief in the closing stages of the First World War, began his military career at the age of 6, for it was then that he spent his time devouring the descriptions of battles in historical works. It was an obsession that was to last him a lifetime.

Lord Kelvin, who became one of the most distinguished of British scientists, lost his mother when he was 6. This tragedy led to an unusually intimate bond between the boy and his father, a mathematics teacher, who trained his son so well that he was able to enter university within four years of his mother's death.

John Stuart Mill, the British philosopher, also benefited (and suffered) from an over-enthusiastic father, with the result that the boy was able to write a History of Rome when only 6½ years old.

Wolfgang Amadeus Mozart was touring the courts of Europe at 6, giving recitals on the violin and the clavier. Minuets written by him at this age are standard learning pieces for many piano beginners.

Jacques Offenbach, at 6, was already playing both the piano and the violin and had written his first song.

Dolly Parton, the American country-and-western singer, the fourth child of twelve in a poor family living in the mountains of Tennessee, was playing the mandolin at 6 and was soon writing and singing her own songs.

Richard Strauss, the German composer, like Offenbach, was playing the piano and violin at 6 and produced his first composition (his 'Christmas Song').

Shirley Temple

Shirley Temple became the youngest person ever to win a Hollywood Oscar, in 1934. The award was given 'In grateful recognition of her outstanding contribution to screen entertainment during the year 1934.' Within a few years she had gone on to top all the other stars as the number one box-office attraction, but by the time she was a teenager her film career was in decline and eventually she turned to politics.

Misfortune at 6

James Barrie, author of *Peter Pan*, was 6 years old when he was deeply shocked by the death of his brother. It was an experience from which he never recovered and he spent the rest of his life trying to recapture the happy years before the tragedy, which is why Peter Pan was the little boy who 'never grew up'.

Charles Baudelaire, the French poet, was an only child whose father died just before his sixth birthday. But this was not the major misfortune of his childhood, because he was happy to be with the mother whom he adored. However, when he was 6½, she married again and he felt she had rejected him. Hating his stepfather, who was a soldier and a statesman, he developed what his teachers called 'precocious depravity', becoming rebellious, debauched, drug-addicted, extravagant, eccentric and generally immoral. Whether he became a great poet despite his trauma at 6, or because of it, is open to debate.

Rudyard Kipling, born in Bombay to colonial parents, was brought to England when 6 and left there in a foster home for five years. His misery during this period was so intense that he wrote of it at length as an adult. Unlike Baudelaire, he did not rebel against the source of his unhappiness, for he was later to be dubbed a 'jingoist imperialist' for his writings.

John Metcalf, the English road-builder known as Blind Jack, lost his sight at 6 following a smallpox attack. Despite this he went on to become an adult of extraordinary energy and activity. He was a musician, a huntsman, a fish merchant, a stagecoach operator, a violinist, a horse-dealer (by touch), a smuggler of brandy and tea, a marathon walker, a racing jockey, and a card-player. For a man with full sight his accomplishments would have been remarkable; for a blind man they were extraordinary. So was his marriage, for he eloped with an innkeeper's daughter the night before her wedding to another man and married her before her original groom could find her. His crowning glory, however, was that between 1767 and 1792 he built 180 miles of road – the best of its time – and was a better judge of the quality of a road surface than any sighted man. He pioneered good roads even before the great McAdam.

James Thurber, the brilliant American humorist, was the same age when he too suffered serious damage to his sight. Fortunately he only lost the sight of his left eye, but his right eye was never strong and gave increasing trouble as he grew older. Tragically, he was totally blind for the last twenty years of his life, despite a series of operations he underwent in his 40s. But the accident which damaged his eyes so badly when he was 6 was also responsible for the charming simplicity of his comic drawings.

K'ang Hsi

Peter Pan

7

Seven is the age when the typical child first starts to challenge its parents. It is the year of sulks and complaints, of alibis and blames. There is an increasing interest in reality, also in riddles and jokes. It is a time for lapsing into quiet moods and for having secrets with close friends. For many children it is also the age when they show the first serious sexual curiosity, and when there is a slight splitting apart into distinct groups of 'the boys' and 'the girls'.

Growth The average weight increases from about 49 lbs at the start of the year to 55 lbs by its close. The average height increases from 48 inches to 50 inches. This makes the average child 71·5 per cent of its adult height (69 per cent for boys and 74 per cent for girls).

The brain-size of the 7-year-old is almost at the adult level. Permanent teeth are appearing rapidly now.

Body Movements At this age the child is well co-ordinated and physically skilful. The hand-eye co-ordination, in particular, shows great improvement and from this point on there are no major thresholds to be passed, merely a strengthening and gradual maturing of all the bodily muscle skills.

Learning The 7-year-old learns to handle simple numerical calculations, to tell the time easily, and to draw a human figure in profile.

Compulsory education has already begun in most advanced countries, but a few are late with this step, including Denmark, Norway, Sweden and the USA.

Social Relations At 7 there is the first sign of true group membership. This is part of the splitting away from the parents and the discovery of new bonds of attachment with companions of the same age. A new sensitivity appears where personal relationships are concerned. The 7-year-old is on the threshold of becoming a social being in its own right, rather than merely an offspring of others. Loyalties that are outside the family are emerging.

According to a Hindu proverb, it is the first five years of life that are crucial in forming the adult personality, but the Jesuits put the age slightly higher. St Francis Xavier: 'Give me the children until they are seven and anyone may have them afterwards.' This attitude – that the first seven years are concerned with character formation and that, once this is complete, the individual can be turned over for formal education – is echoed in verse by Thomas Tusser, writing in the sixteenth century: 'The first seven years bring up as a child/The next to learning, for waxing too wild . . .' This split, between character development and intellectual development, may well be related to the switch from 'offspring' to 'social being' at 7. It is tempting to see the child

spending its first seven years finding out what kind of person it is, from the security of a powerful family attachment, and then, this done, setting off to gain experience and conquer the social world around it, relating less and less to parents and more and more to friends. In reality the shift is much more gradual, but if there *is* a discernible jump in the process it comes here, at the age of 7.

Fred Astaire

Early Start at 7 Fred Astaire, the American dancer, was already touring the vaudeville circuit with his sister Adele as a dancing partner. Bearing in mind that he was still acting at the beginning of the 1980s, it is remarkable that this 7-year-old début was made in the Edwardian period some eight years before the start of the First World War.

The Venerable Bede began his monastic career at this early age. Taken to the Monastery of St Peter at Wearmouth, he was entrusted to the Abbot, and was to remain behind monastery walls for the rest of his life. From this sheltered position he was somehow able to write his *Ecclesiastical History of the English People*. Completed in the year 731, it marked the beginning of English literature and also introduced the BC/AD system of dating events that has been used ever since.

Frederic Chopin, the Polish composer, was playing the piano and precociously produced his Polonaise in G Minor.

José Iturbi, the Spanish concert pianist who became world famous through many appearances in Hollywood films, was also musically precocious – so much so that at 7 he was already teaching pupils three and four times his own age.

Thomas Macaulay, the British historian, who had been reading incessantly from the age of 3, began a compendium of universal history when 7 and followed this with a treatise intended to convert the natives of Malabar to Christianity.

John Stuart Mill, the British philosopher, spent this year of his young life reading, in the original Greek, Aesop's *Fables*, Xenophon's *Anabasis*, and the whole of the works of the ancient historian Herodotus. He was also acquainted with the satirist Lucian, the historian of philosophy Diogenes Laertus, the Athenian educational theorist Isocrates, and the six dialogues of Plato. He also read a great deal of history in English, absorbing it all with his photographic memory.

Margaret O'Brien, the American child actress, won an Oscar in 1944 as the outstanding child actress of the 1940s. She was to that decade what Shirley Temple had been to the 1930s, but she was a more naturalistic performer with great skill and charm, despite her tender years.

Adelina Patti, the coloratura singer, went to America as a child and appeared in concerts in New York at the age of 7.

Misfortune at 7
Ian Dury, a successful English pop star of the 1980s, with his band The Blockheads, managed to succeed in the competitive world of rock music as a lead singer, despite having been badly crippled by polio when 7.

Stendhal, the French author, who described himself as an 'observer of the human heart', adored his mother who died when he was this age. He continued to worship her for a long time afterwards but quickly came to hate his father. His early trauma had a marked effect on his later writings and possibly also on his sex life, since he seemed irresistibly attracted to married women who were as unattainable as his longed-for mother. He died a bachelor.

Émile Zola, another great French writer, was a nervous, sensitive, lisping, short-sighted, delicate, melancholy-faced hypochondriac who, as a 7-year-old child, was sexually assaulted by a servant called Mustapha, an experience which left him with a life-long hatred of all homosexuals.

From *The Clowns* by Fellini

Winston Churchill at 7

Misbehaviour at 7
Winston Churchill was sent to school at 7 and hated it from the moment he arrived there. He was always bottom of the class and was considered a backward child by his father, his governess and by most of his teachers. They were all convinced that he was a dull and untalented child – a fact which can be taken as a source of comfort by other children who suffer similar criticisms.

Federico Fellini, the Italian film director, ran away from his boarding school at this age and spent several wide-eyed days following a travelling circus. It was an experience he never forgot and is a vital clue to the stylistic qualities of his film masterpieces such as $8\frac{1}{2}$, for which he won his third Oscar, *Juliet of the Spirits*, *The Clowns*, and *Roma*, in each of which he somehow manages to turn the whole of life into a magic circus ring.

Sigmund Freud, the father of psychoanalysis, was just 7 when he went into his parents' bedroom and deliberately urinated there, prompting his father to remark 'That boy will never amount to anything', a prophecy as mistaken as the criticisms of Churchill's teachers.

Eight is a bold, expansive age. It is the year that sees the birth of interest in team sports and competitive swimming. It is a year when writing becomes more important and leisure-time reading is more common. The 8-year-old also begins to show interest in collecting things.

Eight is the earliest age at which a child in England can be legally guilty of an offence. A younger child who deliberately kills a baby brother or sister by suffocating it with a pillow (as has happened occasionally) is considered to be incapable of forming sufficient malice aforethought to be guilty of a crime.

Eight is considered to be the critical age for the completion of childhood indoctrination, according to Lenin in a speech to the education commissars in Moscow in 1923: 'Give us the child for eight years and it will be a Bolshevist for ever.' This is longer than the time demanded by the Hindus (five years) or the Jesuits (seven years); communist brainwashing, it would seem, requires a little more effort.

Growth
The average weight increases from 55 lbs at the start of the year to 62 lbs at its close. The average height rises from 50 inches to 52 inches.

In girls, 8 is the earliest age at which breast buds and body hair may appear. Armpit hair may become visible by about 8½ and pubic hair towards the close of the year. But this is by no means the typical year for these pubertal developments. The majority of girls will have to wait several more years for these signs of approaching sexuality.

Social Relations
The 8-year-old is now, for the first time, truly at home as a member of a group. The sexes are growing apart at this stage, with boys forming gangs and displaying gang loyalty. There is a great increase in the sense of humour, and much playful bickering. The 8-year-old is argumentative but is also sensitive to criticism. He or she will challenge parents over issues that a younger child would simply have accepted.

Accession at 8
James III became King of Scotland in 1460 when his father was killed in the siege of Roxborough Castle.

Early Start at 8
Charlie Chaplin first appeared on the stage in a clog-dancing act called 'Eight Lancashire Lads', when he was a tiny, 8-year-old boy. Taught to sing and dance by his parents, who were music-hall entertainers, the rest of his childhood was a misery. His father was soon dead and his mother was often in and out of mental homes, so the small boy spent most of his time in a dreary succession of boarding schools and orphanages, interspersed with periods when he was on the streets.
Sister Juana Inés de La Cruz, the Spanish colonial poet, scholar and nun, was a child prodigy who was only 8 when she composed a dramatic poem in honour of the Blessed Sacrament, and begged her parents to disguise her as a boy and send her to the University of Mexico. Her brilliance was such that before long her fame had reached the Viceroy, at whose invitation she dazzled the court with her knowledge, to the amazement of forty assembled professors.
Joy Foster, who represented Jamaica in the West Indies Table Tennis Championships in 1958, was the youngest ever international player at this sport.
César Franck, the Belgian-born composer, was so musically precocious that he was able to enter the Liège Conservatory at 8. Within a few years he was touring cities giving piano performances. Throughout his life he is reputed to have retained the same personality he had at this early stage of his life, remaining 'sweet and simple' and 'innocent of the ways of the world'.
Hugo Grotius, the Dutch pioneer of international law, was already writing Latin elegies at this age. Within three years he was a student at Leiden University and was so brilliant that he was taken on a diplomatic mission to Henry IV of France when only just into his teens. His legal masterpiece, *On the Law of War and Peace*, was the first great work on international law.
J.B.S. Haldane, the eccentric but brilliant British geneticist and physiologist, became a scientific assistant to his father, J.S. Haldane, also a physiologist,

Charlie Chaplin in *The Gold Rush*

when barely past his eighth birthday.

Ludwig Koch, the German-born pioneer recorder of animal sounds, made his very first recording in 1889, when he was 8. His father had given him an early Edison phonograph and a set of cylinders. On one he made a recording, which still survives, of his pet Indian Shama bird – the earliest animal sound recording ever made.

Pope Leo X, the extravagant Renaissance pontiff who, by his lavish expenditure, made Rome the centre of European culture, while at the same time depleting the papal treasury, began his religious life at the age of only 8, when he received the tonsure, indicating his change of status from lay to clerical.

Carolus Linnaeus, the eighteenth-century Swedish naturalist who revolutionized the classification of plants and animals, was already known by the nickname of 'the little botanist' when he was 8, because his obsession with collecting and studying plants developed so quickly in his childhood.

Franz Liszt, the Hungarian composer and virtuoso pianist, began to compose at this age – the first of more than 700 works which he produced in his lifetime. He had already been playing the piano for three years and had developed an interest in both church and gypsy music.

Arthur Rimbaud, the French poet, who went on to become the homosexual teenage lover of Verlaine, was the child of a broken home. His soldier father left when he was an infant and he was brought up by his mother. At the age of 8 he showed an unusual talent for writing and quickly became a brilliant scholar. This early start led to an early peak, all his important work being completed before he was out of his teens.

Johnny Sheffield, the American child actor, made his first Tarzan film *Tarzan Finds a Son*, in 1939. He was an overnight success in his role as 'Boy', which was the start of a thriving film career.

Natalie Wood, the American actress, who had played her first film role three years earlier, was given her first featured part in the film *Tomorrow is Forever* with Orson Welles and Claudette Colbert in 1946, when she was 8.

Misfortune at 8
Bessie Love, the American star of silent movies, attempted suicide at the age of 8. She was used to being taken to the theatre with her family and enjoyed it so much that when, one night, her mother left her at home, she decided to take drastic action. Her aim was to make a 'terrible mess', a mess that the family would remember for ever. She succeeded in this, using a knife from the dining-room drawer, but happily did not succeed in killing herself and went on to become a star in D.W. Griffith's classic film *Intolerance*.

Life-span of 8
The Son of the Candy Man Killer, Ronald O'Bryan, is not the first 8-year-old to

James III

have been killed by his father, but there is something strangely horrifying about his death. On Hallowe'en night in 1974, his father gave him a sweet laced with cyanide so that he could collect £10,000 on the boy's life insurance. The cold-bloodedness of this act so angered America that the 'Candy Man', as he became known to his fellow prisoners, was refused repeated appeals against his conviction and, with poetic justice, was sentenced to death by poison on Hallowe'en night 1982. This made him the first person in America to be executed by injection (of sodium thiotental).

Johnny Sheffield in *Tarzan Finds a Son*

Nine is the age of widening interests. The 9-year-old collects things almost indiscriminately and makes detailed lists and inventories of possessions. It is a time for 'new crazes' and independent discoveries. There is a pulling away from the parents and a greater interest in being with friends rather than joining in family excursions.

Nine is an age of variable moods – up one moment and down the next; shy one minute and bold the next. There is increasing self-motivation and self-dependence and a characteristic resentment of interruption.

It is still an essentially truthful and honest age – the more devious years are yet to come. The typical 9-year-old expresses open contempt for the opposite sex and eagerly joins own-sex groups for a wide variety of activities.

Growth The average weight increases from 62 lbs at the start of the year to 70 lbs at its close. Up to this age there has always been a slightly higher average weight for boys, but this is the first year when the girls are heavier. This is because the growth spurt that comes with the start of puberty begins a little earlier in girls than boys. By the end of this year the average male weighs 69·2 lbs and the average female 71·6 lbs. In height there is hardly any difference between the sexes at 9: the increase is from 52 inches at the start of the year to 54 inches as it ends. This makes boys about 75 per cent of their adult height, on their ninth birthday, and girls nearly 81 per cent.

Nine and a half is the earliest age at which there is an accelerated growth of the male testes, but this pubertal development will not reach its peak for another two years.

Accession at 9 Edward VI, the only legitimate son of Henry VIII, became King of England in 1547. A gifted child, both mentally and physically, he had little power during his short life, the affairs of state being left to his uncles.

Romance at 9 Lord Byron was introduced to the mysteries of physical sex at the early age of 9 by his family nurse who, although a devout Christian Scottish girl, used to creep into his bed whenever possible to stimulate him sexually. She also allowed him to watch her making love and aroused in him a lifelong obsession with the erotic.

Dante, Italy's greatest poet, was the same age when he set eyes on his beloved Beatrice and 'from that time forward, love quite governed my soul'. Beatrice Portinari became his ideal lady and he never forgot her. Although she was beyond his reach and died very young, he saw in her beauty a revelation of the divine. She became the focus of his whole existence, the 'first of the great inspiring muses of modern poetry'. He dedicated most of his poetry and all of his life to her.

Early Start at 9 Béla Bartók, the Hungarian composer, began to compose small dance pieces at the age of 9 and within a short time was playing in public, with one of his own compositions in the programme.

Georges Bizet, the French composer, was, like almost all great musical talents, born into an intensely musical family and at the age of 9 was already studying composition at the Paris Conservatoire, where he won a series of prizes culminating in the Prix de Rome in his teens.

Michel Fokine, the Russian-born ballet dancer and choreographer, entered the Imperial Ballet School at this age and rapidly distinguished himself, going on to become one of the most influential figures in twentieth-century ballet.

Ruth Lawrence, the English mathematical prodigy, passed her A-level examination at 9 (which is half the normal age for taking this exam) and went on to pass her Oxford entry exam only two years later, gaining top marks. Accepted to start at St Hugh's College in autumn 1983, she was destined to become Oxford's youngest ever undergraduate, at 12. She achieved this without attending school, having been taught at home exclusively by her father.

Franz Liszt, the Hungarian composer, was 9 when he made his first appearance as a concert pianist. His playing so impressed the local magnates that they put up a sum of money to pay for his musical education.

Felix Mendelssohn, the German composer, also made his first public appearance at 9, in Berlin. During his boyhood he wrote numerous compositions including five operas, eleven symphonies, concerti, sonatas and fugues. He went on to reach his creative peak very early, while still a teenager.

Annie Oakley

Annie Oakley, the American markswoman, was already a crack shot at the age of 9, collecting birds and rabbits to feed her large family of brothers and sisters. Her father had died five years earlier and she had suffered extreme poverty as a small child. Determined to correct matters, she started hunting for food and became so successful that she was soon shipping game to market. In no time she had paid off the mortgage and saved her family from ruin. Later she held a shooting contest with a local marksman, beat him by one shot and married him. She then went on tour demonstrating her skills which included slicing a playing card at thirty paces *thin edge on*, and perforating a card twelve times when it was thrown in the air, before it touched the ground.

Anna Pavlova, the Russian ballerina, like Fokine entered the Imperial Ballet School at St Petersburg at 9, and quickly rose to become the world's greatest dancer.

Clara Schumann, the German concert pianist, began her performing career at this age. As in so many cases of precocious talent, her father was a brilliant teacher in her subject.

Ellen Terry by G. F. Watts

Ellen Terry, the English actress, had her first stage part in Charles Kean's production of *A Winter's Tale* at the Princess's Theatre in London on April 28, 1856. From this early start she went on to become the most famous English actress of her epoch, and Henry Irving's leading lady from 1878 until 1902. She continued acting until November 19, 1925 – making her career just five months short of seventy years on the stage.

Frank Wootton, who was English champion jockey from 1909 to 1912, rode his first winner at the age of 9 years and 10 months.

Misfortune at 9
Genghis Khan, founder of the Mongol nation, was only 9 when his father was poisoned by a band of Tartars with whom there had been a continuing feud. He and his family were abandoned by the rest of their own clan, now led by a rival family, and suffered a life of extreme poverty

Genghis Khan

and near starvation. This period of childhood hardship gave him tenacity of purpose, an unbreakable will and a sense of divine mission. It also made him adaptable and in no way impaired his physical strength, for which he was famous. He was soon attracting supporters by the sheer power of his personality and went on to become the Universal Ruler of the peoples of the Steppes.

James Joyce, the Irish writer of genius, was born into a comfortable, middle-class Dublin family in 1882, and was sent to Clongowes Wood College, which he described as 'the Eton of Ireland'. He was happy there until the age of 9, when misfortune struck his family. His father was an alcoholic tax collector who borrowed money from his office, lost his lucrative post and allowed his family to sink further and further into debt. From 9 onwards, Joyce's boyhood was increasingly sordid, with family goods in pawn, debt collectors arriving, and sudden house moves suffered to avoid rent payments and tradesmen's bills. This double childhood – happy before 9, miserable after 9 – had a profound influence on Joyce's adult writing.

Louis XIV, the 'Sun King' of France, who had come to the throne at the age of 4, suffered an uprising against the crown when he was 9, in 1648. This marked the beginning of a civil war during which Louis experienced poverty, fear, humiliation, cold and hunger, for which he never forgave Paris, his nobles or the common people. When he eventually began his personal rule, he took steps to prevent any ministers or factions of the nobility from acquiring any power which might lead to another challenge of his position as monarch. He became his own prime minister and industriously presided over all affairs of state, his childhood lesson well learned.

Ten is the year of intense friendships – though not with the opposite sex. It is also the year when hero-worship becomes important, and loyalty to particular individuals or groups is strong. The 10-year-old is usually affectionate with parents but is beginning to demand a degree of privacy. There is a new poise in this year and greater relaxation. Personal talents begin to show themselves and, if present, leadership qualities become apparent. Abstract thought, logical arguments and reasoning play a larger part than they have done in earlier years. It is also the age when boys become noticeably rougher than girls in physical activities and want to indulge in more hazardous pursuits.

Conveniently linked with this period of risk-taking is a particularly efficient tissue-repair system. Careful studies by a French biologist revealed that a 10-year-old child is *four* times as quick at wound-healing as a 50-year-old patient.

There is another efficiency at this age: the peak of acoustic perfection of the human ear is reached at 10. After this, slowly at first and then with increasing rapidity, the ability to hear high pitched sounds is lost.

Ten is the first age at which a ball can be caught with some skill when thrown from a distance. There are no boy-girl differences in performing this task, but there *is* a difference in throwing. By the age of 10, boys can throw a ball twice as far as they could at 6 and with considerable power and accuracy. They are more accurate in their throwing than girls, which may be a legacy of their prehistoric specialization as the tribal hunters.

Growth The average weight increases from 70 lbs to about 80 lbs (78 lbs in boys, 81 lbs in girls) during this year. The child is now roughly half its adult weight. The average height increases from 54 inches at the start of the year to 57 inches at its close. This makes the males about 78 per cent and females 84 per cent of their adult heights when they reach their tenth birthdays.

At 10 the human brain has reached approximately 95 per cent of its adult weight.

Early signs of puberty are present at 10. One girl in a hundred starts to menstruate at this age, but this will not occur in the average girl for another two or three years. The earliest cases of full breasts or full pubic hair development are seen at about 10½, but again this is very early and is not typical for another three years. The male spurt in height increase starts at 10, as does the first sign of accelerated penis growth.

Life Expectancy at 10 At the turn of the century, in 1901, 10-year-old boys could expect to live for another 51 years; girls for another 54. But by the 1970s this had increased to 60 for the males and 66 for the females.

Tatum O'Neal, *left*

Early Start at 10 **Johannes Brahms,** the German composer, had proved himself to be a musical prodigy as a very small child and by the age of 10 was successfully taking part in chamber music concerts.

Noël Coward, the English playwright and actor, made his first appearance on the London stage at this age. It was his first public performance, in *The Goldfish* at the Little Theatre.

Thomas Edison, the American inventor, was only 10 when he set up a laboratory in his father's basement. It proved the starting point of the greatest career in the history of modern inventions.

Goethe, Germany's most illustrious dramatist and poet, wrote his first play when he was 10. Before he became a university student at 16 he had already learned Latin, Greek, French, Italian and English.

Zacharias Janssen, the Dutch inventor, devised the first compound microscope at Middleburg in 1590, when he was only 10. He later went on to invent the telescope.

Lord Kelvin, the British scientist, entered Glasgow University aged 10 years and 4 months in October 1934.

Jomo Kenyatta, destined to become the first leader of independent Kenya, spent his early years herding his father's goats until, at the age of 10, he became seriously ill with jigger infections in his feet. He underwent successful surgery at a newly established Church of Scotland mission. This was his first contact with Europeans and he was so fascinated by what he saw that he ran away from home to become a resident at the mission. He was one of the earliest examples of a Kikuyu tribesman abandoning the confining traditions of his own culture and striking out as an individualist.

David Livingstone, the Scottish missionary and explorer, was put to work in a cotton mill at 10, to help his family. He had grown up in piety and poverty, reared as one of seven children living in a single room at the top of a tenement building for factory workers on the banks of the Clyde. With part of his first week's wages he bought himself a Latin grammar and began his long climb to adult success as an African missionary.

Tatum O'Neal, the American actress, won an Oscar at this early age, for playing a tough-talking, cigarette-smoking con-artist, alongside her father, Ryan O'Neal, in the film *Paper Moon*. Within three years she had become the highest paid child star in the history of the movies and, although nicknamed 'Tantrum O'Neal' by the press, she nevertheless had more personality and acting skill than many of her adult colleagues.

Paganini, the Italian composer and the greatest violin virtuoso of the nineteenth century, made his first appearance at 10. He became so famous and successful within the next few years that as a teenager he was able to gain complete independence from his father and indulge excessively in love affairs and heavy gambling. At one point he was forced to pawn his violin to pay his gambling debts, but as he grew older he became increasingly wealthy and developed into a legendary personality.

Camille Saint-Saëns, the French composer, gave his first piano recital at the age of 10, yet another example of a precocious musical talent.

Elizabeth Taylor, the London-born actress, made her first film at 10, in 1942. Called *There's One Born Every Minute*, it was quickly followed by *Lassie Come Home* the following year, and a string of successful films over the next forty years.

Elizabeth Taylor

Paolo Uccello, the brilliant Florentine artist, was already an apprentice at this age, working in the studio of Lorenzo Ghiberti, helping to polish that artist's great bronze doors for the Baptistry in Florence. The year was 1407 and he was then known as Paulo di Dono, but he later came to be called Paolo Uccello (literally: Paul the Bird) because he loved to draw birds from nature, the first Italian artist to do this.

Ingres, self-portrait

Misfortune at 10 Jean Auguste Dominique Ingres, the French neo-classical artist, found his education cut short when he was 10. He had been sent to school at 6 but four years later the French Revolution caused disruptions that closed his school and he could no longer be taught. Many 10-year-old boys would have welcomed this closure, but for some reason it upset Ingres and left a lasting mark on his personality for the rest of his life. He had a permanent obsession with studying, to attain the learning he had been denied. For the rest of the world, however, it had the great advantage of driving him into painting at an early age, and turning him eventually into an artist of great stature and importance.

Somerset Maugham, the English writer, was orphaned at the age of 10 and was subsequently reared by an uncle.

Life-span of 10 Louis XVII, the titular King of France, died of scrofula (tuberculosis of the lymph glands) after a period of imprisonment. When his father, Louis XVI, was beheaded as the climax of the French Revolution, the surviving French nobles declared his 7-year-old son Louis-Charles, King Louis XVII in his place. The boy was given to a cobbler to be looked after, and kept as a valuable pawn in negotiations. But after his mother was guillotined, when he was 8, he was thrown into prison and treated so badly that within two years he was dead. The secrecy of his final months gave rise to wild rumours, some saying that he had been murdered and others that he had escaped. For the next few decades more than thirty men claimed to be the escaped 10-year-old, now fully grown and ready to re-establish the French monarchy, but none of them could substantiate their claims.

11

Eleven is the start of the period of maximum conformity to the peer group. This is the age of clubs, gangs, teams, secret societies with passwords, special rules and rituals, initiation rites. Adults exploit this mood with the formation of scout troops, hobby groups, and other forms of premeditated organization, but the most important memberships are those of gangs created by the 11-year-olds themselves. This period of intense 'gangship' lasts only a few years, to be replaced by 'best-friendship' with one special person of the same sex, which prepares the way for the heterosexual pairbond developments of later adolescence.

Eleven is also the age when children become highly critical of their parents and challenge their knowledge. Communication of certain kinds with the parents starts to become inhibited. For example, although sexual curiosity is very high, sexual matters are now discussed largely with friends. Since friends are ashamed to be ignorant on this subject, 11 becomes a peak age for sexual misinformation.

The 11-year-old is less bold with strange adults and may display a new social shyness which irritates parents who are not anticipating it. Eleven is also the age when hero-worship begins to focus on famous, remote figures, rather than local ones. For some children it is a moody year, with rebellion against previously accepted routines, as the child attempts to impose its own individuality on the family unit. As it sees itself more and more as a separate entity, the child also becomes, for the first time, seriously self-critical.

Eleven is the age when a new sex difference appears: boys are interested in competitions and games, while girls are becoming interested in ... boys.

Growth
The average weight increases from 80 lbs to 89 lbs (86 lbs in boys, 91 lbs in girls) during this year. The average height rises from 57 inches to 59 inches. This makes the males about 81 per cent and the females about 88 per cent of their adult height when they reach their eleventh birthdays.

By the eleventh birthday, four out of a hundred girls have started to menstruate. By the close of the year that figure has risen to 21 per cent. Other signs of puberty are also becoming more common. On the eleventh birthday 16 per cent of girls and 10 per cent of boys will have started to grow pubic hair; 14 per cent of girls will show some breast development; 8 per cent of boys will have had their first ejaculation; 4 per cent of boys will have experienced a deepening of the voice.

Early Obsession at 11
Nevil Shute, the English novelist most famous for *A Town Like Alice* and *On the Beach*, was fascinated by aircraft design from an early age. When 11 he played truant from school to spend days at the Science Museum in London studying the early aeroplanes there. He went on to become a successful aircraft designer and made good use of his technical knowledge when he became a full-time novelist, with such books as *No Highway*.

Early Start at 11
Daisy Ashford was 11 when she wrote her book *The Young Visiters* in 1892. When it was eventually published in 1919, under the sponsorship of Sir James Barrie, it brought her sudden and lasting fame as a precocious literary talent. Sidney Bechet, the American jazz musician, had been playing the clarinet for several years when, at 11, his big chance came, deputizing for a member of the famous New Orleans Olympic Band. It was the start of fifty years of performing that would make him one of the most respected names in jazz, both in America and in Europe.

Ludwig van Beethoven

Ludwig van Beethoven also became a professional musician at 11. As the son of an alcoholic whose family were becoming steadily poorer, he was forced to leave school at this age in order to earn a little money. By the age of 11½ he was assistant to the court organist Christian Neefe, and was soon working on his own compositions. But he was not a 'child wonder' like certain other composers of genius, and is said to have wept at his enforced music lessons. His father, so keen to exploit his son's musical talents, lied about the boy's date of birth. Audiences of the concert tour he made when he was 11 were told that he was 9.

William (Buffalo Bill) Cody, the American frontiersman and showman, went to work at 11 as a mounted messenger for a Kansas wagon freight firm, and before the end of his teens had become an accomplished hunter, plainsman, and Indian fighter.

Magdalena Colledge, the British skater, who skated for Britain in the 1932 Olympic Games at the age of 11 years and 24 days, was the youngest ever competitor for her country.

John Evelyn, the English diarist, began his famous diary at this age and continued it for the rest of his long life, providing posterity with an invaluable in-

sight into English habits in the seventeenth century.
Kelvin Grant, English reggae performer, was 11 when, as a member of the group Musical Youth, he saw their best-selling record *Pass the Dutchie* go to number one in the hit parade.

George Frederick Handel, the German-born composer, was this age when he received royal recognition. It was then that he was taken to Berlin, in 1696, where his talents were noticed by the Elector, later King Frederick I of Prussia. He was already accomplished on the violin, organ and clavier, and had written compositions of his own for several years – yet another example of precocious musical talent.

Anita Jokiel, the Polish gymnast, was the youngest ever international performer when she appeared as a competitor at Brighton in 1977, at 11 years and 2 days.

David Low, the political cartoonist, was contributing cartoons to a local newspaper at the age of 11, the start of a distinguished career that earned him the special hatred of Adolf Hitler during the Second World War, and a knighthood in 1962.

Blaise Pascal, the seventeenth-century French mathematician, taught himself geometry at this age, without the aid of books or teachers.

Robert Schumann, the German composer, produced his earliest known composition in 1822, when he was 11, a setting of Psalm 150. He also embarked on plays, writing a comedy and two horror dramas, and poems, including translations of Horatian odes.

John Sousa, the 'March King' bandmaster and composer of military marches, began his career as a violinist in a dance band when he was 11. He went on to compose 140 military marches including *Stars and Stripes Forever.*

Orson Welles, American genius of the cinema, entered a private school at 11, where he startled the drama class by acting in and directing his own interpretations of Shakespeare, Marlowe, Ben Jonson and various modern authors. This would have come as no surprise to his family, however, since he was already well versed in Shakespeare at an age when most other children were only just learning to read.

Revelation at 11

William Penn, the founder of Pennsylvania, experienced a divine encounter when he was this age. After a secure childhood he had suffered the shock, some months earlier, of having his illustrious father, Admiral Penn, thrown into the Tower of London, following a defeat in the West Indies. The family moved to Ireland after this, where a strong religious conviction suddenly developed in him. He went on to become a militant Quaker, for which he was imprisoned four times.

Misfortune at 11

Francis Chichester, the English round-the-world sailor, suffered acutely at this age. He could never get approval for anything he did from his puritanical father, a Devon rector, and it was as if he spent the rest of his life seeking approval

The young Handel

for deeds of extreme, self-imposed hardship.

W.C. Fields, the American comedian, was 11 when he had a violent fight with his street-trader father and ran away from home. As a result he nearly starved, was repeatedly beaten in street brawls, and had to steal to survive. His nose was so badly damaged by one of the beatings that it never resumed its natural shape. These early months of fear and pain left him with a lifelong hatred of mankind.

Sam Goldwyn, the American film producer, was orphaned at 11 and set off alone from his home in Warsaw, travelling across Europe to London. From there he soon made his way to New York, arriving penniless and still on his own. He quickly rose to become the best glove salesman in the city and then moved on to film production, ending up as one of the greatest figures of Hollywood.

Rudyard Kipling, after the misery of a foster home, was sent to an appalling boarding school at 11, where as a sensitive and short-sighted boy, he was subjected to a nightmare of teasing, bullying and beating which haunted him for the rest of his life, and which accounts for the strain of brutality and cruelty in his writings.

Marci Klein, the daughter of Calvin Klein the American fashion designer, was kidnapped at this age on February 3, 1978, in the USA, but was released unharmed when her father paid a ransom of $100,000.

Guy de Maupassant, the French author, suffered the separation of his parents when he was 11 and it left a mark both on his writing and on his personal life. He hated his father for leaving his adored mother and refused ever to become a husband himself. Instead he became intensely promiscuous. His stories contain frequent references to 'ridiculous husbands' and 'lonely fatherless children'.

Jean Jacques Rousseau, the eighteenth-century French philosopher who believed that man is good and that evil is imposed upon mankind by corrupt society and civilization, experienced his own personal corruption at the age of 11. It was at this age that he was spanked by his female teacher, an incident which turned him into a lifelong masochist.

Twelve is an awkward age, poised between childhood and adolescence. It is a time for secret thoughts, sometimes private fears, sometimes private dreams.

The 12-year-old child is biologically more advanced than most parents realize. In earlier epochs 12 would have been viewed as approaching young adulthood, but the extended Western schooling system of modern times makes this age seem much more juvenile. Because of this cultural shift, Nabokov's decision to make his fictional nymphet Lolita only 12 came as a shock to some, but it would not have surprised Virgil, as these lines show: 'In our orchard I saw you picking dewy apples with your mother ... I had just turned twelve years old, I could reach the brittle branches from the ground: How I saw you! how I fell in love! how an awful madness swept me away!'

According to a German proverb, 'Childhood ends at twelve' and one author, at least, feels this is a time for taking stock. Don Marquis: 'If a child shows himself to be incorrigible, he should be decently and quietly beheaded at the age of twelve, lest he grow to maturity, marry and perpetuate his kind.'

Growth

The average weight increases from 89 lbs to 100 lbs by the close of this year (99 lbs in boys, 101 lbs in girls). The average height rises from 59 inches to 62 inches. This makes boys about 84 per cent of their final adult height, girls 93 per cent of theirs. They are almost big enough to be looked up to now by smaller adults, but not quite. That pleasure will have to wait for another year or two.

At most ages between 1 and 100, males are bigger than females, but at puberty this trend is reversed and, for a few years, females are the heftier sex. This trend reaches its peak at 12 years.

Sexually several changes are now seriously under way. Although there are considerable individual variations, 12 is the most likely year for females to develop pubic hair and breast swellings. About one-fifth of young females have started to menstruate by the time they reach their twelfth birthday and by the time the year is over this figure for menarche has risen to one-half.

Among 12-year-old boys, just over one-third have started to sprout pubic hair and just over one-quarter have experienced their first ejaculation. Voices have broken and become deeper in one out of six boys at this age. Although male sexual developments are only beginning to stir, 12 is nevertheless the most likely age for boys to experience their first true arousal from sex-play.

Accession at 12

Pope Benedict IX became head of the Roman Catholic Church at the age of 12, in the year 1032, making him the youngest pontiff in history.

Louis IX of France became King at this age. He was renowned for his love of peace and concern for the poor, which led to him being the only French monarch to be numbered among the saints of the Catholic Church.

Romance at 12

Prince Henry, Duc d'Orléans, later to become Henry II of France, was only 12 when he fell violently in love with the recently widowed lady-in-waiting, Diane de Poitiers, who, despite the fact that she was twenty years his senior, became his mistress. She remained so until his death, by which time she was nearly 60. Their love affair survived his marriage to Catherine de'Medici and the arrival of his children by her, with Diane holding sway at court while the Queen stayed in the background.

Early Obsession at 12

Cassius Clay, later to become the famous heavyweight boxer Muhammad Ali, was caught drawing in class when he was a 12-year-old schoolboy. The teacher found that he had designed a boxing jacket with the words 'World Champion' written on it. It took him only ten more years to make this childhood dream come true.

Albert Einstein, the genius of physics, was also 12 when he recorded his decision to devote his life to solving the riddle of what he called the 'huge world'. His parents, however, did not have high hopes for him. As a small child he had been so late in learning to talk that they feared he was subnormal. He was also slow in physical development, considered himself a weakling and hated games and athletics. His generally poor school record, up to the age of 15, gave no indication of what was to come.

Early Start at 12

Julie Andrews, the relentlessly wholesome singing star of *My Fair Lady* and *The Sound of Music,* made her début on the London stage in 1947, singing operatic arias in the 'Starlight Roof' revue.

Hector Berlioz, the French composer, received little formal education. At the beginning of the nineteenth century when he was a child, war was disrupting formal schooling and he had to teach himself. With help from his father, he worked out the elements of harmony and, at the age of 12, was already composing for local chamber music groups.

George Cruikshank, the nineteenth-century English satirical cartoonist, had no opportunity for formal training but at this age had already produced and sold his first etching.

John Dalton, known as one of the fathers of modern physical science because he developed the atomic theory of matter, was already a teacher at 12, having taken charge of a Quaker School in Cumberland.

Auguste Escoffier, known as 'the king of chefs and the chef of kings', started work in his first kitchen, the hot and noisy setting in which, in later life, he was to create some of the greatest culinary masterpieces of all time.

Gaius Octavius, later to become the great Roman Emperor Augustus, was introduced into Roman public life at this age, when he made his début by delivering the funeral speech for his grandmother Julia.

Thomas Henry Huxley, one of the most famous scientists of his day and who became President of The Royal Society, only received two years of formal schooling. He hated those years intensely, calling his teachers 'baby-farmers', and while his father sank 'in worse than childish imbecility of mind' and eventually died in an asylum, the young Huxley set about educating himself. He was so successful that at the age of 12 he was already reading advanced works on geology and logic, having also taught himself German.

Horatio Nelson joined the navy at this age, in November 1770 and was soon travelling to the West Indies as 'Captain's servant'.

Achievement at 12

Frederic Chopin, the Polish composer, had already been playing the piano in public for several years when, at the age of 12, he was accorded the honour of playing before the Tsar, Alexander I, on that occasion demonstrating a new kind of piano-organ.

Gertrude Ederle broke the record for the 880-yard women's freestyle swimming event at Indianapolis on August 17, 1919. Her time was 13 mins 19.0 secs, making her the youngest world record-breaker in the history of organized sport.

Marcus Hooper swam the English Channel from England to France in 1979 when he was only 12 years and 53 days old, the youngest person ever to achieve this feat.

Sabu, a penniless and illiterate elephant-helper in the stables of an Indian Maharajah, was discovered when he was only 12 and made the star of a major feature film called *Elephant Boy*, which led him to worldwide fame and later to a glamorous Hollywood home where he was to end his days at the age of 39.

Sabu

Brooke Shields, after years of work as a child model, achieved her first major success at this age when she took the controversial film role of the child prostitute in Louis Malle's *Pretty Baby*, in which her virginity

Brooke Shields in *Pretty Baby*

is auctioned to the highest bidder at the brothel in which she lives. Her sensitive and skilful acting in this part is a reminder that the brain of the 12-year-old should not be underestimated, even where sexual matters are concerned.

Misfortune at 12

James Boswell, the biographer of Dr Johnson, suffered an acute nervous illness at this age, which is hardly surprising in view of his family upbringing, with a harsh father and a mystical mother. He grew up with a horror of sin and hellfire, and at one point contemplated self-castration to avoid sexual sinning. Timid and repeatedly suffering from depression, he seriously thought of becoming a monk. Instead, after meeting the ladies of the town, he became a lifelong womanizer whose treatment of whores was often abominable. He suffered from venereal disease nineteen times before it finally killed him.

Charles Dickens faced disgrace and hardship at 12, being taken out of school and sent to a squalid factory to do manual labour. For an intelligent middle-class Victorian boy, this sudden descent to working-class toil was hard to bear and, as a result, he developed a bitter resentment towards his mother. The social disaster in the Dickens family was caused by his father's wild extravagances that ended with him being sent to prison and the rest of the family having to make drastic economies. Charles did, however, turn this misfortune to good use in *David Copperfield*.

René Magritte, the Belgian surrealist, suffered a childhood trauma at 12 when his mother, suffering from nervous depression, jumped to her death in the Sambre River.

Life-span of 12

Edward V of England, born in November 1470, the son of Edward IV, was only 12 when he came to the throne of England in April 1483. Within a few months he disappeared and was never seen again. In popular legend he became one of the 'Princes in the Tower', supposedly smothered by hired assassins.

13

Thirteen is the age of puberty. Although the 13-year-old is not yet a 'typical teenager', he or she can at last lay claim to belong to that special 'age set' and start to emulate in earnest the cultural patterns associated with adolescence – interest in pop music, teenage idols, clothes, magazines and the rest of the commercially fostered 'crazes'.

Thirteen is also the year when 'spots' appear for the first time. The development of sexual characteristics is accompanied by a susceptibility of the sweat glands to clogging. Once plugged up, these glands fall prey to secondary infections and swell into disfiguring pimples. This problem is caused by the dramatic hormonal changes that are taking place, with the perverse result that the spottier the face of the teenager, the more highly-sexed he or she is.

This is the age when children become highly critical of adults. One young rock star commented: 'Every teenage revolution is the same . . . the function of youth is to change laws made by old men for young men that old men would never break.' Parents come under repeated verbal attack. The change in attitude is summed up by Rebecca Richards: 'Oh to be only half as wonderful as my child thought I was when he was small, and only half as stupid as my teenager now thinks I am.' Laurence Peters puts it another way: 'It never occurs to a teenage boy that one day he will be as dumb as his father.'

Thirteen-year-olds do not object to these attacks – such comments merely confirm their views about the stupidity and obsolescence of the adults concerned. They object much more to those adults who continue to kiss and paw them as if they are still cuddly infants. As one 13-year-old American boy put it: 'Sometimes when you meet your relatives you think you're a lollipop.'

In certain cultures, however, the adults recognize formally the changes that are taking place in their young at the age of 13, and mark the occasion with a special ceremony. This is the traditional age for a Jewish boy to go through the religious initiation ceremony called the *Bar Mitzvah*. This rite of puberty is marked by a service in which the boy/man, wearing a prayer shawl and a cap, stands with his father and the rabbi in his synagogue, on the sabbath closest to his thirteenth birthday, and reads aloud the daily passage from the sacred Torah for the first time in his life.

In most advanced countries formal education continues to be compulsory for some time yet, but in Italy it ceases to be legally required at 13.

Growth

The average weight increases from 100 lbs to 111 lbs by the close of this year (112 lbs in boys, 111 lbs in girls). The average height rises from 62 inches to 64 inches, with the boys edging ahead of the girls once again. These figures reveal that, by the end of this year, the girls' growth spurt, which for several years has put them ahead of the boys', is now being overtaken by the newly arrived male growth spurt. After this year the average weights and heights will always be greater for the males.

Thirteen is the age when girls are most likely to start menstruating. In England, the typical girl experiences her first menstrual flow between 13 years 2 months and 13 years 4 months. In the USA it is very slightly earlier than this. In other cultures and at other times the age for menarche has often been higher, due largely to nutritional differences.

Although, in advanced cultures today, four-fifths of all girls have started to menstruate by the end of this year, most of them remain reproductively sterile for about another year, with no eggs being released at the time of the monthly flow.

Other sexual changes at 13: girls are most likely to develop the brown patches of pigmented skin around their nipples at this age; just over half the boys have experienced their first ejaculation, two-thirds have pubic hair by now, and nearly half have undergone some degree of voice-lowering.

Romance at 13

André Gide, the French writer, fell in love with his 15-year-old cousin, Madeleine, when he was 13. He came across her weeping as she knelt in prayer. She told him of the discovery of her mother's infidelity and that it must be kept a secret. From that day he became devoted to her and twelve years later married her, unaware at that time that he was a suppressed homosexual.

Edward Montague, an eccentric eighteenth-century child, ran away from Westminster School at 13 and enrolled himself at Oxford University as a student of Oriental Languages. He found himself lodgings there with a buxom young lady, fell in love with her and made her his mistress, until his mother arrived and retrieved him.

Early Obsession at 13

Robert Fulton, the Irish-American inventor who built the first commercially practical paddle-steamer, the *Clermont*, which offered the earliest regular steamboat service in the world, starting in 1807, had been obsessed with the idea for thirty years. When he was a boy of 13 he constructed paddle-wheels which he applied with success to a fishing boat, and managed to keep this obsession alive long enough to convert it into an adult achievement at the age of 42.

Louis Leakey, the world's greatest field-worker in the hunt for mankind's ancient ancestors, decided to devote his life to this subject when he was 13. A Christmas gift from England, a book called *The Stone Age*, prompted him to go out searching for Stone Age implements near his home at Kabete in Kenya and by luck he found some almost immediately. To his great excitement they were officially confirmed as such by the curator of the Nairobi Museum, and this set the path for the whole of his future endeavours.

Early Start at 13

Paul Cézanne, the French painter and one of the greatest figures in modern art, was a 13-year-old schoolboy in the provinces when he defended a bullied schoolmate against taunts of 'snooty Parisian'. The two boys shared a fascination for poetry and wrote a great deal of it together. This early literary start changed course when his friend returned to Paris and he himself switched to painting. When they met again some years later in Paris, as young men, there was a gulf between them – the painter unwilling to join in the boozing and wenching expected of him by his literary friend. Many years afterwards, the friend wrote of the painter that he was an 'aborted genius' and their relationship, so vital and important at 13, was finally ended. The friend's name was Émile Zola.

Maurice Chevalier, the French entertainer, made his début as a Parisian café singer in 1901 and went on to spend more than sixty years as one of the great international stars of show business.

Thomas Gainsborough, one of the most impressive English artists of the eighteenth century, a warm-hearted and generous man, was also a persuasive 13-year-old. It was at that age that he pleaded successfully with his father to let him go to London to study art, on the strength of his promising beginnings at landscape painting.

Mahatma Gandhi, the Indian religious and political leader who combined shrewdness with saintliness in ideal proportions, made an early start in family affairs, being married at the age of 13. In his memoirs he wrote: 'I can see no moral argument in support of such a preposterously early marriage as mine.' But it was a happy one nevertheless, lasting for more than sixty years.

Betty Grable, the American pin-up girl of the Second World War, whose cheery, jaunty, leggy appeal made her the ideal heart-throb of wartime, appeared in her first film role at the age of 13 in 1930, singing and dancing in a film called *Happy Days*.

Rita Hayworth, a rival Hollywood star, was also dancing professionally at 13. In her case it was in Mexican night-clubs, where she was spotted by a film producer. She went on to become one of the love goddesses of the 1940s, with a string of husbands including Orson Welles and Prince Aly Khan.

Michelangelo left the classroom at 13, in 1488, to become an apprentice in a painter's studio and started on one of the most spectacular careers in the history of art.

Achievement at 13

Bebe Daniels, the American actress, had done more than merely start her career at this age. As a 13-year-old she was already to be seen in *starring* roles opposite comedian Harold Lloyd, in silent films. She made over 200 of these early comedies before graduating to become the leading lady of feature films opposite such figures as Rudolph Valentino. But she is best remembered in Britain for her courage in staying in London through-out the blitz of the 1940s, to make her weekly comedy show *Hi Gang!* with her husband Ben Lyon.

Anne Frank, the author of *The Diary of A Young Girl*, began writing her diary in 1943 and worked at it for two years. In it she wrote one of the most mistaken sentences ever penned: 'No one will ever be interested in the unbosomings of a 13-year-old schoolgirl.' Her diary became the most famous and the most moving personal account of the Second World War when it was published in 1947, two years after she died in a Nazi concentration camp.

Jacqui Hampson, the English swimmer, was the youngest girl ever to swim the English Channel. A butcher's daughter from Weymouth, she reached the French coast at Calais after sixteen hours in the sea, in September 1982.

Mita Klima of Austria was the youngest ever tennis player to compete at Wimbledon. She was only 13 when she took part in the 1907 competition there.

Revelation at 13

Joan of Arc, a simple, exceptionally pious girl who was the daughter of a ploughman, began to hear a voice from God at this age, teaching her self-discipline and instructing her to deliver France from the English.

Misfortune at 13

Henri de Toulouse-Lautrec, the French artist, was 13 when he fell on the slippery floor of one of his aristocratic family's castles. A frail child, he broke his left femur. Within a matter of months he had suffered a second fall, fracturing his other femur. Because of a congenital bone defect, neither leg recovered. They remained stunted while the rest of his body grew normally, making him into a uniquely shaped semi-dwarf. He went to Paris where he mixed with social outcasts – pimps and prostitutes – with whom he could feel some sympathy. He shared their life of dissipation and dissolute pleasures until alcohol, syphilis and other abuses of his stunted body led to his early death at 36 – but not before he had produced vivid masterpieces of the dancers, singers, whores and circus people of Montmartre he knew so well.

Life-span of 13

Saint Agnes was martyred at the age of 13, in AD 304, but the stories of her end are contradictory, and have clearly been embellished down the centuries. It appears that she refused to marry the son of the Prefect of Rome, stating that she was the bride of Christ alone. For this, she was stripped and thrown into a brothel. Only one youth would touch her, and he was struck blind when he attempted to violate her. She then healed him. In another version, it is said that her hair was so thick that it covered her naked body, and that when the youth tried to rape her he was struck dead. In a third version, she was then burned at the stake, but the flames failed to consume her and in the end she had to be beheaded.

14

Fourteen is the age of discretion. In legal terms, a subject is deemed capable of using discretion at this age, and so becomes subject to criminal law. It is also the earliest age at which males develop mature, functional sperm and therefore become reproductively active. Younger males may have been able to ejaculate, but they were not fertile.

This is also the year for the male growth spurt, two years later than that of the female.

Mentally, the 14-year-old is showing greater interest in adult pursuits and habits, but is still derisive where parents are concerned. Mark Twain sums this attitude up: 'When I was a boy of fourteen, my father was so ignorant I could hardly stand to have the old man around, but when I got to be twenty-one, I was astonished at how much the old man had learned in seven years.'

Growth
The average weight increases from 112 lbs to 125 lbs in boys and from 111 lbs to 118 lbs in girls. The average height rises from 64 to 67 inches in boys and from 63 to 64 inches in girls. This makes boys 91 per cent of their adult height on their fourteenth birthday and girls 98 per cent.

Nearly all girls have started to menstruate by the close of this year (92 per cent to be precise). Ninety-five per cent will show breast development and 98 per cent pubic hair growth during this year.

Boys not only reach their peak of growth acceleration, but they too nearly all complete their sexual development: 92 per cent have developed pubic hair, 82 per cent have ejaculated, and 70 per cent have deeper voices, by the year's end.

Life Expectancy
It has been calculated that the expectation of life in the Lower Palaeolithic (The Early Old Stone Age), at birth, was only 14·5 years. This did not mean that nobody lived to a ripe old age, but rather that huge numbers died very early indeed.

> Among pet animals, 14 is the maximum age ever recorded for the guinea pig.

Accession at 14
Edmund the Martyr, ruler of the Anglo-Saxon Kingdom of East Anglia, became King at this age, in AD 855. A kind and virtuous man, he resisted the Viking invasion bravely, but was overpowered and, refusing to renounce Christianity, was tied to a tree, tortured by being shot with arrows and then beheaded, in AD 870.

Elagabulus, the teenage Roman Emperor in the third century notorious for imposing Baal-worship on the Roman world and for his extravagant homosexual orgies, became Emperor at this age.

Emperor Frederick II, who had been King of Sicily since the age of 3, was declared of age at 14 and assumed control of government in the year 1208. Within a matter of months he was married to the much older Constance of Aragon and with her help tightened his grip on the country and soon spread his power farther afield. An unusual claim to fame is his production of the first great study of bird behaviour and falconry.

Ashikaga Yoshimasa, the Japanese hereditary military dictator, was proclaimed Shogun at 14. Ineffective as a Shogun, he was a great patron of the arts, which flourished during his reign. He abdicated when he was 38, in 1473, and during his retirement refined the Japanese Tea Ceremony, developing it into a fine art.

Romance at 14
Lucrezia Borgia, the Italian noblewoman, was this age when she was subjected to the first of her 'romances'. In reality, these were callously arranged marriages for purely political ends and Lucrezia was a more or less innocent pawn in her family's scheming games of power. Her popular reputation is of criminal and moral excesses, but a truer picture has emerged showing her to have been a person of learning, beauty and charity who gained great respect from her subjects.

Early Start at 14
Margaret Beaufort, wife of Edmund Tudor, was only 14 in 1457 when she gave birth to a future king of England (Henry VII) at Pembroke Castle, nearly three months after his father's death.

Christopher Columbus, the son of a weaver, went to sea at this age, and experienced many adventures as a young man that prepared him well for his great voyages later in life. On one occasion he nearly lost his life when his ship caught fire and he had to swim for shore with the help of an oar.

Diana Dors, the English teenage sex symbol of the late 1940s, gained her first film part in *The Shop at Sly Corner*. Because she was cast as the villain's girl-friend, she had to lie about her age, adding three years to make it an acceptable 17.

Ralph Waldo Emerson, the American poet and essayist, entered Harvard University.

William Hazlitt, the English essayist and critic, wrote, at 14, *A New Theory of Criminal and Civil Legislation*, suggested by a dispute about the 'Test Acts'. A touchy, sensitive, paranoid man who had been an intensely shy and solitary boy with a 'very precocious intellect', his mind obviously had matured at an abnormal rate.

Bobby Jones, the American golfer, competed in the US Amateur Championship when he was only 14. He went on to win a number of US Opens and British Opens, becoming *the* golfer of the 1920s.

Pablo Picasso, the Spanish genius of modern art, when 14 and still a boy studying at the School of Fine Arts, painted pictures of beggars, such as *Man With a Cap* (1895), which already displayed a mastery of nineteenth-century realism and a fully adult technique.

14

Ink drawing by schoolboy Picasso

Andres Segovia, the Spanish genius of the guitar, made his public début at the same age.

Late Start at 14
Sean O'Casey, the Irish playwright, had little or no early education and had to teach himself to read at the late age of 14. Although he was to become so famous, he still had to work as a labourer until the age of 30. During the run of his first play *Shadow of a Gunman* at the Abbey Theatre in Dublin, he was labouring during the day, mixing concrete for road repairs.

Achievement at 14
Linda Blair, the American film actress, played a starring role in *The Exorcist* in 1973, when 14. Her chilling performance as a possessed girl was so skilful that despite her age she was nominated for an Academy Award.

Karel Capek, the twentieth-century Czech author, had his first writings published when he was 14. The son of a country doctor, he suffered all his life from a spinal disease for which his writing seemed to be a compensation.

André Eglevsky, the Russian ballet dancer, became Premier Danseur with the Ballet Russe de Monte Carlo at this early age.

Bobby Fischer, the American chess wizard, became an International Grand Master at 14, having already gained the US Junior Championship the previous year. This made him the youngest Grand Master ever in the history of chess. In another fourteen years he was the World Chess Champion, the first American ever to be so.

Hayley Mills, the English actress and daughter of John Mills, won an Oscar for her 1960 performance in *Pollyanna*.

Samuel Palmer, the English painter of visionary landscapes, at 14 was already an exhibitor at London's Royal Academy.

Revelation at 14
Saint Bernadette of Lourdes, the French visionary founder of the cult of Our Lady of Lourdes, was this age when she had a series of visions, between February 11 and July 16, 1858, in which she claimed the Virgin Mary revealed her identity with the words, 'I am the Immaculate Conception'. As a child she was a frail asthma sufferer and caught cholera during the epidemic of 1854. She died young in great agony at 35, after a strenuous defence of the authenticity of her visions. They were eventually declared genuine in 1862 by the Pope and she was canonized in 1933. Three million people annually visit Lourdes, many hoping for a miraculous cure for their ailments.

Misfortune at 14
Hernan Cortés, the Spanish Conquistador, was already dreaming of visiting the Indies when he was 14, but instead he was sent away to study at Salamanca because he was a 'source of trouble to his parents', being not only clever, but also mischievous, quarrelsome, haughty, ruthless and 'much given to women'. He missed an early chance to join a voyage to the Spanish Indies through injuries sustained by falling from a wall when escaping from the house of a married woman, and he missed another important voyage because, as a young man, he had contracted syphilis. Despite these setbacks he had conquered the Aztecs and completed his dream by the age of 35.

Henry VII, King of England, whose father died before he was born and whose 14-year-old mother rejected him when she soon re-married, was rushed into exile at the age of 14 because of the collapse of the Lancastrian cause. He lived at this age in constant fear of betrayal and death, and usually in great poverty. By the time he eventually did become King at 28, he was a tired-looking, anxious man with small eyes, bad teeth and thin white hair. But he had also learned, through his youthful misfortunes, how to survive, which stood him in good stead in his new role as monarch.

Henry VII

Andrew Jackson, later to become American President and military hero, was captured by the British and imprisoned when 14. Shortly after capture, when he refused to shine the boots of a British officer, he was struck across the face with a sabre, an act which left him with a lifelong hostility to Great Britain.

Fifteen is a threshold year, when adolescents stand at the entrance of adulthood and contemplate the exciting world beyond. Gertrude Stein has expressed the age well in her own curious style: 'Fifteen is really medieval and pioneer and nothing is clear and nothing is sure, and nothing is safe and nothing is come and nothing is gone but it all might be.'

In Sheridan's *The School for Scandal* a singer calls out: 'Here's to the maiden of bashful fifteen . . . Let the toast pass, Drink to the lass, I'll warrant she'll prove an excuse for a glass.' The bashfulness of the 15-year-old girl is a key to her emotional condition. She is suddenly aware of her new physical appeal and enjoys the knowledge before discovering quite how to handle it. There is sometimes a great deal of giggling, coy glancing and excited whispering at this age, when females gather in groups. Sexuality also enters the bravado displays of the 15-year-old males. The gangs survive and so do the group loyalties but now the girls are being eyed more seriously. Precocious individuals of both sexes are beginning to break away from their same-sex friends and form into boy/girl pairs, at least for brief spells. Once back in the group, these early forays are usually discussed and compared at length – their details being still to some extent 'group property'.

As a thinking machine, the 15-year-old brain is functioning well and has a special charm in the way it deals with the outside world. David Grayson comments: 'Almost always your lad of fifteen thinks more simply, more fundamentally than you do; and what he accepts as good coin is not facts or precepts, but feelings and convictions.'

Fifteen is also the start of a dangerous period – the deliberate risk-taking phase of the young male. In America the majority of deaths that occur at this point are caused by accidents, with road accidents predominant. This high-risk period, which begins at 15, lasts throughout the late teens.

Many countries make 15 their age for the end of compulsory education, including Australia, Canada, Ireland, Japan, Spain, Switzerland and West Germany.

For film-goers, 15 has become a year of special significance. In England the old 'AA' films, to which 14-year-olds were admitted, were replaced late in 1982 by a '15' category. Such films now require an additional year from their patrons. Other, more explicit films still necessitate an age of 18, implying that 15 is somehow semi-adult. Such distinctions are universally resented by 15-year-olds.

Growth The average weight increases from 125 to 127 lbs for boys and from 118 to 123 lbs for girls, as the year passes. The average height rises from 67 to 68 inches for boys, but still hovers around 64 inches for girls, who have almost reached their adult height. On their fifteenth birthdays, boys are 95 per cent of their adult height and girls are 99 per cent of theirs.

Sexually, this is the year when almost all teenagers have reached maturity and when quite a few of them have begun to test their new equipment. Most parents imagine that their 15-year-old offspring are still sexually innocent, but some of them will already have experienced their first sexual intercourse by the end of this year. The precise figures vary from country to country and also from survey to survey, but it seems that in some advanced nations today as many as a quarter of all females and a third of all males have lost their virginity by the close of this year. This represents a major change; only twenty years ago the figures involved would have been no more than a few per cent.

Physiologically, nearly all girls have started to menstruate (97 per cent) and develop breasts (98 per cent) by the close of this year, and all have developed pubic hair. Boys have reached a similar advanced condition: 97 per cent of 15-year-olds have pubic hair, 92 per cent have had their first ejaculation, and 84 per cent have experienced a voice-change.

Life Expectancy at 15 On the day they reach 15, males can expect to live for another 56 years, females for another 62 years.

Accession at 15 **Edward III** became King of England in 1327 after his mother and her lover contrived to get rid of his father, Edward II, who was deposed and then brutally murdered. While still a teenager, Edward III avenged his father's death by having his mother's lover seized and hanged.
James IV became King of Scotland in 1488, following the death of his father who was killed in a battle with rebellious nobles. The new king had supported the nobles against his own father, but once the throne was his he quickly asserted his mastery over them and, despite his young age, rapidly became an energetic and popular ruler.

Romance at 15 **John Barrymore**, the American actor, was seduced by his stepmother when he was a 15-year-old virgin. It was the beginning of a life of endless lechery in which he made love to countless women but could never really trust one of them.
Charles II, the 'Merry Monarch' of seventeenth-century England, was introduced to sex at the same age, also by an older woman. In his case it was his former governess, Mrs Wyndham. It must have been a rewarding encounter because he showed a strong taste for older women in the years to come. His stated belief was that 'God will never damn a man for allowing himself a little pleasure', and he made sure that he had his share, producing no fewer than fourteen illegitimate children and thoroughly relishing his libertine reputation.
Louis II, King of Hungary and Bohemia from the

age of 9, married Maria of Austria in 1522, when only 15. The two of them pursued a life of such riotous pleasure together that the teenage king was soon disqualified from affairs of state. He died fleeing a battlefield when only a few weeks out of his teens.

Friedrich Nietzsche, the German philosopher, was a 15-year-old schoolboy when a 30-year-old, blonde, sado-masochistic, nymphomaniac countess, disguised as a man, sneaked into his school dormitory and beat him until she was sufficiently sexually aroused to make love to him. She tormented him so much during their ensuing love affair that he himself was driven to flogging her with a riding whip, only to discover that this too aroused her. His next sexual adventure was with a prostitute who gave him syphilis which, added to the terrible migraines and insomnia he suffered most of his life, eventually drove him mad, so that he spent the last eleven years of his life in an asylum.

Early Start at 15

Brigitte Bardot was 15 when she posed for the cover of France's leading women's magazine *Elle* and the striking image of her pouting child-woman face was seen by Roger Vadim, then an assistant film director. They married in 1952 and with his help she became a star in 1956.

George Best, the British footballer, went to Manchester United when he was 15, in 1961. Matt Busby, the manager, was trying to re-build his shattered Manchester United team, following the Munich air disaster in which many of them died, when he took the young Belfast boy on to his playing staff – and triggered one of the most spectacular careers in the history of soccer.

Billie Holiday, the American jazz singer, an illegitimate child whose early life was full of suffering and deprivation, started singing at a bar in New York at this age. She was an instant success and within a few years was making records and heading for stardom.

Edmund Kean, one of the greatest tragic actors of all time, displayed a precocious talent for singing, dancing and acting. The bastard son of a street-hawker mother and an unbalanced father who had killed himself at 22, his childhood was one of wilful vagrancy until, at 15, he became his own master and set out to conquer the stage. Within a few months he had married and started to develop his acting genius, which took him to the peak of his profession despite advancing alcoholism.

Achievement at 15

Louis Braille invented his method of fingertip reading at this age. He had been blind since the age of 3, following an accident. His 'Braille' system was adopted in France in 1854 and later became worldwide.

Tracy Caulkins, the American swimmer, was this age when she won one silver and five gold medals at the 1978 World Swimming Championships.

Lottie Dod was only 15 years and 9 months old when she became the youngest Wimbledon Tennis Champion ever, in 1887.

Andrew Fitzgibbon was 15 years and 100 days old when he became the youngest ever recipient of the VC, on August 21, 1860.

Shane Gould, the Australian swimmer, won one bronze, one silver and three gold medals at 15, when she competed in the Olympic Games in Munich in 1972.

Hugo Grotius, the Dutch pioneer of international law, had his first set of political monologues published at this age, offering a synthesis of the political situation in 1598 – somehow the least likely of precocious achievements.

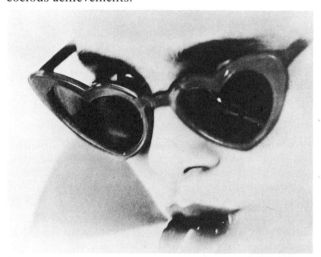

Sue Lyon as Lolita

Sue Lyon, the American actress, made her début at this age, playing the role of the 12-year-old Lolita, in the film of Nabokov's notorious novel. Although she made other films later, her career did not fulfil its early promise.

Yehudi Menuhin achieved the unique distinction of being the youngest entrant in *Who's Who*, other than those there for hereditary reasons.

Mushtaq Mohammad was 15 years and 124 days old when he played cricket for Pakistan against the West Indies in 1959, making him the youngest ever player in a Test Match.

Claude Monet, the French Impressionist, had his first commercial success as an artist at 15 when he sold some of his caricatures, which were already exceptionally well observed and drawn.

Wolfgang Amadeus Mozart wrote his *fourteenth* symphony, which is still in the repertoire of most orchestras, when he was only 15.

Giuseppe Verdi, the Italian operatic composer, although less precocious than Mozart, did see the first performance of his first symphony in 1828, when he too was 15.

Life-span of 15

Edward VI, King of England since the age of 9, died from tuberculosis at this early age without ever enjoying any kingly powers.

16

Sixteen is the age of consent. In England and Scotland this is the age at which a girl's consent to sexual intercourse is valid. Below this age, with or without her consent, having 'carnal knowledge' of her is a criminal offence. A boy may consent to sexual intercourse at any age although homosexual acts remain illegal until both partners reach 21.

Sixteen is also the legal age for driving mopeds, invalid carriages, agricultural tractors and mowing machines, but it is still considered to be too young, in the UK at least, for driving motor cycles or motor cars.

Ecclesiastical Law states that the profession of religious vows starts at the age of 16.

Compulsory education ends at this age in Denmark, France, Holland, Norway, Sweden and the UK.

Most advanced countries demand a higher age for the legal purchase of alcohol, but Austria and Italy allow this at 16 and Holland permits the purchase of non-spirits. Sixteen is also the age at which cigarettes may be bought in the UK.

In the American cinema there was a 1968 ruling allowing 16-year-olds to see 'X' rated films, but this age was raised to 17 in 1970, when such films became more explicit and more violent.

These examples show how confusing it is for teenagers to know precisely when they are 'adult' and in which respects. Because of this delegates at the Liberal Party Conference in England in September 1982 supported a motion calling for 'a uniform 16 age of majority at which young people would become fully independent financially, socially, sexually and politically'. If this were ever made law it would mean that 16-year-olds could, for example, vote, stand for parliament and have a homosexual relationship without reference to parental or guardian control. Whether or not 16 is the best year for 'coming of age' in an advanced society, it would certainly be a great advantage to have all the various legal age limits set at the same point, so that people knew with certainty when they changed from 'minor' to 'major'. To have different ages for different activities destroys the significance of all of them, by eliminating a major rite of passage.

Terence, the Roman playwright, saw 16 as 'The very flower of youth'. Shakespeare agreed, but saw it as the last of the appealing years of youth: 'I would there were no age between sixteen and three-and-twenty, or that youth would sleep out the rest; for there is nothing in the between but getting wenches with child, wronging the ancientry, stealing, fighting.' (*A Winter's Tale*.)

Growth Average weight increases from 137 to 146 lbs in males during this year, and from 123 to 125 lbs in females. Male height rises from 68 to 69 inches, but the females have virtually stopped growing in height at an average of 64 inches. Males at this age have reached 97 per cent of their adult height, females almost 100 per cent.

In sexual development virtually all males and females now have all their secondary sexual characteristics. For example, only 1 in 100 boys have yet to develop pubic hair; only 4 in 100 have yet to ejaculate; and only 7 in 100 have yet to experience a deepening of the voice. Only 1 in 100 girls have yet to menstruate or show breast development. The vital phase of puberty is effectively complete. Biologically one might expect to see a great deal of sexual activity at this age, but culturally attempts are made to hold it in check. A few decades ago, sexual intercourse occurred in only a small percentage of 16-year-olds in the UK. More permissive attitudes in recent years have, however, resulted in a return to a more 'biological' age for first intercourse. A survey carried out in 1976 revealed that 56 per cent of UK females at age 16 had already indulged in sexual intercourse – a dramatic increase following the release of sexual suppression.

Accession at 16
Ivan the Terrible was crowned Tsar of Russia at $16\frac{1}{2}$, in 1547 (Tsar is short for Caesar). He also married the first of his six wives when he was 16.

Nero, the notorious Roman Emperor, came to power at this age when his mother Agrippina poisoned her husband Claudius with mushrooms.

Romance at 16
The Buddha was married at 16 to a 16-year-old girl. His bride was a princess called Yasodhara. Their loving relationship produced a son, but shortly after the baby's birth the Buddha was drawn away from earthly pleasures, stole away in the dead of night and gave himself up to shaven-headed wandering in the search for truth and enlightenment.

Gustave Flaubert, the French novelist, was 16 when he completed *Mémoires d'un fou*, which recounted his devastating passion for an older woman, Elisa, the wife of a music publisher. It was a secret, one-sided romance of which she knew nothing until thirty-five years later, when she was a widow.

Achievement at 16
Clyde Beatty, the American wild animal trainer, was this age when he starred in his first animal act. Over the years he developed this into the most daring act of its kind in the history of the circus. It involved his presence in a cage with a mixed collection of tigers of both sexes, lions, leopards, pumas, hyenas and bears, an almost lethal combination.

Edward, the Black Prince, son of Edward III, father of Richard II, won his spurs at this age at the Battle of Crécy on August 26, 1346.

Albert Einstein, the genius of physics who was a backward schoolboy, had by this age begun to show dramatic signs of improvement. At 16 he wrote an essay which contained the beginnings of his special theory of relativity.

William Morris, the industrialist, was 16 when he opened his small bicycle shop in Oxford in 1893 with a capital of £4. He first repaired bicycles, then raced them, then built them, then designed them himself. Next he designed motor cycles and finally motor cars. In 1912 he opened his car factory at Cowley to make the first cheap British car for the British worker. Morris Motors went on to become one of the most successful car companies in history and he himself became a great benefactor, establishing the Nuffield Foundation and Nuffield College, Oxford.

Domenico Scarlatti, the Italian composer, achieved early success when his first opera was performed in Naples, when he was 16.

Brooke Shields, the American film actress, was only 16 when her film earnings permitted her to purchase a large town house in New York, on East 62nd Street in Manhattan. The house, previously owned by George Barrie, head of Fabergé, cost her £1,100,000.

Sophocles, the author of *Oedipus Rex* and 122 other dramas, only seven of which have survived, was only 16 when he was chosen to lead the choral chant at the celebrations in 480 BC for the Greek victory over the Persians at Salamis.

Cornelius Vanderbilt, the American tycoon, was a poorly educated boy, a farmer's son, who, at 16, bought a ferryboat with which he carried farm produce and passengers between his Staten Island home and New York. Within two years he had bought three boats and went on to become the greatest transport tycoon in the United States, owning several shipping lines and also railway lines. When he died in 1877 he left $100,000,000 and had founded Vanderbilt University.

William Walton, the British composer, received the Carnegie Award for a Pianoforte Quartet in 1918, at this early age.

Yevgeny Yevtushenko, the Russian poet, is a unique figure in the modern-day USSR – a loyal critic. In 1949 at 16 he published his first book of verse, *Prospectors of the Future*. He went on to become famous, both in East and West, for his 1960s poetry readings and created a problem for the Soviet authorities since he was both an outspoken critic of his homeland and at the same time intensely loyal to it. Wisely, he was permitted to speak and to publish, thereby becoming the best 'roving advertisement for the "de-Stalinization" process'. He probably did more for the new Russian image than any other single person. His most famous poem is *Babiy Yar*, in which he attacks Soviet hypocrisy and anti-Semitism.

Misfortune at 16

Charles Darwin, one of the most original and important figures in the history of human thought, described himself as a child as being 'slow in learning', 'incapable of mastering any language' and 'in many ways a naughty boy'. When he left school at 16 he was 'considered by all my masters and by my father as a very ordinary boy, rather below the common standard in intellect'. His father once said to him: 'You will be a disgrace to yourself and all your family.'

Aldous Huxley, the English author, suffered a much more severe misfortune than a few harsh words. When he was at Eton, aged 16, he contracted an eye infection that made him nearly blind. He learned Braille while in this condition but instead of going completely blind his sight slowly returned and two years later he was able to walk alone and read with a magnifying glass. His illness prevented him from studying biology, in the family tradition, and he turned instead to literature.

Alberto Moravia, the Italian author, also suffered an illness at 16. In his case it was tuberculosis, but it had a similar result. Like Huxley, he found his formal education almost impossible and spent his sanatorium years studying French and English and reading Boccaccio, Shakespeare and Molière. He also began to write during this period, and set himself firmly in the direction of his future career.

Lady Jane Grey

Life-span of 16

Lady Jane Grey, who was Queen of England for nine days in 1553, was beheaded in the Tower of London at the age of 16, on February 12, 1554. Her short life was a tragic one. Harshly treated by her parents, she took refuge in learning and spoke and wrote Greek and Latin at an early age. Sent into a miserable marriage at 15 for political reasons, she soon suffered a nervous breakdown. Despite her condition, she was named as his successor by the dying king, Edward VI. Jane fainted when she was given this news, but reluctantly took over the throne on July 10, 1553. She was deposed with great haste and sent to the Tower accused of treason, her place being taken by the rightful heir Mary Tudor.

17

Seventeen is the romantic age. Childhood is over but full adulthood has not yet arrived. In the brief limbo between the two there is time for flights of fancy and wild imaginings.

It is not a year for being serious. As Stevie Smith says: 'To become sensible about social advance at seventeen is to be lost.'

It is a year for idealism and love: 'And all things that were true and fair/Lay closely to my loving eye,/with nothing shadowy between – /I was a boy of seventeen.' (N.P. Willis, 1835). Also, a year of female charm, from Tennyson: 'Claspt hands and that petitionary grace/Of sweet seventeen subdued me as she spoke', to the lyrics of Rock 'n' Roll: 'She was only seventeen/But she caught my eye/At a teenage dance/Yeah – caught my eye/Put me in a trance.'

For some, it is a year of extreme and extravagant gestures. Dostoyevsky cites this suicide note left by a 17-year-old girl, pathetic in its affectation: 'I am undertaking a long journey. If I should not succeed, let people gather to celebrate my resurrection with a bottle of Clicquot. If I should succeed, I ask that I be interred only after I am altogether dead, since it is particularly disagreeable to awake in a coffin in the earth. It is not chic!'

Legally, 17 is the age at which motor cars and motor cycles can be driven in the UK. In the United States it is the age at which entrance to 'adult films' is permitted.

On the sports field, 17 is the 'youngest age' for a variety of achievements, including: Youngest player in an English FA Cup Final at Wembley (Paul Allen of West Ham on May 10, 1980). Youngest player in the World Cup (Norman Whiteside of Manchester United in Spain in 1982). Youngest players in a Rugby Union International (Ninian Jamieson Finlay and Charles Reid – both 17 years and 36 days when they played for Scotland against England in 1875 and 1881 respectively). Youngest player in a Rugby League Cup Final (Reg Lloyd of Keighley on May 8, 1937). Youngest winner of the Golf Open Championship (Tom Morris, Jnr, at Prestwick in 1868).

Growth

The average weight for males increases from 146 to 151 lbs during the course of this year and their height rises from 69 to 70 inches. With this year males have reached 99 per cent of their adult height, and growth for them is coming to an end. For females it has already effectively finished, with the averages standing at 125 lbs and 64 inches.

Life Expectancy

The life expectancy of a slave in ancient Rome was only 17·5 years (compared with the 58·6 years' expectancy of the privileged Roman clergy in the provinces of the Roman Empire).

Accession at 17

Henry VIII came to the throne on April 22, 1509. At the start of his reign he was very popular – quick-witted, intelligent and well educated, in addition to being a powerful athlete, a tireless hunter and an expert dancer. But his formidable mind also had a formidable temper and as time went on he proved himself to have a monstrous ego that demanded constant adulation.

Hussein of Jordan became King at this age and has ever since struggled to keep a balance between his English-educated bias towards the West and the urgings of Arab extremists.

Achievement at 17

Howard Carter, the British archaeologist who was to discover Tutankhamen's tomb – the greatest prize of the ancient world – took part in his first Egyptian dig at this early age, and set the groundwork for his later triumph.

Zsa Zsa Gabor, the Hungarian actress, won the title of Miss Hungary in a beauty contest when she was 17 and was soon on her way to Hollywood and a long career of house-keeping ('When I divorce I keep the house'). When she received valuable gifts from General Rafael Trujillo, a US Congressman referred to her as 'the most expensive courtesan since Madame de Pompadour'.

Judy Garland was just 17 when she appeared in the role that made her world famous – Dorothy in *The Wizard of Oz*. Screened in 1939, the film had originally been intended for Shirley Temple, but instead it became a triumph for Judy and her uniquely emotional singing style. Her performance of 'Somewhere Over the Rainbow' became part of screen history.

Frank Richards, the British author who invented 'Billy Bunter' and the other inmates of Greyfriars School, sold his first story when he was 17 – for five guineas. He was unbelievably prolific, writing about 60,000,000 words in his lifetime, the equivalent of about 800 novels. He once wrote 18,000 words in a single day, wearing his inevitable black skull-cap, dressing-gown, and cycle-clips to keep his legs warm. He made a fortune but lost nearly all of it at the tables in Monte Carlo.

Franz Schubert, the Austrian composer, was this age when he composed his first opera, and the first of his settings for a Goethe poem: 'Gretchen at the Spinning Wheel', for which he is credited with creating the *Lied* or 'art song'.

Jean Seberg, the small-town American actress, was chosen at 17 from thousands of girls to play St Joan for Otto Preminger. The stage-struck girl became famous overnight in a blaze of studio publicity, but the film was not a success. Her achievement was in gaining the part rather than playing it.

Natalie Wood, the American actress, was 17 when she played her most famous role in *Rebel Without a Cause* (1955). Like her co-stars in that film, James Dean and Sal Mineo, she was to suffer, in real life, an early and dramatic death.

Revelation at 17

Joseph Smith, the founder of the Mormon Church, was a farmer's boy who had

had visions as a child and who experienced a dramatic revelation when he was 17. On September 21, 1823, he claimed to have been visited by an angel who told him about *The Book of Mormon* which was engraved on gold plates and was buried on a hill called Cumorah, near Palmyra. He was instructed to dig for it and translate the engravings, both of which he managed to do, though how remains a mystery. The book contained divine revelations, supplementary to the Bible, and formed the basis of the new church which he founded.

Misfortune at 17

Lenin was 17 when his beloved older brother was hanged for conspiring to assassinate the Russian Tsar, Alexander III. As a result, his family was stigmatized for having reared a 'state criminal'. It was this traumatic loss and its aftermath that set the young Lenin firmly on the path to revolution. Thirty years later Russia was his.

Koo Stark was this age when she was persuaded to star in an erotic film called *Emily*. She could never have guessed that this harmless acting role would become a world-wide news item a few years later, when she was taken for an equally harmless holiday to the West Indian island of Mustique by Prince Andrew. The holiday had the Queen's blessing and was a reward for the Prince after taking part in the Falklands Conflict, but the puritans of Fleet Street would have none of it and the couple were hounded at every turn. Even the august BBC showed Koo stripping to the waist in *Emily* on their main newscast.

Life-span of 17

Anastasia, Grand Duchess of Russia, youngest daughter of Tsar Nicholas II, was this age when she and her entire family were murdered by the Bolsheviks. They were taken down into the cellar of the house where they were being held prisoner, shot and bayoneted, the bodies burned and finally thrown down a mine-shaft. These facts were established after the White Russian forces had taken the town where the murders took place, but despite this there were repeated rumours that Anastasia lived on in exile. Several women outside Russia have since claimed her identity, but all the evidence points to her death at 17 with the rest of her family.

Judy Garland in *The Wizard of Oz*

Eighteen is the year of becoming fully adult. Growth is at last completed for both sexes. Sexually it is an intensely active year. It is also the year when there is a strong urge to live away from the parental home for at least part of the time. It is a year for pulling up roots.

Legally, it is the most common age to gain the vote. In the UK until 1969 the minimum voting age was 21, which was then considered the year of 'adult-hood', but then the voting age was reduced to 18, bringing the UK in line with Australia, Belgium, Canada, France, Holland, Ireland, Italy, Spain, the USA and West Germany.

Eighteen is also the age at which alcoholic drinks may be bought legally in Australia, Denmark, Ireland, Norway, Sweden, the UK and West Germany. In North America the age varies from State to State and, in Canada, from Province to Province (18–21 in USA, 18–19 in Canada). In Belgium it is 18 for an unaccompanied person, 14 being sufficient for a child accompanied by an adult. In Holland, it is 18 for spirits but only 16 for weaker drinks. Only three advanced countries have no legal drinking age: France, Spain and Switzerland.

For UK film-goers, 18 is the age at which they are permitted to attend all movies, including the most 'adult' ones.

When hanging was legal in England, 18 was the youngest age at which one could be punished in this way.

Mentally, 18 is an age when males and females become very alike in attitudes and interests. Small boys and girls play with different toys and older males and females diverge in their social patterns, but at 18 there is an almost unisexual mentality. This is caused by the shared urge to break away from parental control. The rejection of parental guidance and the urge to find modes of self-expression unite the 18-year-olds, and their increased interest in being with one another rather than in same-sex groups also helps to make them more alike. Gail Sheehy in *Passages* writes: 'Both he and she are insecure, inexperienced, and as yet undistinguished by the carapace of firm social and vocational roles. Enthralled as much by what they are learning about themselves as by the other one, young lovers gladly lose their egos in each other as if in a warm whirlpool bath.'

For the human male, 18 is the peak of sexuality. It is the year when his testosterone secretion reaches its maximum daily output, after which there is a slow fall-off. This means that the 18-year-old male is not only experiencing his most powerful interest in the opposite sex, but is also capable of responding quickly and repeatedly to sexual stimulation. Many young females claim that males of this age are poor lovers because they are *too* quick, leaving their partners only partially aroused, but the young male's capacity for repeated climax should partly compensate for this shortcoming.

According to major surveys such as the Kinsey Report, the peak of sexuality for the human female comes much later, when she is in her 30s. However, this supposed sex difference has recently been challenged. It is now suggested that the 18-year-old female is also biologically at her sexiest and that what used to increase as she approached her 30s was not sexual passion but the boldness to enjoy it and to admit enjoying it.

Life Expectancy
It has been calculated that 18 was the average life expectancy at birth in the Bronze Age and Early Iron Age in Greece.

> Among animals, 18 is the greatest age ever recorded for a pet rabbit, and it is the age of the oldest horse ever to win a flat race (Marksman at Ashford, Kent, in 1826) or over jumps (Sonny Somers at Lingfield Park in 1980).

Accession at 18
Victoria became Queen of England at this age on June 20, 1837 and was crowned on the 28th. She was a tiny figure, less than five feet tall, but very dignified and composed. She ruled for the rest of the century and produced a large family which, by marriage, created a royal network across much of Europe. This complex arrangement of family ties played an important part in international affairs.

The Accession of Queen Victoria

Romance at 18
Oona O'Neill was only 18 when she married the 54-year-old Charles Chaplin. She gave him eight children and said the marriage was a lasting success because of 'the difference in age between us . . . Provided the partners are suited, such a marriage is founded on a rock.' In their case it lasted lovingly for thirty-four years, until his death at the age of 88.

Achievement at 18
Shelagh Delaney, the English playwright, saw a play by Terence Rattigan when she was 18 and, convinced that she could do

better, sat down and wrote *A Taste of Honey*. A funny-sad story of an illegitimate pregnancy, it received wide critical acclaim and won major drama awards for its teenage author.

Michel Fokine, the Russian ballet dancer, made his début with the Imperial Russian Ballet on his eighteenth birthday, and went on to an outstanding career both as a dancer and a choreographer.

Hedy Lamarr, the beautiful Austrian actress, created a sensation when she appeared naked in the Czech film *Extase*, at the age of 18. The film was made in the early 1930s when the cinema was going through a particularly puritanical phase but, despite her notoriety, she went on to become one of Hollywood's most glamorous leading ladies of the 1940s.

Françoise Sagan, the French novelist, had her first novel published when she was this age. Called *Bonjour Tristesse*, it explored with a new frankness the sexual relationships of the modern generation in post-war France, and was an immediate success.

Defection at 18

Duke Ellington defected from the fine arts to devote his life to jazz. His parents had high hopes that he would pursue a successful career in the art world and he had shown considerable artistic promise during his childhood. He was offered a scholarship to attend the Pratt Institute to study art, but he declined it and spent the rest of his long life developing one of the most original jazz talents in America.

Martina Navratilova announced her defection from her native Czechoslovakia in September 1975, on the final day of the US Open Tennis Championships, just before her nineteenth birthday.

Misfortune at 18

Frederick the Great of Prussia suffered an intensely unhappy childhood when he was a young prince. He was a studious and artistic boy who clashed repeatedly with his boorish, tyrannical, military-minded father. The periodic rows in the royal apartments were so loud that they used to draw crowds on the pavements outside. Even when he was an adolescent, his father would thrash the Crown Prince in public and force him to kiss his boots. The desperate youth decided to run away to England with a friend called Katte, but both were imprisoned by the King, accused of desertion from the army. Katte was condemned to death and the 18-year-old Frederick was forced to watch his friend's beheading. Threatened with a similar fate, he finally submitted to his father's will, and was restored as Crown Prince.

David Hume, the eighteenth-century Scottish philosopher, suffered a nervous breakdown at 18, following a period of intense intellectual discovery and voracious reading. The collapse was probably caused by the conflict between his own passionate interest in philosophy and the demands by his parents that he should follow in the family tradition by studying law, a subject which he found distasteful.

Joán Miró, the great Spanish artist, suffered a similar fate. He, too, experienced a major mental and physical breakdown when he was 18, following a period of two years working as a clerk in an office. His parents took him away to an estate in Montroig for a period of convalescence and finally allowed him to fulfil his wish to attend an art school in Barcelona.

Maurice Utrillo, the brilliant French painter, became an alcoholic at 18. The man who portrayed so evocatively the attractively dingy street-corners of the artists' quarters of Paris, was in and out of sanatoriums all his life. He took up painting as a form of therapy, encouraged by his mother, who had been Renoir's favourite model and whose sexual generosity in her youth meant that Utrillo never knew the identity of his father.

Life-span of 18

Elagabulus, the teenage Roman Emperor whose openly held, homosexual orgies scandalized Rome and whose effeminate, silkclad, face-painted public dancing outraged the military, was murdered by the praetorian guard when he was 18. His body and that of his detested mother, Soaemis, were dragged on hooks through the city of Rome and thrown into the Tiber. His outlandish behaviour as Emperor, devoting himself to the worship of his phallic god, had led to the establishment of his young cousin Alexander as a potential successor. Sensing a threat from the upright and sensible boy, Elagabulus made an attempt on his life and it was this act that finally drove the soldiers into open and murderous revolt.

Tutankhamen, the young Egyptian Pharaoh whose body now resides in the Cairo Museum, was only 18 when he died after a short reign of six years during which time he had no opportunity to show his qualities as a ruler. His main claim to fame is that his tomb, unlike those of other pharaohs, was not disturbed by tomb-robbers and was discovered intact by Howard Carter and Lord Carnarvon in 1922.

The gold funeral mask of Tutankhamen

19

Nineteen is the year of the student, for those who are pursuing a higher education; for others it is already a fully adult year, not only physically, but also socially.

Nineteen is an intensely creative year, bursting with new ideas but still lacking the experience to present them in a mature form. It is also a time of great sexual activity, either in the shape of passionate love affairs, or high-scoring lust. Sadly, it is also an age of sexual naïvety for many, as is demonstrated by the fact that this is a common year for abortions.

Although most countries allow 18-year-olds political independence, in Austria and Sweden the legal voting age is 19.

Jane Austen

Achievement at 19

Jane Austen, one of the greatest novelists ever to write in the English language, was a vicar's daughter who lived a quiet, domestic, studious, happy and uneventful life. She is the proof that it is not necessary for an artist to suffer in order to create great works. At the age of 19 she wrote her first novel, called *Eleanor and Marianne*. The year was 1795, but the book was not published until 1811 when it appeared anonymously. In the intervening years she had changed its name to *Sense and Sensibility*, and it remains today a supreme example, like all her work, of beautifully constructed story-telling and characterization.

Charles Babbage, the father of the computer, was 19 when the idea first came to him of calculating numerical tables by machinery, the vital concept that led to the building of the first calculator and eventually to the first computer. As a Cambridge student he founded the Analytical Society, also when he was 19, because his tutors were so backward and he wanted to lead an attack on 'the dotage of the University'. He had little respect for authority and it was his hatred of The Royal Society that led to the foundation of The British Association.

Josephine Baker, the American-born French dancer, was this age when she took Paris by storm in 1925, in La Revue Nègre. She symbolized all that was exotic, both on stage and in real life. On a fine day she could be seen walking her pet leopard down the Champs-Élysées.

Lorenzo Bellini, the Italian physician and anatomist, was a 19-year-old student at Pisa University when he published a description of the kidney tubules now known as Bellini's ducts. He showed for the first time that the kidney consists of an immense number of tiny canals.

Erté, the master of Art Deco design, was 19 when he left Russia for Paris and became active there as a dress and theatrical costume designer. He became famous for the ornate, fantastic costumes he invented for the Folies Bergère, the Ziegfeld Follies and George White's Scandals. He later spread his talents to include designs for furniture, perfume bottles, playing cards, posters and house interiors.

Evonne Goolagong, the first Australian Aborigine to play international tennis, was 19 when she won the Wimbledon Women's Singles title in 1971.

Colin Maclaurin, the Scottish mathematician and philosopher, became Professor of Mathematics at Marischal College, Aberdeen at this age, on September 30, 1717. Two years later he was elected a member of The Royal Society and a few years after that became Professor of Mathematics at Edinburgh University on the suggestion of Sir Isaac Newton. He is an example of the early maturation of the mathematical ability of the human brain.

John Everett Millais, the English artist, was 19 when he joined with his friends Rossetti and Holman Hunt to form the Pre-Raphaelite Brotherhood. He immediately attracted controversy because of the naturalistic way in which he portrayed religious scenes. The public, used to seeing religious figures classically idealized, was scandalized. Millais was attacked by Dickens for showing the Virgin Mary 'so horrible in her ugliness... that she would stand out as a Monster in the vilest cabaret in France'. All Millais had done was to show her looking anxious because her son had hurt his hand.

Sergei Rachmaninov, the Russian composer and pianist, was 19 when he wrote his Prelude in C Sharp Minor, which gave him world-wide fame. Despite its immense popularity, it became a work that he hated.

Mickey Rooney, the diminutive American actor, became the world's top box-office star when he was this age, starring in the Andy Hardy films of which he made fifteen. He made at least fifty films as a child, and a total, in his career, of over 200, which earned him a fortune of more than $12,000,000. Despite this he went bankrupt at one stage, largely due to huge alimony payments. On the occasion of his eighth wedding he was asked how short he was and replied, 'About seven million bucks, one million for each of my previous wives.'

Helena Rubenstein, the Polish-born cosmetics tycoon, went to Australia at the age of 19 with no plans and little money, but carrying with her twelve jars of face cream made up from an old family formula. It kept her skin smooth even in the hot dry climate and friends started clamouring for more. She wrote to her mother for the formula and was soon manufacturing it. From this early beginning, she built her immense international cosmetics empire.

Revelation at 19

Elizabeth Barton, the Maid of Kent, following an illness, began to have visions and utter prophecies in 1525 when she was just 19. Her forecasts included threats to King Henry VIII about his forthcoming divorce and she was cunningly exploited by enemies of the king. This inevitably led to her arrest and she was eventually tortured into admitting that her prophecies were false. As a result she was executed at Tyburn in London in 1534.

Conversion at 19

Richard Leakey, the son of the famous anthropologist Louis Leakey, had shown a marked reluctance to follow in his father's footsteps. He left school at 15 and set himself up in business, supplying small animals to research laboratories. Then he became a safari guide, taking tourists around his native Kenya. One day, between safaris, he was flying over some outcrops that looked ideal as fossil beds and decided to pay them a visit. Almost

A design by Erté

immediately he and his companion found a valuable *Australopithecine* jaw-bone, which caused a great stir in the world of palaeontology. Richard Leakey was 19 when this discovery was made and he was hooked – from that point on he devoted himself totally to the hunt for man's ancestors, like his brilliant father before him.

Crime at 19

Brendan Behan, the alcoholic Irish author, was in trouble for most of his short life, but especially so when he was 19. Having served two years at a reform school following his arrest in England when on an IRA sabotage mission, he was this age when he was deported to Dublin in 1942. There he was involved in an incident in which a policeman was shot and wounded, and was sentenced to fourteen years for attempted murder. He was released, however, under a general amnesty in 1946 and a few years later made the prison in which he had been kept (Mountjoy Prison in Dublin) the setting for his most famous play *The Quare Fellow*.

Life-span of 19

Joan of Arc was 19 when she was burnt at the stake, six years after hearing the voice of God instructing her to deliver France from the English. To some she was a saint, to others a deluded hysteric, but what was undeniable was that she became a great military leader – a strange and unique achievement for a teenage girl. After a long show-trial to discredit her, she was condemned as a heretic and executed in 1431. Her executioner was deeply disturbed by his task and declared afterwards that her heart would not burn and that he had found it intact amid the ashes.

Evonne Goolagong

Twenty is the age of the young bride. In the United States this is the average age for the 'first marriage' for females. In the UK a major survey revealed that two girls out of three felt that they *should* get married by this age. One out of three went further, saying that if a girl did not get married by 20, she was considered to be 'on the shelf'. However, when questioned in a different way, only one out of three said they would *like* to get married by this age. In other words, 20 is the year of the bride partly because of personal desires and partly because of social pressures. Only half the young brides walk up the aisle because they really want to; the other half do it because they fear there will be no suitable bridegrooms left for them if they delay.

For many people, 20 is the start of a period of adult 'sobering-up' after the giddy self-explorations and romances of adolescence. The soul-searching questions about 'Who am I?' become the more matter-of-fact queries about 'How do I make a start in life?' The concentration on the inner world switches to a serious examination of the complexities of the outer world. Stendhal, in *The Red and The Black* (1830): 'When one is twenty, ideas of the outside world and the effect one can have on it take precedence over everything else.'

With this new direction, the 20-year-old suddenly finds himself or herself in competition with much more experienced adults on *their* terms. The 'special licence' of the teenager has expired. A Japanese proverb sums up this shift: 'At ten years a wonder child; at fifteen a talented youth; at twenty a common man.'

If this is a daunting prospect, there is one great compensation: as a functioning machine, the human brain is at its finest at 20. This is not a modern idea; Montaigne expresses a similar thought in his *Essays* of 1580: 'As for my part, I think our minds are as full grown and perfectly joined at twenty years, as they should be, and promise as much as they can. A mind which at that age has not given some evident token or earnest of its sufficiency, shall hardly give it afterward; put it to what trial you list. Natural qualities and virtues, if they have any vigorous or beauteous thing in them, will produce and show the same within that time, or never.'

Some authors see 20 as the age when there comes the first pang of 'time slipping by', the first realization of life's ephemeral nature. Shakespeare, for instance, in *Twelfth Night*: 'Then come kiss me, Sweet-and-twenty, Youth's a stuff will not endure.' Or A.E. Housman in *The Shropshire Lad*: 'Now of my three-score years and ten,/Twenty will not come again,/And take from seventy springs a score,/It only leaves me fifty more./And since to look at things in bloom,/Fifty springs are little room,/About the woodlands I will go,/To see the cherry hung with snow.'

For parents, suddenly to have a child of 20 comes as something of a shock. John Holmes: 'A child enters your home and makes so much noise for twenty years that you can hardly stand it; then departs leaving the house so silent that you think you will go mad.'

For the mortally wounded, 20 is a terrible age to die: 'Don't cry, I need all my courage to die at 20,' were the poignant last words of Evartiste Galois, a young mathematician killed in a duel in 1832, when he saw the weeping faces of those clustered around his bleeding body.

For mothers, 20 is just about the safest age for avoiding Down's syndrome (mongolism) in a newborn baby. At 20 the chances of giving birth to a mongoloid child are 1 in 2,000. At 35 the risk will have risen to 1 in 1,000 and then up sharply to 1 in 100 at 40.

In most countries 20 is already beyond any 'minimum legal age', but there are a few exceptions: in Denmark, Japan, Norway and Switzerland, this is the legal voting age. And in Japan it is also the minimum age for buying alcohol – the highest such age in any advanced country.

The general sobriety of Japanese youth is reflected in these prison figures: only $2\frac{1}{2}$ per cent of the Japanese prison population is 20 or under. In the USA the figure is 16 per cent, in the UK $28\frac{1}{2}$ per cent, and in Eire 41 per cent.

Life Expectancy at 20
Having reached this age, males can expect to live for another 51 years and females another 57. This is a great improvement on figures for the start of the century. In 1901, males could only look forward to another 43 years, females another 45.

Accession at 20
Lorenzo de' Medici, Lorenzo the Magnificent, assumed power in Florence at this age. A benevolent tyrant, he was well-educated, cultured and athletic. The early years of his rule saw golden days of festivals, balls, carnivals, tournaments and princely receptions. He contributed more than anyone to the flowering of Florentine genius during the second half of the fifteenth century.

Romance at 20
Lady Diana Spencer became the Princess of Wales on July 29, 1981. Her wedding, just four weeks after her twentieth birthday, was the most public culmination of any romance in history, watched as it was by 700,000,000 people in fifty different countries around the world, via satellite TV.

Achievement at 20
Lauren Bacall, the American actress famous for her husky, slinky, sweet-and-sour persona in her leading roles of the 1940s and 1950s, gained her first starring part at this age, opposite Humphrey Bogart in *To Have and Have Not*. She married him the following year and remained devoted to him for the rest of his life, nursing him through his terminal cancer in 1957.
Samuel Colt, the American firearms manufacturer

and inventor, patented his famous Colt revolver in 1835, when he was only 20. Success was slow to come, but he later built the world's largest private armoury and became one of the wealthiest men of his day.

Howard Hughes, the eccentric American tycoon, was 20 when he moved to Hollywood and started ploughing some of his vast wealth into film production, with considerable success. Orphaned in his teens, he had inherited the family oil-drilling business, but with an impulsive, unpredictable nature, he was dissatisfied with a quiet, rich, administrative life, and so created more demanding challenges in the spheres of aviation and show business, before sinking into a sinister reclusive lifestyle in later years.

William Holman Hunt was 20 when he became one of the founding members of the Pre-Raphaelite Brotherhood in 1848. It was he more than any of the others who remained faithful to its original aims and principles. These were: 1. To have genuine ideas to express in art; 2. To study directly from nature; 3. To depict events as they must have happened. This was a rebellion against the false sentimentality of Landseer and other deplorable Victorian artists, and against stilted classical design and composition.

Dante Gabriel Rossetti also became a founding member of the Pre-Raphaelite Brotherhood at this age, along with Hunt and Millais, but when they came under attack a few years later, he retreated into a dreamlike medieval world of myth and legend, abandoning the strict Brotherhood rules.

Jane Russell, the American actress, was 20 when she made her film début in *The Outlaw*, produced by Howard Hughes, and given its first showing in 1941. The film became notorious because of the provocative display of her 38-inch bust and its unusually visible cleavage. Hughes had conducted a nationwide chest-hunt for the star of his movie, a film which today looks almost puritanical, but which was considered nearly pornographic in its day. It made Jane Russell the victim of big-bust jokes for many years and almost ruined the career on which it had launched her.

Mary Shelley, who, in her teens, had become the poet Shelley's second wife, was only 20 when she published her famous novel *Frankenstein* in 1818. A philosopher's daughter, she wrote the gothic tale at Byron's suggestion, but could never have guessed that the story of the man-made monster would become so lastingly popular.

Crime at 20
Salvador Dali, the bizarre Spanish genius, was imprisoned at this age at Figueras and Gerona for political activity against the Spanish government. In the same year, 1924, he had already been suspended from the Madrid School of Fine Arts for twelve months, for inciting the students to insurrection against the school authorities. After his imprisonment he returned to the art school, but was finally expelled permanently for extravagant behaviour in October 1926.

Life-span of 20
Catherine Howard – fair, slim and merry – the fifth and most beautiful of Henry VIII's wives, was the first one with whom he felt in love. This did not save her from an early death, however. He had forgiven her her earlier, premarital affairs, but when he suspected that she was continuing them as his queen, he arranged for parliament to pass a bill of attainder declaring it to be treason for an unchaste woman to marry the king. Two days later he had her beheaded at the Tower of London, in the year 1542.

Charles Keefer, a US Army private, chose a strange way to die at this young age. In the summer of 1982, he took an M-60 US fighter tank, crashed through the perimeter fence of his base in West Germany, drove it at 30 mph through the shopping centre of nearby Mannheim, scattering shoppers, wrecking a tram, fourteen cars, three lamp-posts, power lines and pavements, and then raced on out of the town and on to a bridge over the River Neckar. There he reversed the huge tank over the edge of the bridge and died as it sank in the river below – suicide by drowning on a spectacular scale.

left to right, Jane Russell, Rossetti's Proserpine, Lauren Bacall

Twenty-one was the traditional age of majority in England for many years. Young people celebrated their arrival at what was considered to be 'full adulthood' with special twenty-first birthday parties. The old song: 'Twenty-one today, twenty-one today, never been twenty-one before, now I've got the key of the door' represented the new independence and the final, formal loss of parental control. The front-door key was the symbol of the occasion, with key-shaped cards, cakes and flower arrangements. Speeches were made, toasts were proposed, and there was a strong folklore tradition of 21 being a time of social 'coming of age'.

This 'rite de passage' is still celebrated by some people, but it now lacks much of its earlier impact because of a change in the law, reducing the legal age of majority from 21 to 18. This shift was a result of the Family Law Reform Act of 1969, which came into force on January 1, 1970. From that date, the authority to vote, marry without parental consent, make a will, sign a contract, sue, and perform other such adult actions arrived three years earlier in life. As a result, the great threshold of the twenty-first birthday became something of an anti-climax.

The law being a notoriously untidy institution, however, it is still possible to find some strange remnants of the old 21 age limit. For instance, it is this age, not 18, that is required if you are to supervise a young person 'carrying an air-gun in a public place', or having 'an assembled shot-gun in his possession'. Homosexual acts become legal between consenting males at this age. In the USA there are still twenty-five States that require a minimum age of 21 for the purchase of alcoholic beverages, including California, Illinois and Pennsylvania. And some countries, Puerto Rico for example, still retain the old 21-year minimum for marriage without parental consent. Ecclesiastical Law states that the obligation for fasting begins at the age of 21. But in general 21 is now looked upon as being slightly past the age of assuming responsibility.

For poets and thinkers, 21 seems a proud, strong year brimming with energy and self-assurance. Ben Jonson: 'Twenty-one do I feel/The powers of one-and-twenty, like a tide,/Flow in upon me'; Samuel Johnson: 'Towering in the confidence of twenty-one'; Thackeray: 'Lightly I vaulted up four pairs of stairs/ In the brave days when I was twenty-one'; Pope: 'Long as the year's dull cycle seems to run/When the brisk minor pants for twenty-one.'

Accession at 21 Mohammed Reza Pahlavi was this age in 1941 when he became the Shah of Persia – later the Shah of Iran. Despite his aristocratic pretensions and dignified bearing, he was in reality the son of a donkey-driver. His father had risen to power in the 1920s via a military coup, proclaiming himself the new Shah. When he was forced to abdicate by Britain and Russia, during the Second World War, his son was allowed to replace him because he was more sympathetic towards the Allies. After the war he did a great deal to modernize Iran and transform it into one of the advanced countries, but power went to his head and he became an absolute dictator, employing torture and assassination experts, until he was overthrown by the Ayatollah Khomeini.

Brigitte Bardot

Achievement at 21 Brigitte Bardot, the sex kitten of the 1950s and 1960s, achieved world-wide fame at this age in the film *And God Created Woman*. It was the first film in which she was directed by her husband Roger Vadim, and established her image as a child of nature who was driven by the need for both love and sex. Her flat-faced, almost featureless, beauty retained a childlike quality all through her career, but it made it difficult for her to age gracefully and she wisely retired early, her last film appearing before she was 40.

William (Buffalo Bill) Cody was only 21 when he was put in sole charge of feeding 1,200 railway workers who were building the railroad across America to the West. To carry out this task efficiently he had to slaughter thousands of buffalo and it is claimed that he was such a crack shot that he never missed one of his targets. In eight months he slaughtered

4,280 of these huge animals, which later came to the brink of extinction.

The Marquess of Queensberry, an eccentric Scottish peer, was this age when he introduced his famous 'gentlemanly rules' to limit and control the violence inherent in the sport of boxing. This has resulted in his name being permanently associated with the sport. The Marquess of Queensberry Rules, published in 1865, included the instructions that 1. Gloves must be worn; 2. Rounds must be 3 minutes' duration; 3. There must be 1 minute's rest between rounds; 4. No wrestling allowed.

Yves Saint-Laurent, the French fashion designer, was only 21 when, following Dior's early death in 1957, he was named as the new head of the great House of Dior, a $20,000,000 a year enterprise. A protégé of the firm since he was 17, he immediately set about introducing new lines of his own, starting with the 'little girl look' and the 'A-line'.

Conversion at 21
Martin Luther, a miner's son who became a great religious reformer, was originally intended for the law and had obtained a brilliant degree at Erfurt University. Then, much to his father's distress, he renounced the professional world and decided to enter a monastery. This decision arose from an incident on July 17, 1505 when he was overtaken by a terrible thunderstorm and cried out, 'Help, St Anne, and I'll become a monk.' Luther says in his writing, 'Not freely or desirously did I become a monk, but walled around with terror and agony of sudden death, I vowed a constrained and necessary vow.'

Misfortune at 21
Douglas Bader, the British air ace of the Second World War, lost both his legs in a flying accident in December 1931.

Robert Graves, the English author, was severely wounded in 1916 when he was serving at the front in the First World War. The horror of trench warfare was a crucial experience in his life, leaving him deeply troubled for at least a decade. Subsequently his writing became a form of emotional therapy for him, in a culture which he considered to be sick.

Charles Lamb, the English essayist and critic, was 21 when his sister Mary killed their mother in a fit of madness. He reacted with courage and loyalty, taking on the task of looking after his now recurrently insane sister for the rest of his life. The burden eventually drove him to drink and he died from a drunken fall.

Life-span of 21
William (Billy the Kid) Bonney, the American outlaw, was shot down in the dark by Sheriff Pat Garrett in 1881, at Fort Sumner, New Mexico. A smiling, good-natured, but extremely vicious criminal, he was reputed to have killed twenty-one men, one for each year of his short life.

Eddie Cochran, a budding Rock 'n' Roll star being groomed to rival Elvis Presley, was killed in a car

Eddie Cochran   Yves Saint-Laurent

crash on the way to London Airport at the end of a British tour with Gene Vincent in 1960. His taxi, which was travelling too fast, struck a kerb and spun backwards into a concrete lamp-post, throwing him out on to the road. His hit record at the time of the accident was *Three Steps to Heaven*.

Lord Darnley, consort of Mary, Queen of Scots and father of James I of England, became an embarrassment to his wife as his cultured charm gave way to drunkenness, arrogance, indolence and obsessive jealousy, until he was conveniently blown up with gunpowder at the house in Edinburgh where he was staying. The suspected murderer, the Earl of Bothwell, was acquitted, and married Mary three months later.

Duncan Edwards, the captain of the Manchester United football team, was only 21 when he was killed along with seven of his team-mates in the air disaster at Munich Airport on February 6, 1958. The team had been to Belgrade for a European Cup match and were returning to England when their plane crashed on take-off. Edwards is commemorated in two stained-glass windows at St Francis's Church, Dudley, near his home. On the day of his funeral there were at least 5,000 people outside the church to pay homage to the young man who was clearly destined to have become England's captain, had he survived.

Pochahontas, the American Indian princess who was friendly to the English colonists, died of smallpox at 21. She was only a young girl when she saved the life of a captured English officer who was about to have his head beaten to pulp on her tribe's sacrificial stone. She placed her head on top of his and pleaded with the tribal chief, her father, to spare him. Later she was taken as a hostage, to be used as a negotiating weapon. Treated well, she became converted to Christianity and, although her father paid the ransom for her return, she fell in love with an Englishman, married him and together they went to London where she was presented at court as a princess and a heroine. After seven months in England, she set sail for her homeland, but died just off the coast, never seeing her tribe again.

Twenty-two is the age of fecundity. It is the year when the pregnant mother has the smallest chance of losing her unborn child. Foetus death is about 12 in 1,000, compared with 47 in 1,000 when the mother is 45.

For careers, this is the start of an important period, stretching from 22 to 28, when we concern ourselves with what we *ought* to do with our lives. The previous years have been focussed more on the problem of pulling up roots and gaining independence. Now that newly won independence has to be used and channelled – the future course of life has to be fixed. This is the exhilarating challenge for the ambitious 22-year-old.

Byron records the excitement of being 22 with these words: 'Oh, talk not to me of a name great in story;/The days of our youth are the days of our glory;/And the myrtle and ivy of sweet two-and-twenty/Are worth all your laurels, though ever so plenty.'

For some, who have enjoyed success too soon and too easily, it can, by contrast, be a depressing age with a long, seemingly empty future stretching ahead. Punk-rock star Johnny Rotten had this to say about it: 'I'm only twenty-two and I feel I've seen everything. It makes it very difficult sometimes.' Happily he is the exception rather than the rule.

Life Expectancy The inhabitants of the Neolithic settlement of Khirokitia in Cyprus some 7,000–8,000 years ago had a life expectancy at birth of 22·1 years. Other studies in Rome and in Britain reveal that 22 was also the life expectancy there about 2,000 years ago.

Alfred the Great

Accession at 22 **Alfred the Great** came to the throne at this age, becoming the Saxon King of Wessex in 871. A scholarly man who had never wished for power, once he became King he ruled with great skill, fairness and foresight, excelling as a warrior, an administrator and an educator. He saved England from the Danes and made his kingdom secure by building a series of defensive forts and warships.

Achievement at 22 **Svetlana Beriosova**, the Lithuanian-born dancer became Prima Ballerina of the Sadler's Wells (now Royal) Ballet in 1955 and danced the entire classical repertoire, her interpretation of Giselle ranking among the greatest of all time.

Charles Best, the Canadian physiologist, was the co-discoverer (with Frederick Banting) of insulin in 1921, thereby giving hope to thousands of diabetes sufferers.

Nellie Bly, American ace reporter, sailed from New York on November 14, 1889 to beat the record of Jules Verne's Phineas Fogg journey *Around the World in Eighty Days*. She did it in 72 days, 6 hours, 11 minutes and 14 seconds, travelling by ship, train, sampan, horse and donkey. The last lap, from San Francisco to New York, was by special train which was greeted with brass bands and fireworks.

Ralph Boston, the black American track athlete, was this age when he became the first long-jumper to achieve a leap of over 27 feet – 27 feet 1¾ inches to be precise. He did this in 1961 during a US/Soviet duel in Moscow and managed to extend it by a further three inches in 1963.

Pierre Boulez, the French composer and conductor, became the musical director of a theatrical company, the Marigny Company of Jean-Louis Barrault and Madeleine Renaud, at this early age.

Clara Bow, the Hollywood actress, became the 'It Girl' at 22 in 1927 when she was chosen by Elinor Glyn to play the title role in the film of her novel *It*. It made her the reigning sex goddess of the silent cinema with 40,000 fan letters per week. On screen she typified the emancipated flapper, with bobbed hair, cupid-bow mouth, huge eyes and dimpled knees. Off screen she was equally emancipated and gained notoriety for her sexual generosity. She made love to many men, sometimes more than one at a time. Her conquests included Gilbert Roland, Gary Cooper, John Gilbert, Eddie Cantor, Bela Lugosi and the entire USC football team in after-game parties where nude football was played at her Beverly Hills mansion.

El Cid, the Spanish military leader and legendary hero, was appointed Commander of the Royal Troops and Standard Bearer to Sancho II. His early promotion to such an important post underlines his exceptional military talent, but he was also a wily and iron-willed politician who was ruthlessly self-serving to such a degree that history had to be rewritten to convert him into an acceptably heroic figure.

Cassius Clay, later to become Muhammad Ali, became the World Heavyweight Boxing Champion for the first time at this age, when he defeated the reigning champion Sonny Liston, in 1964. He won the title again in 1974, beating George Foreman, and a third time in 1978 when his victim was Leon Spinks – making him the greatest heavyweight in the history of boxing.

Charles Darwin was 22 when he set off for his great voyage of discovery on *The Beagle*. Although gaining the position of naturalist on the trip did not seem like a major achievement at the time, it turned out to be the most important step he took in his entire life. By the time he returned, five years later, he had already developed the concept of evolution, which was to make such a dramatic impact on our way of thinking about human existence.

Walt Disney made his very first cartoon film when he was 22, in 1923.

Gerald Durrell, the natural history author, led his first animal collecting expedition to the British Cameroons, in 1947. This and subsequent trips all over the world provided material for a series of best sellers such as *The Overloaded Ark*, *The Bafut Beagles* and *My Family and Other Animals*.

Helen Keller, the blind and deaf American author and lecturer, was 22 when she published her autobiography *The Story of My Life*, in 1902.

Sylvia Kristel, the Dutch actress, gained international fame and some notoriety for her title role as *Emmanuelle* in the first soft-porn film to avoid the ugly banality of the typical product of this genre.

Oswald Mosley, the founder of the British Fascist Party, became the youngest MP in the House when he was elected as Conservative Member for Harrow in 1918. As an eager 22-year-old Member of Parliament, he took the 'liberal' or progressive side in almost every issue and saw himself as the champion of the young against the old. He soon abandoned the Conservatives to become a Labour MP and then finally switched his sympathies from Socialism to National Socialism, appearing on the streets as a British Hitler, inciting violence in the 1930s and ending in detention during the Second World War.

Isaac Newton, the English physicist, developed the Binomial Theorem and Differential Calculus in 1665, thanks to the plague which closed down Cambridge

Nureyev with Fonteyn, 1962

University and sent him home, where he had a chance to think.

Mark Spitz, the American champion swimmer, was 22 when he won the astonishing total of seven gold medals at the 1972 Munich Olympics. Nobody in the history of the Games has ever won so many golds at a single Olympics. His events were the 100 metres freestyle and butterfly; the 200 metres freestyle and butterfly; the 4 × 100 metres freestyle relay and medley relay; and the 4 × 200 metres freestyle relay.

Defection at 22

Rudolph Nureyev was a rising star of the Russian Kirov Ballet Company when he defected to the West during a visit to Paris in 1961. His performances with the much older but still brilliant Margot Fonteyn later became world famous.

Life-span of 22

Lillian Board, the British athlete, was nominated Sportswoman of the Year in the last month of her short life. Two years earlier she had won an Olympic silver medal for the 400 metres in Mexico, but then contracted cancer. She continued running in great pain, as late as June 1970, only six months before she died.

Terry Fox, the Canadian marathon runner, met a similar fate at the same age. After losing a leg to cancer, he valiantly continued to compete in long-distance running events, on an artificial limb, before he finally succumbed at 22, in 1981.

Buddy Holly, one of the pioneer figures of Rock 'n' Roll in the late 1950s, died in an air disaster on February 2, 1959. Second only in popularity to Elvis Presley at the time, he had been giving a concert at Clear Lake on February 1, after which he boarded a light aircraft at Mason City Airport with other Rock 'n' Roll performers. The plane crashed only minutes after take-off, killing all on board.

Oswald Mosley

23

Twenty-three is the age of the American bridegroom. In the United States this is the average age for the 'first marriage' for males. This is three years younger than the typical marrying age for American men at the turn of the century.

It is also, surprisingly, the year of the voyeur. The Indiana University Institute for Sex Research founded by Alfred Kinsey reports that the average age for a Peeping Tom is 23·8 years. This seems unusually young in relation to the popular image of the 'dirty old man', but the American researchers explain: 'Peeping is chiefly a young man's avocation, calling for some agility and fleetness of foot. A middle-aged man may have strong voyeuristic desires, but is less inclined to scale fences or run foot races with irate husbands and neighbours.'

To the poets, 23 is already an age to be associated with the passing of youth. Milton, in his sonnet 'On his being arrived at the age of Twenty-three', wrote 'How soon hath time, the subtle thief of youth,/Stol'n on his wing my three-and-twentieth year.' Housman: 'May will be fine next year as like as not;/Oh aye, but then we shall be twenty-four.' And a Dutch proverb sees this age as already being past the wild period of high risk-taking: 'He who is not dead at twenty-three ... will live long.'

Achievement at 23

Akutagawa, the Japanese author, wrote a short story called *Rashomon* in 1915 when he was this age. Through its publication he met Natsume Soseki, the outstanding novelist of his day, who encouraged him to write a series of stories based on twelfth- and thirteenth-century Japanese tales. Two of his tales, *Rashomon* and *In a Grove*, were combined in 1951 as the basis of a film by Kurosawa. It was this film that introduced Japanese cinema to the world and turned Akutagawa's youthful writing achievement into international fame. Sadly he was not able to enjoy this belated success, having killed himself in a fit of depression while still a young man, in 1927.

Bo Derek, the American actress, shot to sudden fame at this age. An unwritten law of Hollywood states that in each decade one American actress of outstanding mammalian proportions shall accept the mantle of Sex Symbol of the Western World. In the 1940s it was a pert and lively Betty Grable; in the 1950s funny and vulnerable Marilyn Monroe; in the 1960s glistening and voluptuous Raquel Welch; then came a hiatus, a vacancy on the throne of the sex goddess, until a Dudley Moore comedy called *10* introduced the next challenger, the shapely and explicit Bo Derek.

F. Scott Fitzgerald, the American novelist, became famous at 23 through the publication of his first novel *This Side of Paradise*. Considered scandalous at the time, it became a best seller and made him rich, although it was far from being his best work.

Werner Karl Heisenberg, the German atomic physicist, was 23 when he discovered a way to formulate quantum mechanics in terms of matrices, for which he was later awarded the Nobel Prize. Although he did not openly oppose the Nazis during the 1930s and early 1940s, he was privately hostile to their policies and acted in such a way that he prevented Germany from developing effective nuclear weapons.

Flora Macdonald, the Scottish Jacobite heroine, was 23 when she helped Charles Edward, the Young Pretender – 'Bonnie Prince Charlie' – to escape from Scotland after his defeat at Culloden in 1746. Pursued by the English, he took refuge in the Hebrides, where Flora was visiting some friends. She allowed him to join her party disguised as a woman – 'an Irish spinning maid, Betty Burke' – and sailed with him to Skye on the pretence of visiting her mother. On parting, the Prince gave her a portrait of himself in a gold locket. Unfortunately for her, the boatmen revealed the truth when they returned to the Hebrides and Flora was arrested and imprisoned in the Tower of London. She was later pardoned and has since been immortalized in Jacobite ballads and legends.

Isaac Newton, the English physicist, developed the Theory of Colours, Integral Calculus, and the Universal Law of Gravitation, all in the year 1666.

Jesse Owens

Jesse Owens, the black American athlete, once described as a 'physical genius', broke four world records in forty-five minutes on May 25, 1935: the 100 yards, the 220 yards, the 220 yards hurdles and the long jump, at the age of 23. The following year at the Berlin Olympics, he broke or equalled twelve Olympic records and won four gold medals, much to the fury of Adolf Hitler.

Saint Francis of Assisi

Karlheinz Stockhausen, the German avant-garde composer and pioneer of electronic music in the 1950s, was this age when he presented *Kreuzspiel* in 1951, his first clean and complete break with the music traditions of the past.

Mikhail Tal of the Soviet Union became the youngest World Chess Champion in 1960 at the age of 23 years and 120 days.

Orson Welles created panic in the United States on October 30, 1938, when he broadcast a realistic report of an invasion from Outer Space. A few minutes after 8 pm a sombre voice announced that Martians had landed and were sweeping across the USA in a series of bloody battles. Although it was in reality a presentation of H. G. Wells's *War of the Worlds*, its newscast style, which included eye-witness interviews and an impersonation of a Presidential message, caused a stampede. New Jersey was said to be the earliest invasion site and cars were soon jamming all roads leading from this area as the inhabitants fled in terror, many of them with wet towels wrapped around their heads as a protection against gas, as instructed by the broadcasters. Restaurants emptied and the US Navy recalled sailors to ships in New York harbour. In the South, women were praying in the streets. The switchboards of papers and radio stations were overwhelmed and there was one attempted suicide. CBS faced a total of $750,000 in lawsuits, but all claims were eventually withdrawn and the ratings of the 23-year-old Welles's radio drama series *Mercury Theatre* soared.

Revelation at 23

René Descartes, the French mathematician who has been called the 'father of modern philosophy', had a vision on November 10, 1619, of a Unitary Universal Science that would link all human knowledge together in an all-embracing wisdom. This became his vocation in life.

Conversion at 23

Saint Francis of Assisi, the Italian saint who became famous for the intense devotion with which he imitated the life of Christ, for his love of nature and his acceptance of poverty, was very different as a young man. In his youth he had an exuberant love life and his worldliness made him the natural leader of the young men of the town. He fought in a war and was taken prisoner. Late in 1205 he experienced a powerful dreamlike state. During this trance, probably caused by fever, he received an instruction to return to Assisi and await a call to a new kind of knighthood. He did so and gave himself up to prayer. Renouncing material goods, he then devoted the rest of his life to the care of the helpless, and became the founder of the Franciscan Order.

Life-span of 23

Dalai Lama Tshangs-Dbyangs-Rgya-Mtsho, the sixth reincarnation of the original Dalai Lama, ruler of Tibet, was not well suited to his sacred position, being a libertine and a composer of romantic verse. The Mongols deposed him and he died en route for China under military escort.

Edmund Ironside, King of England in 1016, was the son of Ethelred the Unready. He was the leader of the struggle against the Danish invaders under Canute and, after a number of battles, divided the country with his enemy – Canute taking the north and he the south. When he died at the early age of 23 it was a simple matter for Canute to take over as ruler of the whole kingdom, and some authorities believe that Edmund was murdered by Canute.

Henri Gaudier-Brzeska, the French sculptor who lived in London and was a member of the avant-garde Vorticist art movement, was promising to become one of the great artists of the twentieth century, but was killed in action in 1915 after joining the French Army to defend his country in the First World War.

Orson Welles

24

Twenty-four is the year of the British bridegroom. The average age for the first marriage for males in the UK is 24·9 years. This is a drop of three years since the turn of the century, but it is still slightly later than that of the typical American male.

It is also the year when those who have experienced a lively adolescence begin to feel their age, seeing another batch of exuberant youngsters springing up beneath them with threatening new fads and fashions of their own. But Oliver Wendell Holmes comments wryly: 'You make me chuckle when you say that you are no longer young, that you have turned 24. A man is or may be young to after 60, and not old before 80.'

Achievement at 24 René Clair, the French film director, wrote and directed his first film *Paris qui Dort*, in 1923.

William Coolidge, the New York born British historian, priest and mountaineer, accomplished the first winter ascent of the 4,158-metre *Jungfrau* in Switzerland, in 1874, one of his 1,750 mountain ascents during a lifetime of systematic exploration of the Swiss, French and Italian Alps.

Jack Dempsey, the American World Heavyweight Champion boxer known as the Manassa Mauler because of his extreme ferocity in the ring, won his title at this age on July 4, 1919 when he defeated Jess Willard in four rounds. He drew record-breaking crowds to his boxing contests and lost only five of his sixty-nine bouts, later retiring to become a successful *restaurateur* in New York. He has been immortalized by having an animal named after him – the 'Jack Dempsey Fish' being a species of Cichlid that is as fearlessly aggressive as its namesake.

Alexandre Dumas, known as Dumas *fils*, the son of the famous writer of the same name who wrote immensely successful romantic historical novels, was himself something of a rebel against his father's lifestyle. The older man had been a libertine and a prodigious lover of many women, while his son was a high-toned moralist. He is best remembered today for his story *La Dame aux camélias*, published as a book when he was 24. It was later turned into a play, *Camille*, and an opera, *La Traviata*, providing a vehicle for many great female performers, from Sarah Bernhardt to Garbo, who revelled in the role of the dying courtesan.

David Garrick, the English actor/producer, was this age when he played Richard III on the London stage with such brillance that the Garrick legend was born overnight. He had not dared tell his parents that he had entered a profession held in such low esteem (in 1741), until his triumph as Richard gave him the courage to break the news to his father and start him off on one of the great theatrical careers of all time.

Audrey Hepburn, the skinny, radiant, gazelle-like actress, won an Oscar for her performance as a young princess in *Roman Holiday*, in 1953. She was well-suited to the part, being in real life the daughter of a baroness.

Christopher Isherwood, the 'autobiographical novelist', an outspoken homosexual, a religious (Hindu) pacifist and an anti-patriot, took his most important step in 1928, when he was 24, by setting up home in Berlin. He lived there for five years absorbing the decadent atmosphere and gaining the material for his best writings, especially *I am a Camera*, which was later turned into the internationally acclaimed film musical *Cabaret*.

The Marquis de Lafayette, the French nobleman who became the 'hero of two worlds', fighting with the Americans in their struggle for independence against the British, was only 24 when he participated in the climactic defeat of General Cornwallis at Yorktown, on October 19, 1781. His inspired leadership in the field of battle and his ability to rouse the French to send military aid were instrumental in bringing the American Revolution to a successful conclusion – a remarkable feat for one so young.

Jack London, the American author, grew up in poverty, became a wandering hobo and was arrested for vagrancy in New York, before joining in the Klondike gold-rush. Ill health forced him to abandon this and he settled down to become a writer, having by now accumulated a rich store of personal experiences. His first book appeared in 1900 when he was 24. Called *The Son of the Wolf*, it was the first of fifty which he completed in the next seventeen years.

Friedrich Nietzsche, the German philosopher, showed such promise as a student that he was appointed to a professorship at Basle University in 1869, at 24, the appointment being made even before he had graduated.

Isaac Pitman, the inventor of shorthand, was this age when his book *Stenographic Sound* was published in 1837, introducing the concept of shorthand based on sounds.

William Pitt the Younger became the youngest ever Prime Minister of England at the age of 24 years

Audrey Hepburn in *Roman Holiday*

and 205 days on December 19, 1783. A delicate child who was educated at home, he grew into a shy, lonely bachelor who made little personal impact on his colleagues. Political know-alls gave his leadership two to three weeks before collapse, but it lasted for seventeen years, his political wisdom more than compensating for his personal deficiencies.

John Travolta, the American actor, became an overnight sensation at this age and started the international craze for disco-dancing, thanks to his starring role in the film *Saturday Night Fever*.

Misfortune at 24

Bob Dylan, one of the most original talents in modern pop music, who came to stand for the thoughtful rebelliousness of the 1960s, suddenly disappeared in 1966 and vanished from the public scene for four years. The official explanation was that he had been almost killed in a motorcycle accident and required the time to recover, but this has always been doubted. The true reason remains a mystery. When he reappeared he was a changed man, his voice having altered its tonality and his lyrics having become sentimental and even religious.

Götz of the Iron Hand, the Imperial German Knight, had his hand shot away while taking part in a siege in 1504 and had an iron one made as a substitute. With this he managed to survive to a ripe old age in a violent and turbulent era and was romanticized in legend as the German Robin Hood.

Crime at 24

David 'Son of Sam' Berkowitz was this age when he terrorized New York in 1976-7. An apparently random, psychotic sex-killer, he attacked thirteen girls, six of whom died, before he was caught and confined to an asylum for the criminally insane. He acquired his nickname when he wrote to the *New York Post* signing himself 'Son of Sam'. Sam, it turned out, was a neighbour through whose dog he received his orders to kill. By the summer of 1977, the FBI had been called in and the 5,000 New York members of the Mafia were also searching for him, but he was finally trapped when he left his car badly parked while killing a girl, and was traced through his parking ticket.

Charlotte Corday was also this age when she assassinated the Jacobin leader Jean Paul Marat. Horrified by the excesses of the Reign of Terror, the quiet, convent-educated girl decided to rid France of his influence and sought an interview with him. He granted her this while he was taking a bath, which made it easy for her to draw a concealed knife from under her skirt and stab him through the heart. She went bravely to the guillotine four days later.

Leonardo da Vinci was 24 when he was arrested for committing sodomy with a beautiful young male prostitute called Jacopo Saltarelli. He and three other Florentines had been accused of the crime anonymously and he was imprisoned for two months while he was subjected to a thorough interrogation. He was

finally released without a trial, the charges having been dropped, so the incident remains a mystery, although it has been pointed out that there is no trace of a woman anywhere in his life, before or after the arrest.

Myra Hindley, the sullen-faced blonde typist who was a willing aide to the savagely sadistic Ian Brady in the 'Moors Murders', was tried and convicted when she was 24. She allowed herself to be photographed, using delayed action, copulating with Brady before assisting him with his brutal child tortures and killings. Sentenced to life imprisonment, she is said to have reformed, while Brady has 'deteriorated'.

David Garrick as Richard III

Lee Harvey Oswald, the presumed assassin of President John Kennedy, was 24 when he supposedly fired the fatal shots from the sixth-floor window of the Texas School Book Company in Dallas on November 22, 1963. Oswald was cornered and arrested in a nearby cinema about an hour after the assassination. Two days of police questioning followed before he was removed from the police building to be taken to jail. As he left under guard, Jack Ruby stepped forward and shot him at close range, in full view of television cameras.

Life-span of 24

Clyde Barrow, the Clyde of Bonnie and Clyde, was a juvenile thief who met the 19-year-old waitress Bonnie Parker in 1930. Two years later they set off on a robbing and killing spree which ended on May 23, 1934 when they were trapped in their car and lawmen fired a total of 187 bullets into their bodies.

James Dean, the restless, tormented star of *Rebel Without a Cause*, died in a high-speed crash when driving his Porsche sports car to Salinas to compete in a racing event. Although he took the lead in only three films, the others being *East of Eden* and *Giant*, he achieved greater international fame than most Hollywood performers can hope for in a lifetime of acting. After his death he became a cult hero with a massive following of fanatics who saw each of his films many times over, as if attending religious services.

25

Twenty-five is the age of physical maturity, a condition which arrives much later than sexual maturity. At 25 muscular strength reaches its peak, after which it shows a slow decline.

This is also the year when a final, almost imperceptible growth process is completed – the fusion and strengthening of the bones. The end of body development gives some slight justification for Dante's definition of 25 as 'The end of childhood', although to most people this seems ridiculously late.

More common among poets and authors is the suggestion that 25 is somehow the ideal age, the age we all want to be, both when we are younger and when we are older: Thomas Wolfe: 'A young man of twenty-five is the Lord of Life. The very age itself is, for him, the symbol of his mastery . . . like an ignorant fighter, for he has never been beaten, he is exultant in the assurance of his knowledge and his power.'

Some see it as an age by which time creativity must have shown itself. Wordsworth: 'A poet who has not produced a good poem before he is twenty-five we may conclude cannot and never will do so.' Others see it as an age when it becomes too late for children to relate to their parents. Howe: 'After a man's children reach twenty-five, he has none. No parent was ever very comfortable with a child after it had reached twenty-five.' And George Orwell makes an amusing observation concerning the symbolic figures that appear in comic cartoon drawings, especially saucy seaside postcards: 'Sex appeal vanishes at about the age of twenty-five. Well-preserved and good-looking people beyond their first youth are never represented. The amorous honeymooning couple reappear as the grim-visaged wife and the shapeless, moustachioed, red-nosed husband, no intermediate stage being allowed for.'

Tennyson begrudgingly admits that 25 is a peak year, in this negative statement: 'Ah, what shall I be at fifty/Should nature keep me alive,/If I find the world so bitter/When I am but twenty-five?' If it seems strange to consider this as a 'bitter' age, then it is worth looking at the suicide figures. Twenty-five is the start of the peak period, not only for physical strength and energy, but also for intense mental disturbance in females. Insincere suicides (attempts that are not meant to kill), called 'para-suicides', are at their most common among girls of this age. These phoney suicide bids are really cries for help, tricks to create concern, and they reveal that, at the very moment when the successful are at their most boisterous, the less successful feel their shortcomings most acutely, and are driven to desperate measures. It is, therefore, in addition to everything else, an age of contrasts.

Life Expectancy
The life expectancy at birth for males living in Gabon is only 25 – the lowest figure for any country in the world today.

> Among the animals, 25 is the longest life-span recorded for any species of antelope.

Accession at 25
Elizabeth I of England became Queen at 25 on November 17, 1558. An erect figure with a pale, aquiline face and red hair, with a great interest in elaborate costumes and jewels, she was a popular arrival on the throne whose strong personality soon gave her the same sort of absolute authority enjoyed by her father, Henry VIII, before her.

Elizabeth II of England also came to the throne at this age on February 6, 1953, although she was not actually crowned until she was 26, on June 2. She first heard the news that she was Queen when staying at the Treetops Safari Lodge in Kenya and flew home immediately to attend her father's funeral and assume the monarch's role.

Romance at 25
Madame du Barry took the fancy of King Louis XV at this age, but could not become the official royal mistress because she was not married to a French nobleman. The King solved this problem by arranging a nominal marriage between her and his brother, after which she lived on the estates given to her by the French monarch and became a great patron of the arts. Despite her grand lifestyle, she was of humble origins, having been a Paris shop assistant after growing up as the illegitimate child of a dressmaker. It was her great beauty alone that elevated her, via several rich lovers, to the King's bed.

Achievement at 25
Roger Bannister, the British athlete, ran the first four-minute mile, at Oxford on May 6, 1954, when he was 25.

Theda Bara, the great screen vamp of the silent cinema, rose to fame when she appeared in *A Fool There Was* in 1915. An overnight sensation, she was billed as a woman of mystic powers, born in the Sahara desert as the love-child of a wild French artist and an Egyptian girl. In Hollywood she lived a lifestyle as exotic as those depicted in her films, wearing indigo make-up, surrounding herself with death symbols such as skulls and ravens and being served by Nubian slaves in rooms filled with incense. In reality, she was a tailor's daughter from Cincinnati and her screen name was nothing more than an anagram of 'Arab Death'.

Mrs Beeton, whose *Book of Household Management* became an immediate and lasting success, was educated in Heidelberg and trained as a pianist before marrying a publisher and starting to write cookery articles for him in one of his magazines. These articles were so well received that it was decided to convert them into a major book covering all aspects of cooking and housekeeping. The finished work, running to 1,112 pages, is always thought of as representing the distilled wisdom of a mature, elderly Victorian lady and it comes as a shock to discover that it was instead the work of a girl of 25.

Hilaire Belloc launched his literary career at this age with his 1896 volume called *The Bad Child's Book of Beasts*.

Lawrence Bragg, the English scientist, was the youngest ever winner of a Nobel Prize. He won it in 1915 for his work on the X-ray analysis of crystal structure. It was awarded to him jointly with his father for work they had done together two years earlier. In Bragg's own words: 'It was a glorious time, when we worked far into every night with new worlds unfolding before us in the darkened laboratory.'

Al Capp, the American cartoonist, invented the character Li'l Abner in 1934 when he was 25, his drawings portraying American foibles through the medium of an eccentric mountain community.

André Courreges, the Paris fashion designer, was pushed into a 'safe' career as an engineer by his butler father, but at 25 he rebelled and achieved his goal to become a designer. He quickly rose to the peak of his profession and his early engineering experience proved to be of great value, giving his futuristic, geometric fashion styles a marked originality.

Charles Dickens became the most popular writer in Britain at 25, with the publication of his comic novel *Pickwick Papers*, which was serialized in twenty parts, in 1837.

Marlene Dietrich was 25 when she enjoyed her great success in *The Blue Angel* in 1935, under the direction of Josef von Sternberg.

George Gershwin had his first major Broadway success at this age, in 1924, with *Lady Be Good*.

Charles Lindbergh made the first non-stop solo flight across the Atlantic at 25.

Henry Luce, the American publisher, having worked his way through college by waiting on table, launched a new magazine on March 3, 1923, at the age of 25, having successfully borrowed $86,000 from friends to do so. He and a young friend paid themselves salaries of $30 a week. The friend died young, but Luce went on to become a multi-millionaire. The name of the magazine was *Time*.

Norman Mailer wrote *The Naked and the Dead* when he was 25, in 1948, while still enrolled at the Sorbonne in Paris. It was hailed as the finest novel to come out of the Second World War.

Richard Sheridan, the English playwright noted for his comedies of manners, wrote *The School for Scandal* when he was 25, in 1777.

Peter Ustinov, the British actor, author and raconteur, directed his first feature film at the age of 25 – *The School for Secrets* – in 1946.

James Watson, the American geneticist, was this age when he and Crick broke the genetic code and uncovered the structure of DNA – the double helix.

Pearl White, the American actress famous for being tied to railway tracks and hanging from clifftops, was 25 when she starred in a Pathé serial called *The Perils of Pauline*, in 1914. She became the most famous actress of her day.

Fay Wray, the Canadian film actress, was this age when she became immortalized as the screaming, writhing, leg-waving love object of the pathetic giant ape King Kong. She made no fewer than eleven feature films in that year, and has been working in the cinema from 1923 to 1980, but all else was overshadowed by her spectacular part in *King Kong*.

Misfortune at 25
Bonnie Prince Charlie (Charles Edward, the Young Pretender) was decisively defeated at the Battle of Culloden Moor on April 16, 1746, when he was 25. He escaped to France and spent the rest of his days wandering Europe, drunk and debauched, trying unsuccessfully to revive his cause.

Retirement at 25
Johnny Sheffield, the American actor who played 'Boy' in the *Tarzan* films, retired at this age after seventeen years in the cinema and made a new career in real estate.

Life-span of 25
Aubrey Beardsley, the English illustrator, died of tuberculosis at 25, having contracted the disease at the age of 6. A master of morbid eroticism, his dying request was 'I am imploring you – burn all the indecent poems and drawings', a plea that happily was ignored.

Brian Jones, one of the original members of the rock group The Rolling Stones, was found dead in his swimming pool on July 3, 1969. The verdict was misadventure, but he had been becoming more and more deeply involved in the drug scene and it is possible that his death was self-inflicted. He had quit the Stones only twenty-four days before his death and a final telegram he had sent read 'Don't judge me too harshly.'

John Keats's short life was haunted by the tragic death of loved ones – first his father who was killed in a fall from his horse, then his mother who died of tuberculosis, nursed by her devoted son, then his beloved brother Tom, also from tuberculosis and also nursed by John. Finally, he himself succumbed to the same disease dying at the early age of 25 and robbing nineteenth-century England of one of its greatest poets.

Baron Manfred von Richthofen, the German aviator who became the most famous air ace of the First World War, nicknamed 'The Red Baron' because he always flew in a bright red Fokker, was shot down in 1918. He had destroyed eighty enemy planes during his legendary career and had Goering as a member of his élite squadron. When he was killed, his plane crash-landed in one piece and his body was photographed still sitting in his cockpit, so that copies of the picture could be dropped over German lines to demoralize the troops.

Twenty-six is a serious age, when adult responsibilities are treated with more respect. Social awareness increases and there is a greater interest in political matters. In the words of Hilaire Belloc: 'At the age of twenty-six/They shoved him into politics.'

Life Expectancy
In the Roman Empire between the first and fourth century AD, there was a life expectancy at birth of only 26·6 years, with men (30·4) faring better than women (22·9). There was virtually no change in this figure by medieval times when, in the ninth century AD, the life expectancy at birth was 26·7 years.

> Among the animals, twenty-six years is the record longevity for any deer and also for any marsupial.

Romance at 26
Madame de Montespan became the mistress of Louis XIV at this age, retaining his favour for thirteen years and giving him seven children. When she was chosen by the king, her husband, the Marquis de Montespan, displayed his resentment, for which he was exiled.

Prince Philip, the Duke of Edinburgh, married Princess Elizabeth in November 1947. Earlier that year he had become a British subject after renouncing his rights to the Greek and Danish thrones.

Achievement at 26
John Alcock, the English aviator, made the first non-stop flight across the Atlantic, accompanied by Arthur Whitten-Brown, on June 14-15, 1919. They won a prize of £10,000 from the *Daily Mail* and were knighted at Windsor a week later.

Jean Anouilh, the French playwright, achieved his first major success, in 1937, with his play *The Traveller Without Luggage*. His main concern was to discuss, through drama, how far people must compromise with truth to obtain happiness.

Learie Constantine became the first West Indian cricketer in England to achieve the cricket 'double' of 1,000 runs and 100 wickets in a single season.

James (Gentleman Jim) Corbett, the American boxer, became world champion at 26 in 1892, and held the title for five years. Before his arrival in the ring, boxing had been considered a vulgar sport, but he brought a new dignity to the contests and made it socially acceptable.

Jacques Cousteau, the French ocean-explorer, was largely responsible for the development of the aqualung in 1936. All sub-aqua work since then owes him a great debt.

Joe Davis became world snooker champion at this age, in 1927, and held the title until his retirement in 1946.

Walt Disney invented Mickey Mouse when he was 26, and also made his first *sound* cartoon film.

William Fargo, one of the founders of Wells Fargo, was this age when he organized the first express company to operate in the west of the USA.

Dawn Fraser, the Australian swimmer, became the first woman swimmer to win gold medals at three consecutive Olympics (1956, 1960, 1964). Her achievement in 1964 was amazing, bearing in mind that she had been injured in a car crash that year.

Hans Geiger, the German physicist, invented the Geiger Counter in 1908 when he was a 26-year-old assistant to Ernest Rutherford at Manchester University. Geiger Counters became essential tools in prospecting for uranium and other radioactive elements.

Hannibal became the Commander-in-Chief of the Carthaginian army at this young age and prepared to attack Italy with 40,000 troops and 38 war elephants.

Henry the Navigator, who made huge advances in the science of marine navigation and cartography in the fifteenth century, and whose work led to the foundation of the Portuguese empire, was 26 when he was made Grandmaster of the Order of Christ, sponsored by the Pope. As a result, all his ships carried a red cross on their bows and he was obliged to dedicate himself to a chaste and worthy life. The finances made available from the Order were primarily to assist voyages which would lead to the conversion of foreign pagans to Christianity.

Amy Johnson, the English aviator, was the first woman to fly solo from England to Australia. She achieved this in May 1930 when she was 26. The journey took nineteen days and was only one of many record flights she made, helping to turn her into the 'air heroine of the 1930s'. The extraordinary feature of her epic Australian flight was that previously she had flown only as far as from London to Hull. Despite this, she took off, piloting a tiny Gypsy Moth called *Jason*, on what was then an extremely hazardous journey even for an experienced international pilot.

Margaret Mead, the American anthropologist, produced her first book, *Coming of Age in Samoa*, in 1928. A perennial best seller, it was the first of twenty-three books on similar subjects which she published during her lifetime of tribal research.

John Osborne, the angry young British playwright,

Nero   Samuel Pepys

had such a huge success with his play, *Look Back in Anger*, in 1956, that it haunted him for the rest of his career, the label of 'angry young man' clinging to him long after he had ceased to be young. He did, however, remain angry and in the 1960s referred to television as a medium run by 'dim, untalented little bigots'.

Samuel Pepys, the famous London diarist, began keeping his diary on January 1, 1660, just before his twenty-seventh birthday, and went on recording in it until May 31, 1669, when he had to stop because of failing eyesight. His honesty and acute observational powers make his great work, consisting of 1,250,000 words, an invaluable historical record of the events and customs of his day.

Barbra Streisand, the American singer and actress with the ski-slope nose and the powerful voice and personality, won an Oscar at this age for her first film – *Funny Girl* – in 1968.

Orson Welles, the tall, brooding genius of Hollywood, also won an Oscar for his first film – *Citizen Kane* – in 1941. Full of brilliant innovations, it created a furore, both for its originality and the clash it caused between the young 26-year-old Welles and the immensely powerful newspaper magnate William Randolph Hearst, who was caricatured as Kane in the film.

Henry Wood, the English conductor, was 26 when he gave the first Promenade Concert at the Queen's Hall, London in 1895. He continued to do so for fifty years, dying shortly after his jubilee year in 1944.

Crime at 26
John Wilkes Booth, the American actor, was this age on the evening of April 14, 1865, when he entered the President's Box at the Ford Theatre in Washington and shot Abraham Lincoln in the head. Leaping on to the stage, Booth shouted out the words he had used many times before when, playing Brutus, he had killed Caesar: '*Sic temper tyrannis*', adding 'The South is avenged!' With this he escaped from the back of the theatre and made off on a fast horse. He was not caught for twelve days, but was eventually trapped and shot in a barn by troops.

Nero, the Roman Emperor, was 26 when he was suspected of starting the great fire of Rome in AD 64 and of singing to his lyre while the city burned.

Life-span of 26
Claude Duval, the French-born highwayman, was captured when drunk, and hanged at Tyburn for a series of daring robberies. He was famed for his gallantry towards his female victims and his epitaph read: 'Here lies Duval: Reader, if male thou art,/Look to thy purse; if female, to thy heart./Much havoc has he made of both; for all/Men he made stand, and women he made fall.'

Fatimah, the Shining One, daughter of Muhammad, the founder of Islam, died at this young age. In later centuries she became the object of veneration for many Muslims.

Jean Harlow

Jean Harlow, the original platinum blonde actress, who dyed her pubic hair to match the hair on her head, became a major Hollywood star in the 1930s, but could not find private happiness. Her troubled personal life and her three 'marriages of inconvenience', as she called them, seem to have stressed her into an early death, from uremic poisoning, in 1937.

Baby-Face Nelson, the American gangster, died in a hail of bullets at this age, killing his FBI killers in the process. Only 5 feet 4 inches tall, he was so wild that few gangs would have him, his lunatic tantrums putting valuable operations at risk. After Dillinger he became Public Enemy No. 1, but was angry that the price on his head was lower than his predecessor's.

Gram Parsons, the American rock star, died at 26 while rehearsing in the desert outside Los Angeles in 1973. His body was hijacked and burnt at the Joshua Tree National Monument – an act of loyalty by his friend and manager, Phil Kaufmann, fulfilling a request made by Parsons. Since no autopsy was performed, the cause of death was never established, but Parsons was heavily involved with the drug scene. Once, when warned that his drug abuse could kill him, he replied 'Death is a warm cloak, and old friend ...'

Otis Redding, the influential black American singer, famous in the 1960s as the first to bring black soul music to white audiences with any success, was killed in an air tragedy on December 10, 1967, when his plane crashed into an icy lake near Madison, Wisconsin.

Twenty-seven is the ideal age for becoming a mother. This is the year when a mother is least likely to suffer the death of her new-born baby. It may seem strange that this is five years older than the ideal age for being pregnant, but there is a simple explanation for this. The super-fecundity of the 22-year-old woman is not matched by her experience at giving birth. She is far more likely to be a first-time mother than a 27-year-old. The latter is superior to her in producing live babies because there is a much greater chance that her system has been through it all before.

The most amusing quotation on this young and wonderfully vigorous age is by F. Scott Fitzgerald who, in 1920, wrote: 'She was a faded but still lovely woman of twenty-seven.' This is a classic example of the relativity of being 'young' or 'old'. Fitzgerald was 24 at the time he wrote those words, an age at which, for a man, a 27-year-old woman seems elderly!

> Among the animals, 27 is the greatest life-span ever recorded for any species of rodent – a Sumatran Crested Porcupine in the Washington Zoo having lived to the impressive age of 27 years and 3 months. It is strange to think that a mere rodent can outlive any antelope or deer (see 25 and 26 respectively).

Accession at 27

Cosimo de' Medici, Duke of Florence from 1537 to 1574, rushed to the city when he heard of the assassination of his cousin, Alessandro. Regarded as inoffensive, he was warmly welcomed and the government was handed over to him, but he soon showed himself to be a ruthless despot, secretly liquidating his opponents throughout his reign. He was, however, an excellent administrator, with a passion for efficiency and organization.

Achievement at 27

Jean-Paul Belmondo, the athletic French film actor with the pug-ugly appeal, achieved overnight stardom as the new tough guy of the French cinema in 1960 with his cheeky-macho performance in Godard's first feature film *Breathless*.

Robert Burns, the national poet of Scotland, published his first poems, in 1786. They were immediately popular and from that point on he lived the dual life of a tenant farmer and a literary lion.

John Calvin, the French theologian, published his masterwork, *Institutes of the Christian Religion*, in 1536, making him one of the most important Protestant reformers of the sixteenth century.

Bette Davis, the iron-willed, domineering leading lady who punches her words out in intense performances that have made her a Hollywood original, was awarded her first Oscar for *Dangerous*, in 1935, in which she played an ageing alcoholic.

Sergei Eisenstein, the Russian film pioneer, made his cinema masterpiece, *The Battleship Potemkin*, in 1925.

Yuri Gagarin, the Russian cosmonaut, became the 'Columbus of the Interplanetary Age' or 'The First Swallow in the Cosmos', to use Khrushchev's phrases, when, on April 12, 1961, in *Vostok I*, he orbited the earth for 108 minutes – the first man to see with his own eyes the spherical shape of our planet. Afterwards he became a good ambassador for Russia, taking tea with the Queen and visiting many other countries, always to rapturous welcomes in both East and West.

Yuri Gagarin

Henri Giffard, the French inventor, flew the first (non-rigid) airship, near Paris, on September 24, 1852. It was steam-powered, propeller-driven and with directional control. (It was another half century before Zeppelin first went up in *his* airship.)

Graham Greene, the English novelist obsessed with evil, produced his first famous book when he was this age, in 1932. Called *Stamboul Train* in England, it was renamed *The Orient Express* in the USA, and was later made into a successful film.

Jacob Grimm, the elder of the Brothers Grimm, was this age when he started work on their famous collection of folktales now known as *Grimms' Fairy Tales*, in 1812. This led to the birth of the serious study of folklore.

Lillie Langtry, the celebrated beauty and actress who was born on Jersey in the Channel Islands and became known as 'The Jersey Lily', created a scandal in 1881 when she became the first society woman to go on stage. She later became the mistress of Edward VII.

Guglielmo Marconi, the Italian physicist who invented a successful system of radio telegraphy, managed to transmit signals across the Atlantic for the first time, in 1901, the signals being sent from Cornwall to Newfoundland.

William Morris, the English designer, craftsman, poet and socialist, founded with friends the firm of Morris, Marshall, Faulkner and Co., in 1861 – an association of 'fine-art-workmen' – creating furniture, fabrics, stained glass, wallpaper and other such products – which attempted to put art into everyday objects and destroy the distinction between the artist and the craftsman.

William Morton, the American dentist, was the first to demonstrate to the medical profession the use of ether in surgery. He used it successfully to extract a tooth on September 30, 1846 and a month later repeated the process with a hospital patient undergoing an operation for a tumour. He is credited with gaining the medical profession's acceptance of surgical anaesthesia, for which we all owe him a great debt, but he was unsuccessful in his lengthy attempt to patent the use of ether, since he was not the actual discoverer, but merely the implementer of its use in practical terms.

Jean Baptiste Racine, the French dramatist, scored his first great success with *Andromaque* in 1667, in which he presented what was to become his favourite theme – the tragic folly and blindness of passionate love.

George Sand, the French romantic novelist and early champion of women's rights, abandoned her husband, the Baron Dudevant, and ran off to Paris in her mid-20s to become a writer. When her first novel was published in 1832, *Indiana*, she was 27. It was a daring work glorifying free love and brought her immediate success. Like most of her novels, it followed closely the themes of her own life.

Life-span of 27

John Alcock, the English aviator who had just become a famous record-breaker after his transatlantic flight, was only 27 when he took off on a much simpler trip from England to France, to demonstrate and exhibit a Vickers Viking amphibious aeroplane. He came down in a mist twenty-five miles from Rouen and the plane crashed forward on to its nose. He died later in Rouen hospital.

Bix Beiderbecke, one of the legendary figures of early jazz, died in comparative obscurity. But a book about his life by Dorothy Baker, called *Young Man with a Horn*, created enormous interest in him and a huge posthumous following. He was important as the first white player to be a true innovator in jazz music.

Lyman Bostock, one of the highest paid baseball players in the USA, was shot to death at 27 by the estranged husband of the woman he was escorting on September 23, 1978.

Rupert Brooke, the English poet idolized as a young hero of the First World War, died of blood poisoning on a hospital ship off Skiros in 1915. He was buried in an olive grove on the island and one of his most famous poems could have served as his epitaph: 'If I should die, think only this of me;/That there's some corner of a foreign field/That is forever England.'

Peter Collins, the English racing driver, was 27 in 1958 when he was killed in the German Grand Prix, just as his career was about to reach its peak.

Rupert Brooke   Robert Burns

Jimi Hendrix, one of the wild men of the rock scene, who was known to play his guitar with his teeth, to set fire to it on stage, and to behave so outrageously that he was occasionally banned from appearing, drowned in his own vomit in 1970. His death took place in the apartment of Monika Danneman in London and the official cause of death was given as 'inhalation of vomit following barbiturate intoxication'. He had been jailed once for fighting with a member of his group and again for taking drugs in Canada. His drug problem had grown steadily worse, culminating with the tragedy in London.

Janis Joplin was a female counterpart of Hendrix. Loud, hoarse-voiced, uninhibited and frantically sexual, she was a blues singer from Texas who was found dead in her Hollywood hotel room at this same age, also in 1970. The cause of death was an overdose of heroin. Her foul language, hard drinking, free sex, and bizarre clothes made her the symbol of the more desperate rebels of the late 1960s. Her death, like that of Hendrix, was a result of the intense pressures of the Rock 'n' Roll world of that period.

Janis Joplin

Twenty-eight is the year of the divorcee. Studies in the USA have revealed that for the past fifty years, 28 has been the most common age for wives to leave their husbands, or be left by them. This gives a factual basis to the popular concept of a 'seven year itch', since most young couples will have been married for about eight years by the time the wives are 28 – allowing seven years to acquire the itch and a year to scratch it and create the emotional chaos that finally leads to separation.

In career terms, this is the year when there is a strong urge to make some sort of dramatic change or switch in direction. For those who resist the temptation of novelty, there is a deepening of the commitment to the old way of life. In other words, 28 is an age for *stepping out* of one's rigid career channel, or *stepping up* one's involvement in it. It is a year for intensifying one's life. The 20s have been spent establishing a personal identity and now it is time to do something with that identity.

> The greatest age a spider can expect to reach is 28 – the record having been achieved by a large tarantula.

Accession at 28

Henry VII, after exile in Brittany, invaded England in 1485, killed Richard III at Bosworth, claimed the throne and was crowned on October 30.

Muammar Al-Qaddafi, the puritanical, idealistic, fanatically anti-western leader of the coup that overthrew King Idris I in 1969, became the Prime Minister of oil-rich Libya in 1970 at the age of 28. Although a 'political' appointment rather than an accession, it made him a ruler more absolute and powerful than any contemporary monarch. Now called President of Libya, he remains one of the most dangerous of world leaders, capable of provoking a war at any time.

Achievement at 28

Charles Atlas was a 98-pound weakling when he was beaten up by a thug and vowed never to suffer the same experience again. By the age of 28 he had built himself up into a massively muscle-bound figure and was voted 'The World's Most Perfectly Developed Man' by the magazine *Physical Culture*. He became the pioneer of body-building and famous for it around the world.
Sebastian Bauer built and tested his first submarine, staying below the surface for seven and a half hours in 1851 and emerging in the midst of a funeral service for himself and his crew. The machine was called *Le Plongeur-marin*.
Elizabeth Blackwell was this age in 1849 when she became the first woman to obtain the degree of MD from a medical school in the USA. She went on to establish a New York hospital staffed entirely by women, and founded a full course of medical training for them there.
Napoleon Bonaparte conquered Egypt in July 1798. His aim was to cut off England's commercial route to India and he achieved this with a force of 280 transport ships, 55 warships, 38,000 troops, 1,200 horses and 171 guns. But just before his twenty-ninth birthday, Nelson destroyed his fleet in the Battle of the Nile and stranded him there for over a year.
Noam Chomsky, the American linguist, was 28 in 1957 when he published his controvercial concept of human language in *Syntactic Structures*, arguing that there is a universal grammar, or language structure, which is inborn in humans and biologically determined at a genetic level. In other words, humans are specially programmed as language-learning animals. Chomsky is the greatest figure to emerge in the rather sterile science of modern linguistics. Sadly he, like his colleagues, suffers from an inability to express himself clearly in simple English, an irony that makes it almost impossible for an outsider to penetrate the dense, technical jargon.
Arthur Conan Doyle, the British doctor and author, was 28 when he introduced the eccentric Sherlock Holmes to the reading public, in a tale called *A Study in Scarlet*, in 1887. The most successful detective invention of all time, Holmes was based on one of Doyle's teachers at the medical school in Edinburgh, a man who was noted for his deductive reasoning. Doyle was knighted, not for creating Holmes and Dr Watson, but for his own medical work in the Boer War.
Gabriel Fahrenheit was 28 in 1714 when he invented the mercury thermometer and introduced the Fahrenheit Scale – a scale so useful because of its fine calibration that it has defeated all attempts to introduce the Centigrade Scale into British thinking.
André Garnerin, the French aeronaut, became the first man to use a parachute from a great height, with repeated success. On October 2, 1797, he gave his first exhibition of parachuting, in Paris, when he jumped from a height of 2,000 feet. Later he made a descent from 8,000 feet in England. His white canvas parachute was umbrella-shaped and approximately twenty-three feet in diameter.
Christiaan Huygens, the Dutch mathematician, astronomer and physicist, invented the pendulum clock, in 1657.
James Prescott Joule, the English physicist, formulated the First Law of Thermodynamics in 1847. The law of conservation of energy, it states that energy can be converted from one form into another, but cannot be destroyed. His name was later given to a unit of energy. By a curious coincidence, the German physicist **Rudolph Clausius** was also 28 when he formulated the Second Law of Thermodynamics in 1850. This states that heat cannot pass from a colder body to a warmer body but only from a warmer body to a colder body. This heat-transfer principle may seem obvious, but it was the basis for the development of that most valuable human invention – refrigeration.
Billie Jean King was 28 when she became the first

28

Billie Jean King

professional female athlete to be paid more than $100,000 in a single year (1971). In that year her record total prize money was $117,000.

Sir Walter Raleigh was this age when he became the favourite of Queen Elizabeth I and began to acquire lucrative monopolies, properties and influential positions.

Georges Simenon, the creator of Inspector Maigret, was 28 when he published a novel under his own name. It was to be the first of over 200. In addition, he published 300 further books under seventeen different pen-names, making him one of the most prolific authors of all time. He was even more prolific sexually, if we are to believe his own statements on the subject.

Igor Stravinsky saw his ballet score for *The Firebird* performed at the Paris Opera on June 25, 1910, just one week after his twenty-eighth birthday, giving him a major success.

Johnny Weissmuller, the American Olympic champion swimmer who won a total of five gold medals in the 1924 and 1928 Games, made his first Tarzan film, *Tarzan the Ape Man*, in 1932, when he was 28. It was the first of twelve, after which he was demoted to 'Jungle Jim' films as a White Hunter. Later he became a successful business man in Cali-

fornia, selling swimming pools. A whole string of actors took over the famous Ape Man role, but to most film-goers Weissmuller remains the definitive Tarzan.

Crime at 28   **Ian Brady,** the sadistic British murderer, was a sullen, withdrawn child with a record of juvenile theft. He developed private fetishes and read all he could about de Sade and Adolf Hitler. With his female accomplice, Myra Hindley, he dressed in black leather and indulged in flagellation fantasies. Then their fantasies became reality and they abducted, tortured and killed five (possibly eight) children, taking pornographic pictures of them and making tape recordings of their pleadings and screams under torture. They were caught in 1965 and sentenced to life imprisonment early in 1966, when Brady was 28.

Angela Davis, the American black militant, was one of the FBI's most wanted criminals, accused of complicity in the George Jackson affair. She became the focus of a world-wide 'Free Angela Davis' campaign, but was held in solitary confinement for many months before public opinion finally forced a trial. It lasted for thirteen weeks after which she was acquitted of all charges by an all-white jury on June 4, 1972, when she was 28.

Life-span of 28   **Caligula,** the depraved Roman Emperor, whose cruelty and sexual perversity knew no limits, was the victim of a conspiracy in AD 41, and was murdered by a tribune of the guard at the Palatine Games.

Stephen Crane, the American novelist whose book, *The Red Badge of Courage* (1895), was a pioneering work of naturalistic war fiction, describing the true horror of events in the American Civil War, died of tuberculosis and malaria in 1900.

Ruth Ellis was the last woman to be hanged in Britain. She was 28 when she was executed in 1955 for shooting dead her young lover. She acted in a fit of jealousy and anger, but the fact that it was a crime of passion did not save her.

Mike Hawthorn, the British motor-racing champion, was killed at 28, not on the track, but in his private car on the Guildford by-pass, in 1959.

Neville Heath, like Brady a sadistic British murderer, was a lady-killer in both senses. Suave, handsome and charming, the ex-RAF pilot had the phone numbers of 177 girls in his little black book. The ones who became his victims were spread-eagled, tied, whipped, bitten and slashed. He was coolly nonchalant at his trial in 1946 and his final request before being hanged, at 28, was for a glass of whisky. His last words were 'And make that a large one.'

Edie Sedgwick, the 'Queen' of Andy Warhol's 'skin-flicks' in the 1960s, who became addicted to cocaine and heroin, died of a drug-overdose at this early age.

Twenty-nine is a year of new vitality. Both males and females begin to feel too restricted by their old way of life and try to find ways of breaking loose from it or expanding it. Subconsciously this may be due to the anticipation of becoming 30 and the feeling that this somehow marks the end of youth. So 29 becomes a last shout of youthfulness, before the roaring 20s are gone for ever.

This is the year in which the domestic seek the wild experience and the wild seek the domestic experience, both striving to broaden their horizons while they still enjoy the flexibility of youth.

Life Expectancy Few countries have a life expectancy at birth that is this low. The one exception is Chad, in Africa, where 29 is the figure for males, even in these modern times.

> Among the animals, 29 is the age for the oldest known budgerigar and the oldest known domestic dog – a Queensland Heeler called Bluey that lived in Australia from 1910 to 1939 and died aged 29 years and 5 months.

Achievement at 29 **Alexander Graham Bell**, the Scottish-born American inventor, made the first telephone in the world in 1876, at this age. He later gave it to Sir James Murray, the editor of the *Oxford English Dictionary*, who had first introduced him to the mysteries of electricity and whom he called the 'Grandfather of the Telephone'. The machine ended up in the Murray attic in Oxford, where it remained for many years until, during the wartime occupation of the house by billeted soldiers, it was brought down and burned in the fireplace to help warm the house.

Erskine Caldwell, the American novelist, achieved fame at this age in 1932 with his book *Tobacco Road*, a blunt, savage portrayal of the lives of poor white Southerners.

Claudette Colbert, the French-born American actress who became the leading sophisticated-comedy heroine of the 1930s and 1940s, won an Oscar at 29 for her 1934 performance opposite Clark Gable in *It Happened One Night*.

Denis Compton, the English cricketer, still holds the record for the greatest number of *hundreds* in a season. He was 29 in 1947 when he achieved this, with no fewer than eighteen centuries.

El Cordobés, the virtuoso Spanish bullfighter, was 29 in 1965 when he fought 111 corridas, breaking the previous record of 109 established in 1919 by Juan Belmonte. In August 1965 Cordobés killed sixty-four bulls – a record for a single month – and was paid $600,000 for doing so, an extraordinary achievement for an illiterate orphan who started life as a petty thief.

Günter Grass, the German poet, novelist and playwright, produced his first novel at this age – *The Tin Drum* in 1956 – and with it became the literary spokesman for the generation that had grown up in Nazi Germany and survived the war.

Hannibal was 29 when he crossed the Alps and proved himself to be one of the greatest military leaders of antiquity.

Elton John became the Chairman of Watford Football Club at 29 in 1976 and set about scaling the heights of the English Football League, successfully taking his club from Fourth to Third to Second to First Division glory.

Wassily Kandinsky, the Russian artist who was one of the great pioneers of abstract art, was 29 when he achieved his goal of becoming a full-time painter. Offered a professorship in jurisprudence at an Estonian University, and a safe, respectable academic life, he turned it down and took a train to Germany to start his difficult life as an avant-garde artist.

Kirkpatrick Macmillan, a Scottish country blacksmith, invented the first practical bicycle driven by pedals, in 1839, at Courthill in Dumfriesshire. He made a few for sale, but then his idea was copied by another Scot, Gavin Dalzell, who is often wrongly given the credit for having invented the bicycle.

Pelé, the soccer genius, scored his thousandth goal on November 19, 1969. The event occurred at the giant Maracana Stadium in Rio, where a huge crowd invaded the pitch, Pelé's jersey was torn from him and replaced with a silver shirt bearing the number 1,000. Traffic was brought to a halt throughout Brazil as the news spread via radio, and dancing crowds filled the streets. Pelé was carried around the stadium shoulder-high until he fled the field weeping, his place in the game taken by a substitute.

William Penn, the Quaker founder of Pennsylvania, acquired a large tract of land in America, in 1674, and set about developing a colony there organized according to his religious beliefs and his political principles.

Harold Pinter, the son of a Jewish tailor in London's tough East End, who matured into one of the most original playwrights of the twentieth century, found fame with his second play *The Caretaker* when he was 29, in 1960.

Gordon Richards, the English champion jockey, was this age when he achieved a remarkable racing record – twelve winners in a row. He won the last race at Nottingham on October 3, 1933, then all six races at Chepstow the following day, and the first five of the six on the next day as well. During his career he was champion jockey twenty-six times and won a total of 4,870 races before he retired.

Arthur Sullivan, the English composer of comic operas, was 29 when he initiated his immensely successful partnership with librettist W. S. Gilbert with a performance of *Thespis, or the Gods Grown Old* on December 26, 1871 at the Gaiety Theatre in London. It opened a new era in light opera in England, dominated by Gilbert and Sullivan for nearly thirty years.

Frederick Walton, the English inventor of lino-leum, saw the first piece produced at his factory in Staines one night in the autumn of 1863, when he was 29. The following year he built a lino factory and went into large-scale production.

James Watt, the Scottish engineer, invented the Condenser Steam Engine in 1765. It was four times as efficient as the older type and heralded the indus-trial revolution which changed the face of the globe.

Frank Whittle, the English aeronautical engineer who became the father of jet air travel, carried out his first successful test-bed run of a jet engine not long before his thirtieth birthday, on April 12, 1937. Although it was the German Heinkel who achieved the first jet flight (in 1939) it was Whittle's engine that was used as the prototype for all the jet engines flying today. In 1948 he was awarded £100,000 for his work – work which has transformed international travel.

Henry Williamson, the English author, was 29 when he published his famous animal story *Tarka the Otter*, in 1927.

Misfortune at 29

Nijinsky, the Russian dan-cer of genius, developed schizophrenia at 29 and was forced to retire from dancing. The leading figure in Diaghilev's Ballets Russes in Paris, who brought a new vitality to male ballet dancing with his incredible agility and athleticism (he once performed ten entre-chats in a single jump), he was imprisoned in Hun-gary during the First World War, accused of spying for Russia. Forbidden to dance, he suffered severe anguish which eventually led to a complete mental collapse. He spent most of his last thirty years of life in a Swiss mental institution.

Crime at 29

Titus Oates, the English conspir-ator, was a Protestant priest who invented a Popish Plot, saying that King Charles II was going to be killed, London burnt, and all Protestants massacred. He did this in 1678 when he was 29 and it so terrified London that thirty-five innocent Roman Catholics were executed for treason. He was later found guilty of perjury, flogged and imprisoned for life.

Life-span of 29

Fred Archer, the famous English jockey who won the Derby five times, the Oaks four times and the St Leger six times, and who was champion jockey for thirteen years in a row, took his own life at Newmarket when he was delirious with typhoid fever.

Mrs Beeton, famous for her *Book of Household Man-agement*, died at the age of 29, a few days after the birth of her fourth son.

Anne Boleyn, the second wife of Henry VIII and mother of Queen Elizabeth I, lost her head to the axe on May 19, 1536 at this early age. Unable to produce a male heir for Henry, she was accused of adultery and incest and promptly beheaded. Eleven days later Henry married Jane Seymour.

Kitty Fisher, the most successful whore in history, died of tuberculosis after six years of immensely lucrative prostitution. Charging 100 guineas a night in the middle of the eighteenth century, when a common soldier earned only a few pence a day, she quickly became rich and, according to Casanova, once wore over 100,000 crowns' worth of jewels when attending a ball. She spent more than £12,000 in one year, a vast amount for the time in which she lived. Among her clientele were Admirals, Generals, Lords, Earls and a Duke. The Duke outraged her by handing her only a £50 note on the morning after their night of love, so she responded by placing it on her bread and butter and eating it for breakfast. She retired and made a good marriage before succumbing to tuber-culosis.

Carole Landis, the American actress who was the blonde star of many Hollywood films in the 1930s and 1940s, killed herself at 29 with an overdose of sleeping pills, allegedly over the collapse of her love affair with Rex Harrison, who at the time was married to Lilli Palmer.

Christopher Marlowe, the Elizabethan dramatist who wrote *Tamburlaine the Great* and *Dr Faustus*, was stabbed to death in a tavern brawl in 1593. His death is surrounded in mystery and rumours of es-pionage and counter-espionage, but the story told at the inquest suggests that he quarrelled with a com-panion over the tavern bill, drew a knife and had it turned upon himself.

Gunnar Nilsson, the Swedish racing driver, was cut down at 29, not by a speeding car, but by cancer, in 1978.

Percy Bysshe Shelley, the controversial English romantic poet, was drowned in a boating accident at sea on July 8, 1822, when a sudden squall struck his yacht *Don Juan*. By a coincidence, his first wife Har-riet also died of drowning, but that had been some years earlier, in London, and as an act of suicide.

Mrs Beeton   Anne Boleyn

30

Thirty is a contradictory year. For optimists it is the start of a phase of satisfying maturity; for the pessimists it marks the end of youthful pleasures.

Among the optimists – Hervey Allen: 'The only time when you really live is from thirty to sixty'; Baltasar Gracian: 'At 20 man is a peacock, at 30 a lion ...'; Aristotle: 'The body is at its best between the ages of thirty and thirty-five'; Hesiod: 'Marry in the springtime of thy life, neither much above or below the age of thirty'; Glenn Fry (an American rock star): 'The great thing about being thirty is there are a great deal more available women. The young ones look younger and the old ones don't look nearly so old'.

Among the pessimists – Emerson: 'Men and women at thirty years ... have lost all spring and vivacity, and if they fail in their enterprises, they throw up the game'; Chaucer: 'Better than old beef is tender veal: I want no woman thirty years of age'; Montaigne: 'Of all the great human actions I ever heard or read of, of what sort soever, I have observed, both in former ages and our own, more performed before thirty than after ...'; Mick Jagger: 'A lot of people start to fall to bits at thirty ... quite honestly once you are able to reproduce you're over the hill.'

Some of the pessimists refer specifically to the physical decay that supposedly sets in at 30. In the *Vogue Body and Beauty Book* we read: 'The face ages progressively. As a rule, at thirty personality lines usually appear.' ('Personality lines' is Voguese for 'wrinkles'.) Dr Leonard Williams, in his book *Middle Age and Old Age*, has no time for gentle euphemisms: 'As matters stand at present, the ordinary man begins to "rot" at 30 years of age. The physical signs are not difficult of detection. They are writ large on the abdominal contour of the all-too-numerous victims.' In her book *Futures*, Shirley Conran assaults 30-year-old males even more savagely: 'They start to go bald ... they don't look after their skin so it looks as if someone has been cleaning the car with it for the past twenty years. Then the neck disappears in a fleshy roll and lots of them start to look five months pregnant. As far as losing their looks is concerned, men have as much to fear as women.'

It is not clear why certain people wish to exaggerate the gradual ageing process in this way, unless of course they are selling ways of retarding it. It is certainly true that at this age a little more care has to be taken with the body than was necessary in earlier years. The young couple who can enjoy an all-night party at 20 will have much heavier hangovers at 30. James Howell summed this up in 1659: 'Every one is a fool or a physician to himself after thirty'; and there is an old proverb which states: 'A man, as he manages himself, may die old at 30 or young at 80.' So, with a little care, we can ignore the more brutal of the pessimists and look upon 30 as a splendid age for starting a phase of mature creativity.

In some spheres 30 is looked upon as the age for accepting major responsibility. Ecclesiastical Law, for instance, states that a bishop must have completed his thirtieth year.

Life Expectancy Having reached this age British males can expect to live for another 41 years and females for another 47. This is a great improvement on the figures for the turn of the century, when males could expect only another 34 years and females another 37.

Achievement at 30 **Hans Christian Andersen,** the son of a poor Danish shoemaker, published his first book of fairy-tales in 1835, including such stories as *The Tinderbox*, *Little Claus and Big Claus* and *The Princess and the Pea*.
George Balanchine, the Russian choreographer, transplanted European ballet to the United States and founded the School of American Ballet in New York in 1934.

Marlon Brando in *On the Waterfront*

Marlon Brando won an Oscar for his compelling performance in *On the Waterfront*, in 1954. A pioneer of method-acting in films, he introduced the stumbling rhythms of the natural speech patterns of an inarticulate man – the first time that this had been attempted. The incoherence of some of his statements irritated many, but marked him out as one of the great originals of Hollywood.
Bertolt Brecht, the German dramatist, had his first major success with *The Threepenny Opera* (1928), in which the chaos, greed and violence depicted are identified with what the author sees as the cruelty and selfishness of capitalism.
Dave Brubeck, the jazz virtuoso, formed the Dave Brubeck Quartet, in 1951, and created the most original sound in modern jazz.
Luis Buñuel, the Spanish film director and one of the few true geniuses of the cinema, created his surrealist masterpiece *L'Age d'Or*, in collaboration with Salvador Dali, in 1930. He went on to devote his entire creative energy to attacking the Church and established authority, for which he has been called traitor, anarchist, atheist, pervert and iconoclast. Despite the fact that his films have often been banned, nothing has diverted him from this obsessive mission.
William Congreve, the English Restoration dra-

matist, was so furious at the poor reception of his play *The Way of the World* in 1700 that he retired to live as a country gentleman. Yet this play has become considered as his greatest work and is still frequently performed today.

Demosthenes, the greatest of the Greek orators, made his first major speech at this age. In his youth he had suffered from a speech defect and set about a rigorous self-improvement scheme, shutting himself away in an underground study and shaving off half his hair so that he could not go out in public. He then practised speaking with pebbles in his mouth and recited verses when running out of breath. Despite this, his first speech was greeted with laughter and he had to earn a living by writing speeches for other people – until at the age of 30 he finally made his breakthrough.

Allen Ginsberg, the American poet, was 30 when he published his major work *Howl*, in 1956. It was the most important poem of the 'Beat Generation' and dwells in particular on his revulsion at the materialism and insensitivity of post-war America.

John Glenn was this age when he became the first American astronaut to orbit the earth, on February 20, 1962, in the space capsule *Friendship 7*.

Heinrich Hertz, the German physicist, the father of radio and television, was 30 when he discovered radio waves in 1887. His discovery led to Marconi sending the first wireless signals through space, seven years later. It also led to the important science of radio-astronomy.

Jack Johnson broke the boxing 'colour line' when he became the first black Heavyweight Boxing Champion of the World, defeating Tommy Burns by a knock-out in Sydney on December 26, 1908.

Gene Kelly, the man who brought virility and masculinity to dancing in America, was already 30 at the time of his screen début in 1942 opposite Judy Garland in *For Me and My Girl*. The success of the film made him the only serious rival to Fred Astaire in the history of Hollywood.

Peter O'Toole was 30 when he gained international fame for his portrayal of Lawrence of Arabia, in 1962.

Omar Sharif, the Egyptian actor, was the same age when he, too, became an international star, thanks to his role in the same film.

Henry Stanley, the American journalist and explorer, was also this age when he reached Ujiji in November 1871 and uttered the immortal words 'Dr Livingstone, I presume.'

Crime at 30

Al Capone, known as 'Scarface' because of three jagged scars he had acquired after insulting a man's sister when acting as a bouncer at a brothel in Brooklyn, committed one of the most famous of all crimes on February 14, 1929: the St Valentine's Day Massacre. The 30-year-old Capone, who had been ruling the whole of the Chicago underworld since 1925 and was earning $5,000,000 a year, was the subject of an assassination attempt by a rival gang. His revenge was to dress his own killers as policemen and use them to mow down all seven members of the rival group, in a garage on N. Clark Street. After this slaughter he ruled supreme until he was jailed for income tax fraud.

Life-span of 30

Emily Brontë, author of *Wuthering Heights*, lived a short, lonely and tragic life, suffering from crippling shyness. She caught a cold when attending the funeral of her opium-addicted, alcoholic brother Branwell, in 1848, and two months later succumbed herself to tuberculosis.

Mama Cass (Eliot), of the first hippie group, The Mamas and the Papas, died in London in 1974 when she choked on a sandwich.

Manolete, the arrogantly dispassionate Spanish bullfighter, was gored to death at this age in the bullring, in 1947.

Jean Paul Marat, the French revolutionary, was murdered in his metal bathtub by Charlotte Corday who pulled a six-inch butcher's knife from a sheath hidden in her stylish summer dress and plunged it deep into his heart, on the evening of July 13, 1793.

Nero, the blond, paunchy, spindle-legged, bull-necked Roman Emperor who fancied himself as a charioteer, musician and poet and who made attendance at his recitals compulsory, was also killed by a sharp knife – in his case a dagger thrust into his throat. He had become Emperor in AD 54 while still a teenager and soon began to show his true qualities – vanity, cowardice, timidity and a dilettante interest in the arts. His only good quality seems to have been that he ran an efficient administration. As he grew older, he became increasingly cruel. First he poisoned his half-brother, then he tried to poison, crush and drown his mother, in three separate, unsuccessful attempts on her life, before having her openly murdered; next he had his wife executed and later kicked to death his mistress (who had unwisely become his second wife). He also forced his teacher to kill himself. His greatest cruelties, however, were reserved for the Christian scapegoats whom he blamed for the great fire of Rome. These he had crucified, burned to death as living torches, or sewn up in animal skins and devoured by dogs. All this he achieved in a short life of thirty years, before dying in desperate flight from a Rome that had finally turned against him, in AD 68. His dying words were: 'What an artist the world is losing!'

Eva Perón, the ex-actress wife of the Argentinian president, came from a humble background to a life of furs and diamonds, as a kind of 'National Cinderella'. When her husband fell from power in 1945, she roused the workers to paralyse the country until he was reinstated. Then she worked for women's suffrage, help to the aged, child welfare and the relief of poverty before suddenly disappearing from public view and dying a few weeks later at the age of 30.

Thirty-one is the age when the young are supposed to stop trusting you. Jerry Rubin, the noisy American activist of the 1960s, employed 'Never Trust Anyone Over Thirty' as one of his main rallying slogans. It is curious how many people have expressed the feeling that, once past the magic age of 30, they are moving into a cold climate. Matthew Arnold, in a letter written in 1853, remarks 'I am past thirty and three parts iced over.' And a little earlier, in 1834, Emerson writes in his journals: 'After thirty, a man wakes up sad every morning excepting perhaps five or six, until the day of his death.' There is also Swift's cutting comment 'I swear she's no chicken; she's on the wrong side of thirty, if she be a day.'

If 31 is on the wrong side of 30 for some, it is eagerly enjoyed by many others as one of the most active and energetic of all adult years.

Life Expectancy The expectation of life at birth for males living in Togo and females living in Upper Volta, in Africa, is amazingly no more than 31 years, even in the second half of the twentieth century, which is little above that expected in Ancient Rome 2,000 years ago.

> Among the animals, 31 years and 5 months is the record longevity for any species of bat – the record-maker in question being a 'flying fox' which died at London Zoo on January 11, 1979.

Accession at 31 **Robert the Bruce,** King of Scotland and hero of Scottish legend, was crowned at Scone on March 25, 1306, and went on to defeat the English at the Battle of Bannockburn, driving them from Scotland. At one point when he was a fugitive and his fortunes were at a low ebb, he is reputed to have derived hope and patience from watching a spider perservering at spinning its web.

Achievement at 31 **Arnold Bennett,** the English author, published his first novel, *A Man from the North*, in 1898, the success of which encouraged him sufficiently to abandon his job as a solicitor's clerk and devote himself to full-time writing for the rest of his life, during which he completed thirty novels presenting the drab underside of Edwardian England.

Isambard Kingdom Brunel, the English engineer, designed the first transatlantic steamer, the *Great Western*, in 1837.

James Cagney, the short, jaunty and most imitated of Hollywood actors, became famous in 1931 with his role as a vicious gangster in *The Public Enemy*, when he squashed a grapefruit into the face of 'gangster's moll' Mae Clarke.

Daphne du Maurier, the English novelist, gained international fame for her novel *Rebecca* in 1938, which was made into an equally famous film by

Alfred Hitchcock.

Thomas Edison, the greatest technological inventor in the world, who surprisingly had no schooling whatever, patented his sound-recording invention – the phonograph – in 1878.

Isambard Kingdom Brunel

Aleksei Leonov, the Russian astronaut, was the first man ever to climb out of a ship in space. He let himself out of the airlock on March 18, 1965, when he was 110 miles above the Crimea and stayed out in space for ten minutes, manœuvring and taking films.

Édouard Manet, the French artist, was 31 when his major work *Déjeuner sur l'herbe* was rejected by the French Academy and exhibited at the Salon des Refusés, where it aroused great enthusiasm from the group of young painters which was later to become the nucleus of the Impressionists.

Charles Stewart Parnell, the Irish Nationalist leader who was the political champion of Home Rule for Ireland, was chosen as President of the Home Rule Confederation in 1877, thereby becoming the most conspicuous figure in Irish politics at 31.

Misfortune at 31 **Douglas Bader,** the legless British air hero of the Second World War, had already shot down fifteen enemy aircraft when he collided with a German plane over France and was captured and held prisoner in August 1941.

Ludwig van Beethoven was tempted to take his own life at 31 because, in 1802, he realized that his deafness was both permanent and progressive. He

31

wrote: '. . . for six years I have been a hopeless case, made worse by ignorant doctors, yearly betrayed in the hope of getting better, finally forced to face the prospect of a permanent malady whose cure will take years or even prove impossible.'

Johnny Cash, the American country-and-western singer and composer who was raised in a strict Baptist family and became the top recording artist in his field in the 1950s, took to drugs in the 1960s and was caught crossing the Mexican border in 1963 with a guitar-case full of pep-pills. This arrest at the age of 31 was followed by sickness, a car crash, an overdose suicide attempt and divorce, before he was rescued and reformed. Turning to religion, he joined the Evangelical Temple in Nashville.

Curt Jurgens, the German actor who appeared in over 100 films and became an international star of the cinema, playing dignified blond Germans with firm authority but with warmth and humanity, was sent to a concentration camp for 'political unreliables' by a special order of Dr Goebbels, in 1944.

Robert Mitchum, like Johnny Cash, was 31 when he was arrested and jailed on a narcotics charge. Fortunately his drug associations did not damage his career, nor were they responsible for his sleepy eyes and his characteristic slow drawl. These features, he claimed, were a legacy from a heavy blow received during his early boxing days.

Life-span of 31
John Bonham, drummer with the world's most successful Heavy Metal rock group Led Zeppelin, drank himself to death, after which the group split up. The coroner's report gave the cause of death as a massive overdose of vodka and orange consumed during a twelve-hour drinking spree.

Cesare Borgia, the son of one of the most licentious popes ever to rule the Roman Catholic Church, Pope Alexander VI, and one of his many mistresses, Vannozza dei Cateanei, was brilliant in war, daring, ruthless and energetic, but was also pathologically ambitious and had complete disregard for any laws. When his father died, his days were numbered, the new pope being an enemy of the Borgias. He was imprisoned and escaped after two years but was shot dead during a skirmish near the Castle of Viana.

Commodus, the pleasure-loving, irresponsible Roman Emperor notorious for his brutal misrule, was strangled on New Year's Eve, AD 192, by a champion wrestler, on the orders of his mistress, much to the gratitude of the Senate.

Keith Moon, like John Bonham, was a rock drummer who killed himself with an overdose at the age of 31. Moon, the drummer with The Who pop group was infamous around the world as the wild man of rock, who demolished hotel bedrooms and drove expensive cars into swimming pools. His extravagant behaviour was looked upon as a defence mechanism by his close friends, who regarded him with great affection.

Georges Seurat, the French Neo-impressionist painter and the leading exponent of Pointillism – the scientific use of tiny spots of unmixed colour which merge at a distance to create the desired tones – died young of infectious angina. He was the master of the scientific rationalization of Impressionist theories, but his strictly controlled compositions were, for many, too rigid and disciplined.

Rudolph Valentino, the Italian-born film idol, whose female fans fainted in the aisles in the early 1920s, died of peritonitis from a perforated ulcer in 1926. As a teenager he had begged for coins on Paris street-corners. Later he was a dish-washer in New York. Within a few years of arriving in Hollywood he was a slick-haired, eye-flashing leading man, dominating the cinema screen. In private he had a strange sex life, his first marriage never being consummated, and his second turning into a disaster with his wife banned from the sets where he was working. After his second divorce, the studio hired detectives to keep 'The Great Lover' away from strange women, in case it was reported that he had given a poor sexual performance that might harm his public image. Three months before his death he was attacked in the press as a 'painted pansy', and there were headlines such as 'All men hate Valentino' which doubtless added to the stresses of stardom and hastened his death. At his funeral there was a near-riot with people trampling over one another to view the body, and women who had never met him committed suicide in the grief of the moment.

Thirty-two is a productive, industrious age. The shock of entering the 30s is past and it is a time for settling down to the serious business of making a success of life. Exceptions to this rule are those rare individuals who have experienced a juvenile success story. For them it is a time of danger, when they are faced with becoming young 'has-beens' or of attempting to force an extension of their juvenile image. Some come to terms with their early 'over-the-hill' condition and manage to enjoy their new freedom from the stress of immature stardom. Ringo Starr, in the lyrics of a post-Beatles song, comments: 'I was in the greatest show on earth, for what it was worth; Now I'm only 32 and all I want to do is boogaloo'.

Life Expectancy

The life expectancy at birth for males living in Gambia and Upper Volta is still only 32.

Romance at 32

Prince Charles became engaged to Lady Diana Spencer in February 1981 and married her on July 29.

Achievement at 32

Kingsley Amis, the English author, published his first and immensely successful novel *Lucky Jim*, with its new-style, disgruntled hero, in 1954.

Melville Bissell, the American inventor, patented the carpet sweeper in 1876 and turned it into a world-wide business success.

Amelia Bloomer, the American reformer who campaigned for 'rational dress' for women, invented 'bloomers' in 1850. They were loose 'Turkish' trousers, gathered at the ankles and worn under a below-the-knee skirt. Although her innovation failed and was much mocked, the name survived to be used as a term for any loose-fitting knickers.

Benjamin Britten, the British composer, achieved a great success with his opera *The Rape of Lucretia* in 1946, which marked the inception of the English Opera Group, with Britten as artistic director, composer and conductor. This gave rise to the Aldeburgh Festival which became one of the most important of English musical occasions.

Chester Carlson, the American physicist, brought the science of electronics to the art of printing, with his 1938 invention of Xerox copying. He called the process electro-photography or xerography and the very first electrostatic copy was made in a rented room behind a beauty parlour in Astoria. Amazingly he found difficulty in marketing his revolutionary dry-copying process, but eventually succeeded and became a multi-millionaire.

Fidel Castro, the Cuban revolutionary, came to power at 32 on January 1, 1959, when he ousted Batista and ended his corrupt dictatorship.

Sean Connery, the Scottish actor who was the son of a truckdriver and a charlady and who had hardly been noticed in his early films, shot to international stardom in 1962 as James Bond, in *Dr No*, the first of the immensely successful cinema versions of Ian Fleming's spy thrillers.

Raymond Dart, the South African anthropologist, became famous for his 1925 discovery of a skull that became popularly known as the 'missing link'.

Eugene Dubois, the Dutch anatomist, was also 32 when he discovered his own 'missing link', an ancient jaw-bone found in 1908 in Java which became known as 'Java Man'.

Epicurus, the Greek philosopher, founded his first school in 310 BC. The Epicurean philosophy he taught, however, was rather different from its popular image today. There was no excessive wining and dining – water was the usual drink. His aim was to achieve happiness through a life of simple virtuous pleasures, avoiding pain in both mind and body, and ignoring political activity and public involvement. His school admitted women as well as men and sounds remarkably like an early 'commune', except that there were slaves for the menial tasks.

William Halstead, a pioneer of scientific surgery, developed, by self-experimentation, efficient local anaesthetics in 1885. His innovation was not without risk: injecting cocaine into his body as part of his experimental procedure left him with a heavy drug addiction that lasted for two years.

Thor Heyerdahl, the Norwegian explorer, crossed the Pacific on a fragile raft, the *Kon-Tiki*, made of balsa logs tied together with rope. He left South America in April 1947 and arrived in Polynesia, 4,300 miles away, in August. His bravery and curiosity has dramatically revised our ideas about the possible marine mobility of early man.

Allen Lane, the British publisher, left orthodox publishing in 1935 and with a capital of £100 founded Penguin Books, becoming the pioneer of cheap paperback publishing that revolutionized the reading habits of the western world.

Joseph Lockyer, the English astronomer, discovered in the sun's atmosphere a previously unknown element which he named Helium, in 1868.

left, Amelia Bloomer; *right*, Sean Connery

32

Thomas Malthus, the English economist and demographer, published his famous *Essay on the Principle of Population* in 1798, in which he showed that a population always tends to outrun its food supply, that poor countries breed faster than rich countries and that mankind can only be bettered by imposing limits on its reproduction.

Erich Maria Remarque, the German-born author, captured the horror of war in his 1929 novel *All Quiet on the Western Front*, the most realistic treatment of the subject at that time. Written in four weeks, it became a major international success and was made into one of the greatest of all war films. The book was publicly burned by the Nazis.

James Stewart, the gangling, drawling star of Hollywood films, who specialized in roles depicting the ordinary man in extraordinary circumstances, won an Oscar in 1940 for his part in *The Philadelphia Story*.

Orville Wright, the American pioneer aviator, made the first self-powered aeroplane flight in history on December 17, 1903, at Kitty Hawk, North Carolina. The flight, against a 27-mph wind, lasted for just twelve seconds, but it introduced a new phase in human transportation.

Orville Wright's pilot's certificate

Misfortune at 32

King Farouk, the last monarch to rule Egypt, was sent into exile following an effortless coup by his army officers. Renowned for his gluttony, promiscuity and kleptomania, the 21-stone king suffered from bouts of impotence, but attempted intercourse with 5,000 women in his lifetime. When he was deposed his possessions were found to include huge amounts of pornography and warehouses full of stolen goods, including the ceremonial sword and medals filched from the corpse of the Shah of Iran as it passed through Cairo, and Winston Churchill's heirloom gold watch.

Lucky Luciano, the American racketeer, suffered the misfortune of being 'taken for a ride' when he was 32. Rival gangster Legs Diamond had him beaten, shot, tortured and left for dead. Amazingly, he survived, acquired the nickname of 'Lucky', and within a year had wiped out the entire Diamond gang to become the undisputed boss of New York crime. He destroyed the old-style mobsters and became the pioneer of 'organized crime', infiltrating big business, until he was eventually jailed on vice charges and later deported to Italy.

Crime at 32

Lizzie Borden was arrested at 32 for killing her parents on the morning of August 4, 1892. Her mother was killed in her bedroom at Fall River, Massachusetts, her head chopped nineteen times with a hatchet. Her father, napping on a sofa downstairs, was killed with ten hatchet blows. Local children developed a skipping rhyme: 'Lizzie Borden took an axe/And gave her mother forty whacks/When she saw what she had done/She gave her father forty-one.' Although she became world famous as a murderess, in reality she was acquitted of the crime. She was a charity-worker and do-gooder and remained so all her life. She was buried in the family plot alongside her parents, whose true murderer was never found.

Life-span of 32

Alexander the Great died in the year 323 BC. Although only 32, he had in his short life conquered Persia, Syria, Phoenicia, Egypt, and lands to the East as far as India. With an army that today would occupy only about one-third of Wembley Stadium, he travelled some 21,000 miles and dominated an empire roughly the same size as the United States.

John Dillinger, the American gangster dubbed 'Public Enemy No. 1' by the FBI, was shot and killed outside a cinema in Chicago in 1934.

Brian Epstein, the man who masterminded The Beatles, was found dead from a drugs overdose in London in August 1967.

Bruce Lee, the archetypal kung-fu hero from the slums of Hong Kong was, in fact, born in San Francisco and was a philosophy student at Washington University who played minor roles in *Batman* on television before finding fame in 'Chop-suey Westerns'. He quickly developed a devout, global following of martial arts enthusiasts. He died mysteriously after taking a pain-killer for a headache in 1973 when working on a new film, and his more fanatical fans still believe that foul play was involved. He was replaced in Hong Kong films by lookalikes called Bruce Li, Bruce Le and Bruce Lei.

Richard III, King of England, died in battle at Bosworth Field. He was a controversial monarch whose villainous reputation stems more from Shakespeare's play than from history. In reality he seems to have been a just legislator, an able administrator and a courageous soldier.

Charles Rolls, the salesman to Royce's engineer, not only founded Rolls-Royce, but was also a pioneer aviator who was killed when his tail-plane collapsed as he was making a steep descent on to Bournemouth aerodrome, in July 1910. He was the first Englishman ever to be killed flying an aeroplane.

33

Thirty-three is the age at which Byron was already beginning to feel drained of his youthful drive. In his diary, the entry for January 22, 1821 reads: 'Through life's road so dim and dirty,/I have dragged to three and thirty:/What have these years left to me?/Nothing, except thirty-three.' Another sour note is added by French revolutionary Camille Desmoulins. When asked his age by the French Tribunal on April 3, 1794, he replied: 'I am thirty-three – the age of the good Sans-Culotte Jesus: an age fatal to revolutionists.' Sans-Culotte was the name given to the revolutionaries and he was implying that, in his rebellion against authority, Jesus was 'one of them'. The parallel was accurate in at least one respect: like Jesus he was quickly executed.

Despite these gloomy associations, 33 is for many a time of 'young maturity', when the world offers challenges rather than threats, when nothing seems impossible and when no time-span limitations have yet crossed the minds of the successful and the well-adjusted. For those less fortunate, who have become fixated on extreme youth or who have burned themselves out quickly, then Byron's moan will, of course, be more appropriate.

Life Expectancy
Thirty-three was the estimated life expectancy in England in the Middle Ages. And as late as the year 1900, it was the life expectancy for new-born, non-white Americans, fully sixteen years less than that for white Americans. By 1967 this difference had been reduced to only five years and was still shrinking as the status of non-white Americans continued to rise.

In the Central African Republic, however, the figure for life expectancy at birth for males remains at 33 even today.

Achievement at 33
Alexander Calder, the American sculptor, invented a new art form in 1931 when he was 33 – the *mobile*. These moving sculptures were standing or hanging, brightly coloured metal constructions, the parts of which swung or dipped as the wind blew them, making a changing pattern of shapes. A big, burly, bearlike man, he graduated not as an artist but as a mechanical engineer.

Lewis Carroll, an English mathematician whose real name was Charles Dodgson, took three little girls, the daughters of his college Dean, for a trip on the River Thames on the afternoon of July 4, 1862. He rowed them from Oxford to Godstow where they stopped for a picnic on the river-bank. It was there that he told them the tale of *Alice's Adventures in Wonderland*, which was destined to become the most famous children's story in the world. He was persuaded to publish it in 1863, under a pseudonym, at the age of 33.

Ninette de Valois, the Irish-born ballerina and teacher, founded what is now the Royal Ballet in 1931. She helped to create a distinct style of dancing and, with Marie Rambert, was the co-founder of the British Ballet.

Amelia Earhart, American aviator extraordinary, was the first woman to cross the Atlantic Ocean single-handed. Her record-making solo flight took place in May 1932.

Jane Fonda, the American actress, won the first of her two Oscars for her 1971 performance as a prostitute in *Klute*. After an unpromising start with pretty pin-up roles in the 1960s, she suddenly blossomed into the finest film actress of the 1970s. Increasingly left-wing politically, she has often offended the establishment, but they are helpless in the face of her great talent.

Clark Gable also received his first Oscar at this age. It was for his comedy performance in *It Happened One Night*, a role he did not want and which was given to him as a punishment for his objecting to 'brutish' typecasting. The film, a minor project, also won the 1934 awards for Best Actress, Best Film, Best Director and Best Screenplay.

Jean Genet, outstanding French novelist, playwright, male prostitute and burglar, produced his masterpiece *Our Lady of the Flowers* while in prison. It was published in 1944 when he was 33, but he was again convicted of burglary and given a life sentence. Cocteau and Sartre pleaded successfully for his reprieve and he went on to become a major figure in the Theatre of the Absurd.

Edmund Hillary, the New Zealand mountaineer, was this age when he and Sherpa Tensing were the first climbers to set foot on the summit of Mount Everest, at 11.30 am on May 29, 1953. Since then more than a hundred people have managed the same feat.

Boris Karloff, the heavy-browed English actor with the sinister lisping voice, became famous at 33 for his classic role as Frankenstein's monster. It was the

Alice by Tenniel

start of a long career as Hollywood's top horror star, rivalled only by Bela Lugosi.

Auguste Lumière, the French inventor of the 'Cinématographe', presented what is generally considered to be the first ever 'movie' on December 28, 1895, at the Grand Café, Boulevard des Capucines in Paris. The film was called *Lunch Break at the Lumière Factory* and its showing marked the beginning of cinema history and established the 33-year-old Lumière as the father of the cinema.

Martin Luther, the German Protestant leader, horrified by the luxury and corruption of the papal court, finally rebelled against Rome on October 31, 1517, when he fixed to the door of All Saints Church, Wittenberg, an announcement of his condemnation of church abuses, which caused a major religious split and precipitated the Reformation.

Martin Luther

Arthur Miller, the American playwright, achieved his major success with his play *Death of a Salesman* in 1949.

Claude Monet, the French artist, exhibited an oil sketch of the harbour at Le Havre, under the title *Impression: Sunrise*, in 1874. It was from this that the Impressionist movement was christened, when a journalist, making a sarcastic attack on the painting, headed his article 'Exhibition of the Impressionists'.

Alfred Nobel, the Swedish chemist, was this age when he invented dynamite in 1867, the basis of his fortune which has since financed the annual Nobel Prizes.

Julius Reuter, the German-born founder of the famous international news agency which grew to become the largest network of information services in the world, began in a small but imaginative way. His first great success was in 1849 when, at the age of 33,

he organized the transmission of stock exchange prices between Aachen and Brussels by a systematic use of pigeon post.

Crime at 33

O. Henry, the American author, was indicted for embezzlement of bank funds in February 1896. He fled to Honduras but returned when he heard of his wife's fatal illness. His trial was postponed until after her death and, on conviction, he was given the lightest of sentences which was eventually shortened to three years three months for good behaviour. It was while working in the prison hospital that he found time to write a series of short stories employing a characteristic style with a surprise twist in the ending. These were published under the pseudonym of O. Henry (his real name being William Porter) and when he was released from jail and moved to New York he soon became famous as one of the most popular short-story writers in America.

David Stein, the French art forger, was jailed in January 1969 at 33 because of a stupid error. He had successfully and with great skill produced 400 fake pictures by modern artists including Picasso, Chagall, Modigliani, Renoir, Cézanne, Van Gogh and Dufy. Moving to New York he opened the Gallery Trianon to sell his forgeries and quickly made a million dollars. But when an art dealer demanded authentication Stein forged the documents too quickly for them to have been posted from France and he was trapped.

Retirement at 33

Henry Armstrong, the American who was the only boxer to hold three world championships simultaneously, retired from the ring in 1945 and became a Baptist Minister.

Life-span of 33

Eva Braun, the mistress of Adolf Hitler who became his wife shortly before their joint suicide in 1945, was a photographic saleswoman when she first met him. Her function as the Führer's companion seems to have been to provide him with domestic comfort rather than erotic pleasure.

Pretty Boy Floyd, the American outlaw, was gunned down in a field by the FBI in 1934. He was a poor boy who robbed a bank to feed his 17-year-old pregnant wife living on his father's dirt farm. While he was in jail his father was murdered by a neighbour and, once free, Floyd tracked the man down and killed him. After this he became a full-time, machine-gun-carrying bank-robber, until he was cornered by Melvin Purvis and shot.

Dick Turpin, the English highwayman, was an unsavoury criminal involved in smuggling, burglary, highway robbery and murder in early eighteenth-century England. Sentenced to death for horse-stealing he did at least die bravely, leaping into the air from the gallows cart before it was drawn away. He was converted into a romantic figure in the 1834 novel *Rookwood* by Ainsworth.

Thirty-four is seen by some as the 'last year of youth', because it is followed by the traditional half-way mark of 35. This thought seems to have focussed Carl Werner's attention on the age 34 in his poem *The Questioner*: 'I called the boy to my knee one day,/ And I said: "You're just past four;/Will you laugh in the same lighthearted way/When you've turned, say, thirty more?"/Then I thought of a past I'd fain erase –/More clouded skies than blue –/And I anxiously peered in his upturned face/For it seemed to say: "Did you?"'

In America, according to Gail Sheehy in *Passages*, 'Thirty-four is the average age at which the divorced woman takes a new husband. By this time an average of thirteen years has elapsed since her first wedding day. She will have another try at building a partnership to fulfill her need for intimacy.'

Life Expectancy In Gambia the average life expectancy for females is still only 34.

> Among the animals, this is the greatest age that can be expected for any small cagebird. The record is held by a 34-year-old cock canary called 'Joey' owned by Mrs K. Ross of Hull. It lived from 1941 to 1975.

Achievement at 34 **David Bushnell**, the 'father of the submarine', was an American inventor who, in 1776, built a unique turtle-shaped vessel designed to be propelled under water by an operator who hand-turned its propeller. The craft was armed with a torpedo or mine to be attached to the hull of an enemy ship. Several attempts were made against British warships during the American Revolution.
Lon Chaney Jnr, the central figure of many Hollywood horror films, as Frankenstein's monster, Dracula, the Wolfman and the Mummy, created his most impressive role in 1940, as the half-witted Lenny in *Of Mice and Men*.

Clint Eastwood in *A Fistful of Dollars*

Clint Eastwood, the American actor, had his first major success with *A Fistful of Dollars*, the 'Spaghetti Western' that started his international career as one of the highest paid actors in the world. In real life a child of the depression who grew up in poverty, he became the archetype of a new-style hero – a tough loner, brutal, expressionless, resilient and yet oddly soft-spoken.

George Gallup, the American pollster who gave the world the 'Gallup Poll', founded the American Institute of Public Opinion in 1935.

Oliver Hardy, the American actor, teamed up with Stan Laurel to create the greatest comedy duo in the history of the cinema.

Thomas Hardy, the foremost regional English novelist, achieved his first public success with *Far from the Madding Crowd*, in 1874. It was sufficient to encourage him to devote himself entirely to writing from this age onwards.

Ben Jonson, the early English dramatist, also had his first great popular success at 34, with *Volpone* in 1606. Second only to Shakespeare in Elizabethan times, he was eight years younger than his great rival, whom he always managed to out-talk in 'wit-combats' at London taverns. A less happy victory occurred when he killed a fellow actor in a duel. Although he managed to escape hanging, he could not avoid branding, but with some help from his friends this was done with a cold iron.

Arthur Koestler, the Hungarian author and critic, drew on his own experiences as a prisoner under sentence of death in the Spanish Civil War, when he produced his most widely read work *Darkness at Noon* in 1940, in which an old Bolshevik is compelled to confess to treason despite his innocence.

Pierre Larousse, the French encyclopaedist whose ambition was to teach everyone about everything, opened his famous publishing house in 1852, which was later to produce his monumental *Grand Dictionnaire Universel* in fifteen volumes.

Austen Henry Layard, one of the great Victorian archaeologists, who astonished the world with his amazing discoveries in Mesopotamia at Babylon and Nineveh, completed his field-work in 1951 at the age of 34 and steadfastly refused to return to the East, saying '... the climate always disagrees with me, and I can find neither books nor society. I should like to get into Parliament in England ...' which he did with great success.

Edward Lear, one of the world's greatest natural history illustrators, has always been better known for the 'nonsense poems' which he wrote for the grandchildren of the Earl of Derby, whose private menagerie he was illustrating. His first *Book of Nonsense* appeared in 1946 when he was 34. The youngest of twenty-one children, he was an epileptic who never married despite his great love of children.

Arnold Palmer, the American golfer, was the first person to win the US Masters Tournament four times, a feat which he achieved in 1964 when he was 34. He was also the first golfer to win a million dollars in tournament prize money.

George Pullman, the American inventor of the Pullman Sleeping Car for use on railways, saw his first car, called *The Pioneer*, go into action in 1865.

Pullman Dining Car

Arnold Schoenberg, Austrian-born composer who became an American citizen, was the most extreme innovator in twentieth-century music, introducing the concept of 'atonality' in 1908.

Scipio Africanus the Elder, one of the finest soldiers of the ancient world, achieved his greatest victory in 202 BC when his Roman Army finally destroyed the forces of Hannibal at the Battle of Zama, outside Carthage.

Conversion at 34

John Wesley, the English religious leader and founder of Methodism, experienced a dramatic conversion at 8.45 pm on May 24, 1738, when 'I felt my heart strangely warmed. I felt I did trust in Christ alone for salvation, and an assurance was given me that He had taken away my sins.' From this point onwards he broke away from the established church and based his way of living on the *'method* laid down in the New Testament'. He began to hold open-air meetings, preaching a total of 40,000 sermons, travelling on horseback sixty to seventy miles a day. He appealed to the working classes who were bored by the fossilized rituals of the orthodox church.

Misfortune at 34

Captain William Bligh was cast adrift in an open boat on April 28, 1789, by the mutineers who had taken over his ship, the *Bounty.* After a voyage of more than 4,000 miles, he reached Timor on June 14. This misfortune made little difference to his career and he survived two further mutinies to become a Vice Admiral.

Crime at 34

Fatty Arbuckle, the huge comedian who was one of the most popular figures of Hollywood's early silent days, had his career cut short by a scandal in 1921, when he was tried for manslaughter. Although he was acquitted, the scandal ruined his professional life. The incident occurred at a drunken party given by him at the St Francis Hotel in San Francisco. He allegedly took a young starlet, appropriately named Virginia Rappe, into a separate room, tore her clothes off, raped her and then assaulted her sexually with a champagne bottle. He emerged doing a drunken dance as his guests stared down at the blood-covered girl. She died three days later of a ruptured bladder. The jury who acquitted him posed for photographs with the great star before dispersing.

Charles Manson, self-styled hippie messiah, was also 34 when he committed his most brutal crime in 1969, despatching his gang of drop-out 'slaves' to Roman Polanski's Beverly Hills mansion to slaughter all the occupants as savagely as possible. Five people died that night and the following night he made them repeat their ritual slaughter, killing an elderly couple at another large house. Loyalty was obtained by drugs, group sex and Manson's strange, compelling personality. When he was eventually brought to court, it was the longest murder trial in American history – nine and a half months. All the defendants including Manson were convicted and jailed for life.

Life-span of 34

Maureen Connolly, 'Little Mo', the popular American tennis star of the early 1950s whose career was tragically cut short by a riding accident in 1954, died young, of cancer, in 1969.

Georges Danton, the French revolutionary leader who played a major role in overthrowing the monarchy and then governing the country, became increasingly critical of the excesses of the Reign of Terror. He defied the Committee of Public Safety by crying 'Better a hundred times to be guillotined than to guillotine!' Their response, in 1794, was to give him what he considered the better alternative. At his execution, his remarkable last words to the executioner were: 'You will show my head to the people, it is well worthwhile.'

Yuri Gagarin, the Russian cosmonaut and the first man in space, died in an ordinary aircraft crash in March 1968.

Henry V, King of England, a popular warrior and able monarch who made his country the strongest in Europe, died of camp fever at the height of his power in the year 1422, during a siege.

Joe Orton, the English playwright and exponent of savage black farce, who used the stage as a platform for his derisive raging against society, was beaten to death with a hammer by his male lover, Kenneth Halliwell, who then took his own life, in 1967.

Mungo Park, the Scottish explorer of Africa, was killed on his second trip to the River Niger. His canoe was attacked by natives and he drowned while trying to escape, in the year 1806.

35

Thirty-five is the half-way mark to the Biblical 'three-score-years-and-ten'. For some this makes it an exciting peak moment in life, bursting with creative energy. For others it becomes a sombre time when they first contemplate the thought that, in the rise and fall of life, they are about to enter the fall. Carl Jung emphasizes this when he speaks of 'over thirty-five' as being 'the second half of life', as if at 35 we move into the final act of a two-act play. Samuel Johnson provides a note of urgency: 'For, howe'er we boast and strive,/Life declines from thirty-five./He that ever hopes to thrive,/Must begin by thirty-five.'

Schopenhauer adds his weight to this anxiety-making: 'The intellectual powers are most capable of enduring great and sustained efforts in youth, up to the age of thirty-five at the latest; from which period their strength begins to decline.' And an early Victorian poem called simply *Thirty-Five* cries out: 'One backward look – the last – the last!/One silent tear – for youth is past!'

As far as human potential is concerned, the facts do not support this pessimism. Many people continue to grow in mental stature in the years after 35, some of them for many years past it, despite the ageing of the body. But it has to be admitted that, for some, 35 becomes an age of crisis. In America it is the most common age for the 'runaway wife'; also the age when a wife is most likely to be unfaithful – perhaps connected with the fact that it is the age when the average mother sends her last child off to school. And for childless career-women, 35 is the year when they are most likely to stop and contemplate their dwindling years of efficient fecundity. But for the majority of people, like astronauts Eugene Cernan and William Anders who went to the moon at 35, it is the age when youthful vigour and mature experience find their perfect balance.

Life Expectancy

Today a British male who has reached 35 can expect to live for another 36·6 years; a British female another 42·1 years. This is a great improvement on life in eighteenth-century Britain when the life expectancy at birth was only 35 years. Nowadays such a figure at birth is found only for females in Chad.

Accession at 35

Edward I, King of England, was crowned at Westminster in 1274. Before his coronation he was known for his arrogance and violent temper, but as King he mastered his anger to become a great law-maker and administrator – the father of parliament.

Napoleon was crowned Emperor of France in the presence of the Pope, at the Cathedral of Notre Dame in Paris, in 1804. The Pope only presided over the ceremony – Napoleon crowned himself and then Josephine.

Achievement at 35

Blondin, the most famous of acrobats and tightrope walkers, made his first tightrope crossing of Niagara Falls in 1859.

Blondin at Niagara Falls in 1859

Robert Boyle, the British scientist, published in 1662 'Boyle's Law', as it became known: 'So long as the temperature remains constant, the volume of a body of gas varies inversely with the pressure upon it.' He was a pioneer of the experimental method, on which all modern science is based.

Hernan Cortés, the Spanish Conquistador, completed the conquest of the Aztec Empire in 1521, with a small force of only 508 soldiers.

Gustave Flaubert, the French realist novelist, published his masterpiece *Madame Bovary* in 1857. The French government brought him to trial on the grounds of the novel's alleged immorality and he narrowly escaped conviction.

Walter Gropius, the German-born architect, founded the 'Bauhaus' in 1919, a progressive school of arts and crafts where staff members included such giants of modern art as Kandinsky and Klee.

James Harrison, the Scottish engineer, emigrated to Australia, where in 1851 he developed the first refrigerator. The invention was initially used by the brewing industry at Bendigo and for the freezing of meat for shipment.

Victor Hasselblad, the Swedish inventor, developed a revolutionary camera for the Swedish Air Force in 1941. It was the world's first $2\frac{1}{4} \times 2\frac{1}{4}$ inch single lens reflex camera with interchangeable lenses and magazines. It became known as the 'Rolls-Royce of cameras' and was used by the United States to take the first pictures of the moon.

Franz Lehar, the Hungarian composer of operettas,

achieved a major success with *The Merry Widow*, first staged in 1905.

Evangelista Torricelli, the Italian inventor, mathematician and scientist, invented the barometer in 1643.

Misfortune at 35

Victor Brauner, the Romanian surrealist artist, had his left eye put out by a bottle thrown by fellow surrealist Oscar Dominguez, at a studio party brawl in Paris in 1938. By truly surrealist coincidence, a painting produced by Brauner in 1931 showed himself with one eye crushed and bleeding, and another in 1932 depicted him with one eye pierced by a sharp instrument bearing the initial D.

Horatio Nelson, by another coincidence, was also 35 when he lost the sight of his right eye in 1794. A shot from the enemy threw gravel up into his face with such force that it destroyed the eye. Years later he was able to exploit this misfortune when attacking Copenhagen, by raising his telescope to his blind eye to avoid seeing the order to retreat.

Roman Polanski, the Polish-Jewish film director, who, as a boy, had been used for target practice by brutal German soldiers and whose parents had been taken to a concentration camp when he was 8, was to encounter further horrific violence in his life, a few days before his thirty-sixth birthday. On the night of August 8-9, 1969, the Charles Manson gang attacked his house in Beverly Hills, California, and killed all the occupants including his beautiful young wife, actress Sharon Tate, who was eight months pregnant. Sharon Tate begged for her life because of the baby, but she was beaten, hanged, shot, and stabbed sixteen times. Polanski was in London working on a script at the time or he would undoubtedly have been killed with the others.

John Ruskin, the major Victorian art critic, suffered the humiliation of having his six-year marriage annulled because his young wife was still a virgin. Kept in ignorance of sexual matters by his puritanical mother, he had been horrified to discover that, unlike marble statues, women have pubic hair between their legs. This so disgusted him that he had been unable to consummate the marriage.

Crime at 35

Alfred Dreyfus, the French-Jewish army captain, was accused of selling military secrets to the German military attaché in 1894. He was convicted and sent to serve a life sentence on the notorious Devil's Island. But the legal proceedings were based on insufficient evidence and a major controversy arose. He was later pardoned, given the Legion d'Honneur and made a Major. A crime was committed when he was 35, but it was against him, not by him.

Guy Fawkes, the English soldier and key figure in the Gunpowder Plot to blow up Parliament, was arrested on November 4, 1605, when the plot mis-

fired. A man of great courage and cool determination, he had been enlisted by a group of conspirators who wished to assassinate King James and his ministers as a retaliation against increased religious oppression. More than twenty barrels of gunpowder had already been secreted in the cellars beneath Parliament before the attempt was discovered. Fawkes was tortured on the rack until he revealed the names of his accomplices, and he was executed a year and a day later, on November 5.

The Kray Twins, Ronnie and Reggie, who dominated London's underworld in the 1960s with a 150-strong gang known as 'The Firm', were jailed in 1969 for life sentences of thirty years with no possible parole. Immensely rich and powerful from protection rackets and other criminal activities, they were Capone-imitators and Nazi-admirers whose beatings and gangland killings gave them a reputation unique in British crime.

Life-span of 35

Saint Bernadette, the French woman who had a vision of the Virgin Mary and spoke with her, became a nun and served as a nurse in the Franco-Prussian War, before dying of tuberculosis in 1879.

Crazy Horse, the Sioux Indian Chief who fought and killed many US soldiers, eventually became their prisoner. Confined at Fort Robinson he was killed trying to escape.

Ned Kelly, the famous Australian outlaw, a robber and killer who had been on the run for years, prepared for his final showdown by stealing some iron and manufacturing a cumbersome, heavy suit of armour with a cylindrical helmet. He stood and shot it out with the police until his leg wounds weakened him and he was caught and hanged in 1880. His last words were: 'Such is life.'

Jayne Mansfield, the American actress who was a tragic caricature of Marilyn Monroe, died even younger than her famous model, in a car crash near New Orleans on her way to appear in a television show. With her enormous breasts, her tiny waist, her little girl voice and bland, blonde head, she was like a cartoon figure come to brief life.

Amedeo Modigliani, the Italian painter of brilliantly original, elongated nudes and portraits, lived a short, tortured, miserable life. Sickly, half-starved, and always poor, he eventually resorted to alcohol, cocaine and hashish. His only exhibition in his lifetime was closed by the police as obscene. He died in 1920 from tubercular meningitis and the tragedy was made complete by the suicide of his distraught teenage art-student mistress who, pregnant with his child, jumped to her death from a fifth-floor window on the day of his funeral.

Wolfgang Amadeus Mozart, the Austrian musical genius, died penniless in 1791 from heart failure following rheumatic fever and blood-letting, leaving his widow in a state of nervous collapse and with huge debts.

Thirty-six has the distinction of being the start of the human female's sexiest period of life, according to the Kinsey Report. For the American female, at least, the period from 36 to 40 is the phase when the 'Total Sexual Outlet' reaches its highest 'Active Incidence'.

Life Expectancy

If you lived in seventeenth-century London, you had only a 16 per cent chance of surviving to 36. The main reason was the enormous number of deaths in infancy or at birth due to the lack of hygiene in dirty cities. Today the life expectancy figures have risen dramatically, but there are still some Third World countries where infant mortality remains alarmingly high. Life expectancy at birth for females in the Central African Republic, for instance, is still only 36 years.

> Among the animals, 36 is the very longest life-span a domestic cat can enjoy, the oldest known pet cat being a tabby called 'Puss' who lived from 1903 to 1939.

Accession at 36

Queen Elizabeth, the present Queen Mother, unexpectedly became Queen of England in May 1937, following her brother-in-law's unique abdication, and found herself destined to become the mother-figure of a country about to be plunged into war.

Achievement at 36

Edgar Rice Burroughs wrote his first Tarzan story, *Tarzan of the Apes*, in 1914. He went on to produce a total of thirty Tarzan books and saw them translated into fifty-seven different languages and made into many films.

Gordon Cooper, the American astronaut, was faced with the unscheduled challenge of piloting his spacecraft back to earth manually, following the breakdown of the automatic control system at the end of the sixth manned space flight launched by the US, in May 1963. His success was such that he landed only five miles from the recovery ship.

Humphry Davy invented his famous Miner's Safety Lamp in 1815. It saved countless lives in the appalling working conditions underground.

James Frazer, the Scottish anthropologist, published the first edition of his monumental work on magic and religion in the primitive world – *The Golden Bough* – in 1890.

Sigmund Freud managed to persuade his teacher, Josef Breuer, to begin work on their joint paper called 'On the Psychical Mechanism of Hysterical Phenomena' in 1892. This was the starting point of Freud's unique contribution to science – the study of psychoanalysis.

Adolf Hitler was also busy at the age of 36, for it was then that he published the infamous *Mein Kampf*, written during a nine-month stay in jail, following the abortive Munich 'Beer Hall' *putsch* of 1923.

Burt Reynolds, the American actor with the non-chalantly macho screen image, set a new trend by being the first male film star to pose for a nude centrefold photograph in a glossy magazine – the American *Cosmopolitan*, in April 1972.

Revelation at 36

Mohammed Ahmed, the Sudanese Muslim fanatic, declared himself the *Mahdi*, or Messiah, on June 29, 1881, because it was revealed to him that God had appointed him to purify Islam and to destroy all governments that defiled or oppressed it. Welding the faithful into an unconquerable military machine he made himself master of almost the whole of the Sudan.

Misfortune at 36

Frank Sinatra, the scrawny crooner from Hoboken, New Jersey, suffered a major setback. In 1952 his vocal cords suddenly haemorrhaged and he was dropped by MCA. But he was a tenacious fighter and begged a part in the film *From Here to Eternity*, agreeing to work for a nominal ($8,000) sum. Overnight he became an Oscar-winning actor with Hollywood at his feet, and later his voice returned to make him a dominant figure in both popular music and the cinema.

Andy Warhol, the American Pop Artist, was shot and nearly killed by Valerie Solanis, an embittered member of an anti-male hate group. Warhol's Manhattan studio, known as 'The Factory', was busy turning out bizarre films with titles such as *Flesh*, *Bitch*, *Heat*, *Kiss* and *Trash*, and Solanis was one of his 'Factory' rejects.

Retirement at 36

Greta Garbo, the Swedish genius of the cinema, went before the cameras for the last time in 1940. Her final film, *Two-faced Woman*, was a disaster when it was released in 1941 and she promptly announced her retirement, disappearing to become a press-shy, semi-recluse for over forty years.

Adolf Hitler   Andy Warhol

Life-span of 36 Cyrano de Bergerac, the notoriously long-nosed French satirist and dramatist, died at this young age, his most interesting works being published after his death. They were fantastic stories of imaginary journeys to the moon and the sun, remarkably advanced for the middle of the seventeenth century, and which influenced both Swift and Voltaire in their free-thinking materialism and their ridicule of authority and religion.

Georges Bizet, the French composer, died suddenly, shortly after completing what was to be the most famous of his eight operas, *Carmen*.

Lord Byron, the English poet and sexual athlete, died of a formidable combination of malaria, malnutrition caused by a fight against obesity, a severe chill, venereal disease and, finally, the disastrous attentions of his doctors, with their savage bleeding treatment. Three months before his death, on the occasion of his thirty-sixth birthday, he wrote a poem entitled 'On this day I complete my thirty-sixth year': 'My days are in the yellow leaf;/The flowers and fruits of love are gone;/The worm, the canker, and the grief/Are mine alone!'

Caravaggio, the most violent man ever to be possessed of artistic genius, died of malaria two months short of his thirty-seventh birthday. After being arrested and thrown in jail at a port near Rome, he was released angry, alone and penniless. He stormed off along the swampy shore on a sullen trek to the nearest town. By the time he had arrived he had contracted the disease and was soon dead, robbing Italian art of one of its most innovative painters.

General George Custer, the American cavalry officer, was slain by Indians at Little Bighorn. Whether he was a courageous fighter or a foolish glory-seeker is still a subject for debate, but it is certain that as a leader of men he was a disaster, since every one of the 250 he took into this last battle was killed.

Jesus Christ, whose major act of rebellion was to replace a God of Vengeance with a God of Love, was adored by the people and hated by the ruling classes, making his punishment and death at an early age inevitable. Scholars now agree that he was born in the year 6 BC, giving him an age of 36 at the time of his crucifixion.

Bob Marley, the Jamaican pioneer of reggae music, died in Florida in 1981 after a seven-month fight against cancer and was given a lying-in-state in the Jamaican capital, Kingston, before his burial.

Marilyn Monroe was found dead in mysterious circumstances in 1962 at the age of 36 years and 2 months. Officially her death was reported as a suicide from an overdose of barbiturates, but some investigators have sensationally insisted that she was murdered. They claim that the medical evidence rules out suicide but that this was simulated by cold-bloodedly injecting her with barbiturates. They feel she was killed because of her supposed sexual liaisons with President John Kennedy and his brother Rob-

Henry Purcell   Marilyn Monroe

ert, the Attorney General. Whether she was eliminated to silence her and conceal these affairs, or whether she was murdered to draw attention to them and thus discredit the Kennedys is a matter still debated by those who see in her death yet another Great American Cover-up.

Henry Purcell, one of the greatest English composers, was amazingly prolific in his short life, composing 200 songs, 70 anthems, 6 operas, theatrical music, chamber music, and a great deal of church music before his early death in 1695.

Gilles de Rais, a brilliant French warrior who fought alongside Joan of Arc, was one of the wealthiest nobles in Europe, with a personal retinue of 200 knights, and 30 priests to pray for him. In private, however, he was a depraved monster, his greatest pleasure being the sexual torture of small children. Basing himself on the Roman Emperor Caligula, he killed between 200 and 800 young boys and girls before he was arrested in 1440, tried and garotted.

Robespierre, the French revolutionary leader, was guillotined following a suicide attempt in which he wounded himself with a pistol-shot in the jaw. The mastermind behind the Reign of Terror, he was eventually accused of setting himself up as a dictator, was turned on by the people and executed amidst a cheering mob.

Gene Vincent, the pioneer American rock singer, died almost forgotten in 1971 following a long slide into alcoholism. The original greasy-haired, black-leather-jacketed rocker of the 1950s, he was Capitol Records' answer to Elvis Presley – tough, working-class, threatening – the archetypal Rock 'n' Roll Hell's Angel.

Antoine Watteau, the romantic French artist, suffered from tuberculosis for many years, but managed to introduce a new style of court painting, replacing the heavy formality of previous works with lyrical lighthearted, frothy, amorous, escapist scenes, before dying of his ailment at the peak of his career in 1721.

Thirty-seven is the ideal age for a man to marry, according to Aristotle. In the *Politica* he goes on to say that the perfect age for the bride is 18, and defends this age difference by pointing out that the couple would then experience their sexual decline at the same time – he at 70 and she at 50.

Certain modern authors comment apprehensively on this age. They see 37 as the starting point of the 'mid-life crisis' and a peak period for anxiety and personal upheaval. Gail Sheehy in *Passages* remarks: 'The age of 37 kept coming up as a prominent death line among artistic and highly industrious people ... the artist may burn out creatively or literally die.' Paradoxically, however, 37, like the late 30s generally, is a peak period for human creativity and achievement. Either way, it heralds a period of change. As Sheehy points out: '... of those artists who physically and creatively survive this unanticipated crucible, there are few whose work does not undergo a decisive change. The reactions vary from an agonizing eruption to a smoother transition, just as they do in the general population.'

Life Expectancy
Life expectancy at birth for males in Angola, Ethiopia, Guinea-Bissau, Madagascar and Yemen is still only 37.

Accession at 37
Mary I, Queen of England, daughter of Henry VIII and his first wife, Catherine of Aragon, came to the throne in 1553. A plain, forceful, bluff, hearty and initially popular woman, she became known as 'Bloody Mary' because her restoration of Catholicism led to savage persecution of Protestants.

Romance at 37
Richard Burton, the Welsh actor, became involved in a highly publicized romance with his co-star Elizabeth Taylor during the making of the multi-million-pound film epic *Cleopatra*. When the film appeared in 1963 it echoed the off-screen romance between Burton and Taylor that destroyed both their marriages, but converted Burton from an accomplished stage actor into an international film star.

Richard Burton in *Cleopatra*

Paul Gauguin, the French painter, left his wife and family to devote himself totally to his great love – not another woman, but his painting. Although he had been a successful stockbroker, his passion for his art was such that he decided to abandon completely his bourgeois lifestyle. As a result, he suffered discomfort, poverty, ill-health and an early death, but in the process produced some undeniable masterpieces.

Achievement at 37
Alfred Adler, the Austrian psychiatrist, published his famous paper on inferiority and its psychological compensation, in 1907, leading to the popular concept of the 'inferiority complex'.

John Logie Baird, the Scottish pioneer of television, gave the world's first demonstration of the televising of moving objects on January 26, 1926, to members of The Royal Institution at his Soho laboratory in London. Within a few years the BBC had launched the first television service, using Baird's system.

Alan Bean, an American astronaut, spent nearly eight hours walking on the moon's surface in November 1969 during the *Apollo 12* mission.

Lord Beaverbrook acquired a controlling interest in the *Daily Express* in December 1916 and launched himself as one of the great press barons of the century. Politically powerful, he was one of only two people to sit in the British cabinet during both the First and the Second World Wars. The other was Winston Churchill.

Ingmar Bergman, the Swedish film director, shot to international fame in 1955 with his prize-winning film *Smiles of a Summer Night*. He was the son of the chaplain to the Swedish royal family, a stern father who often locked the young Ingmar in a dark closet as a punishment for minor misdeeds. These early experiences had an important bearing on his mature work, his films repeatedly dwelling on philosophical torment and the alienation of the individual.

Louis Blériot, the French aviator, made the first ever ocean-crossing in an aeroplane on July 25, 1909, when he flew across the English Channel from Calais to Dover in a small monoplane of his own design.

Francis Crick, the British molecular biologist, with Watson and Wilkins discovered the structure of DNA – the Double Helix – in 1953, a major step forward in genetics for which they were later awarded the Nobel Prize.

Sergei Diaghilev, the ballet impresario, founded the Ballets Russes in 1909 in Paris. A homosexual Russian aristocrat, he was a domineering man who revitalized ballet by integrating the ideals of other art forms with those of dance.

Antonin Dvořák, the Bohemian composer, reached a creative peak with his *Slavonic Dances* in 1878, which brought him to the attention of a wider public.

Bob Hope, the British-born American comedian, achieved international fame in 1940 when he teamed up with Bing Crosby and Dorothy Lamour for *The Road to Singapore*, the first of seven immensely successful 'Road' films.

Maria Montessori, the Italian educational innovator, opened her first children's school in 1907. The 'Montessori Method' emphasized that young children learn best in an atmosphere of physical freedom and self-help where they can develop their initiative.

Jacques Piccard, the Swiss ocean-explorer, made the greatest ocean descent ever on January 23, 1960, in the Challenger Deep of the Marianas Trench in the Pacific, in a Swiss-built bathyscaphe called *Trieste*, reaching the sea-bed 6·78 miles down.

Lord Reith, the Scottish administrator, established the British Broadcasting Corporation on January 1, 1927. Formidable, towering, stern, unbending and pious, he gave the BBC an integrity, authority and political independence that acted as a model for other broadcasting stations around the world.

John Steinbeck, the American author specializing in studies of human dignity struggling to survive social ills, produced his masterpiece, *Grapes of Wrath*, in 1939.

Barthelemy Thimmonnier, a French tailor, patented the first sewing machine put into practical use, in 1830.

Misfortune at 37

Edward Kennedy, the American politician, drove his car off a bridge on Chappaquiddick Island. He managed to escape from the sinking car himself, but his young companion, Mary Jo Kopechne, drowned. The tragedy occurred on July 18, 1969 and has dogged Kennedy's career ever since.

Life-span of 37

Marie Antoinette, Queen Consort of Louis XVI of France, was by her extravagant behaviour a major cause of the downfall of the French monarchy. Her frivolous lifestyle and apparently callous indifference to the suffering of the poor ('If they have no bread, let them eat cake'), made her unpopular and she became the main target for hatred. She was imprisoned and eventually sent to the guillotine in 1793.

Robert Burns, the national poet of Scotland, died, not of drink as is popularly supposed, but of rheumatic heart disease in 1796. A hard-working farmer's son, he had no idea of becoming a poet until he fell in love.

Nell Gwynne, an illiterate child of the London streets, was put to work as a barmaid in her mother's bawdy house in Covent Garden. With a slender, shapely body and a reckless, high-spirited personality, she charmed her way on to the London stage where she became a leading actress and eventually the mistress of Charles II, whose dying words were 'Let not poor Nellie starve.' She gave him two sons, but died young of 'apoplexy and paralysis'.

Nell Gwynne

Amy Johnson, the English aviator and air heroine of the 1930s for her many record long-distance flights, was drowned in January 1941 after parachuting into an icy Thames. The incident occurred when she was ferrying a plane for the Air Ministry during the Second World War.

Sal Mineo, one of the young stars of the film *Rebel Without a Cause*, grew up in a tough, violent district, portrayed troubled youths in films, and tragically died in character – stabbed to death in an alleyway near his Hollywood home.

Alexander Pushkin, the Russian writer, was killed in a duel at St Petersburg, at the height of his creative powers, in 1837. His marriage to the beautiful Nathalie Goncharova was not a happy one and he lost his life fighting a French nobleman whom he suspected of being her lover.

Raphael, one of the three major figures of the High Renaissance of Italian art (along with Leonardo and Michelangelo) was born on Good Friday 1483 and died, suddenly and unexpectedly, of a fever, on Good Friday 1520, his thirty-seventh birthday.

Ethel Rosenberg, an American spy working for the Russians, was executed in 1953. With her husband Julius, she obtained atomic secrets from her brother who was at the Los Alamos nuclear research station. The brother gave evidence against them which led to their execution – a rare sentence for spies in peacetime.

Count Claus Schenk Von Stauffenburg, a German Staff Officer, was the leader of the unsuccessful attempt to kill Hitler in 1944. On July 20, 1944, he attended a meeting at Hitler's headquarters at Rastenberg and left a briefcase behind containing a bomb. It exploded, but only injured Hitler, and the Count and 150 conspirators were executed.

Vincent Van Gogh, the Dutch genius, shot himself in July 1890 when suffering from intense loneliness and fear for his sanity. He had endured a life of poverty, selling only one of his total output of 800 paintings and 700 drawings.

Thirty-eight is the age of the inventor. More discoveries, inventions and innovations occur at this age than at any other. This appears to be the result of the passing of enough years to add self-confidence to experience, but not so many years as to create rigid thinking.

Life Expectancy Life expectancy at birth for females in Madagascar, Togo and Yemen is only 38 years.

Accession at 38 Frederick the Great, Emperor of Prussia, became ruler in 1740 and although he described the crown as 'a hat that lets the rain in' and himself as 'the first servant of the state', he was in fact an absolute, if enlightened, despot, craving above all things fame and glory, both of which he was to achieve during his reign of nearly half a century.

Achievement at 38 Neil Armstrong, Commander of the *Apollo* mission that put man on the moon, was just two weeks short of his thirty-ninth birthday when he spoke the historic words 'That's one small step for a man, one giant leap for mankind', as he set foot on the moon's surface at 2.56 pm on July 21, 1969.

Rudolf Diesel, German engineer, invented the diesel engine. Although he had been working on his 'Black Mistress' for years, he was not successful in producing a first working model until 1896.

Havelock Ellis, English pioneer of sex research, published the first volume of his major work *Studies in the Psychology of Sex*. The year was 1897 and inevitably he ran into difficulties, including a court case in which the book was attacked as 'a pretence, adopted for the purpose of selling a filthy publication'.

Gabriel Fallopius, the Italian anatomist, discovered what are now known as the fallopian tubes, connecting the ovaries to the uterus; also the semicircular canals of the inner ear, and several major nerves of the head and face. In addition he named the vagina, placenta, clitoris, palate and cochlea. All his discoveries were published in his only major work, *Observationes Anatomicae*, in 1561.

Charles Goodyear discovered the vulcanization of rubber by accident, when he dropped some rubber mixed with sulphur on to a hot stove, in 1839. It was through a development of this discovery that he was able to make possible the commercial use of rubber and it should have made him a rich man, but he died a pauper after being imprisoned for debt, following endless legal wrangles over patents. Others made the millions.

Eugene Ionesco, the Romanian-French playwright, presented *The Bald Prima Donna* in 1950 which established him as the audience-shocking leader of the Theatre of the Absurd. He hated Brecht and all political plays, insisting that 'committed theatre leads straight to the concentration camp'.

Edwin Herbert Land, the American physicist, invented the polaroid camera, which incorporated a revolutionary one-step process for developing and printing, in 1947. Known as the Polaroid Land Camera, it produced a finished print in sixty seconds.

Joseph Lister, a shy, unassuming, gentle genius who believed himself directed by God, pioneered the introduction of antiseptics for hospital surgery, a step which saved countless lives. He first introduced his new method, using carbolic acid, on August 12, 1865, when he was 38 and he was responsible for reducing the mortality rate during surgery to one-third of its previous level.

Georg Ohm, the German physicist, formulated what is now known as Ohm's Law ($C = E/R$) in 1827. In his honour, the practical unit of electricity, the ohm, was named after him. His Law therefore reads amps = volts/ohms.

Linus Pauling, the American chemist, published in 1939 *The Nature of the Chemical Bond, and the Structure of Molecules and Crystals*, one of the most influential textbooks of the century.

James Quick, the American haematologist, developed the Quick Test for assessing the clotting ability of a patient's blood, in 1932.

William Henry Fox Talbot, who retired early from politics to experiment with photography, read a paper to The Royal Society in London on January 31, 1839, entitled *Some Account of the Art of Photogenic Drawing*. In this he described for the first time how to use photographic paper to produce negative prints from which positives could be made. Despite rival claims, this was effectively the moment of the birth of modern photography.

Sherpa Norgay Tensing, the Nepalese mountaineer, with Edmund Hillary, reached the summit of Mount Everest, the world's highest (29,028 feet) peak, at 11.30 am on May 29, 1953. This was the first time the mountain had been climbed.

Charles H. Townes, the American physicist, invented the MASER (Microwave Amplification by Stimulated Emission of Radiation) in December 1953, which quickly led to the development of the LASER, using visible light.

Misfortune at 38 Amelia Earhart, the American aviator, disappeared on July 2, 1937, near Howland Island in the South Pacific. Already a veteran of record-making flights, she had set off to fly around the world, with an alcoholic Irishman, Fred Noonan, as her navigator. After completing two-thirds of the trip her plane vanished and there has since been much speculation about what happened to her, but the facts remain unknown.

Horatio Nelson lost his right arm on March 24, 1797, during an attack on Santa Cruz in Tenerife. A bullet shattered his elbow as he was about to land

Amelia Earhart

World War, he succumbed to influenza. In addition to his poetry, he wrote a surrealist play, in 1917, and this was the first use of the term 'surrealism'.

Charlotte Brontë, the author of *Jane Eyre*, married late, in June 1854, and died as a result of her pregnancy, eight months later, in March 1855.

Caryl Chessman, the American murderer, was executed in the gas chamber of California State Prison, San Quentin, in 1960, only minutes before the news of a reprieve reached the governor. In eleven years, ten months and one week on death row, he had won no fewer than eight stays of execution, a record for any prisoner.

Mario Lanza, the American operatic tenor who attempted the impossible – to make opera popular – became grossly overweight, which suited his voice but not his film image. He adopted such a stringent diet that, along with alcohol and sleeping pills, it killed him at this early age, ending his life with a coronary thrombosis.

Sonny Liston, the American World Heavyweight Boxing Champion, was found dead at his Las Vegas home in 1971. A juvenile delinquent who had been jailed for five years for robbery with violence, he was taught to box by the prison chaplain. Using brute strength rather than style, he forced his way to the top, made and lost a fortune and was soon being arrested for drunken driving and carrying an offensive weapon.

Federico García Lorca, the Spanish poet and dramatist, was shot without trial by the Nationalists, at Granada, during the early days of the Spanish Civil War.

Louis XVI, King of France, was guillotined in 1793. A weak-willed, indecisive monarch, dominated by Marie Antoinette, he committed himself to a policy of subterfuge and deception, but managed to redeem himself slightly by dying with dignity at his execution in the Place de la Révolution.

Tom Mboya, the Kenyan politician, was assassinated while shopping in Nairobi, in 1969. He had succeeded with the anti-White slogan 'Scram out of Africa' and was determined to undermine white power in Kenya, using (by his own account) the confidence he had gained as a student at Oxford University to do so.

Felix Mendelssohn, the precocious German romantic composer, lived at a fever-pitch of intensity until his beloved sister Fanny died in May 1847. He was completely undermined by this, his energies deserted him and he was dead himself by November.

Joseph Smith, the American founder of the Mormon Church, was repeatedly persecuted and ridiculed during his short life. He and his followers were forced to move several times to new locations. Eventually he overstepped himself and, in the year of his death, announced that he was going to run for the presidency of the United States. He was arrested on charges of conspiracy and a mob attacked the jail and lynched him in 1844.

and lead the shore attack. His arm was amputated on board *Theseus*, but it was badly done, trapping a nerve and leaving him in terrible agony.

Lana Turner, the leading pin-up blonde of the Second World War, endured a major family tragedy in 1958. Her lover, underworld hoodlum Johnny Stompanato, was threatening her life when her distraught 15-year-old daughter lost control and stabbed him to death.

Crime at 38
Benedict Arnold, the American traitor, dutifully served the cause of the American Revolution until 1779, when he shifted his allegiance to the British. In April of that year, aged 38, he married a young woman loyal to the British crown and within weeks was making secret overtures to the British Headquarters. As an informer he was soon forced to flee and spent the rest of his days unhappily in England.

Klaus Fuchs, the German-born British spy, was tried and found guilty of passing secret information to the Russians from the Harwell Atomic Research Centre, in 1950. He was sentenced to fourteen years' imprisonment.

Donald Maclean, the British spy for the Russians, was 38 when he fled to Moscow in 1951 with Guy Burgess after being warned by Kim Philby that he was under investigation by the Foreign Office.

Life-span of 38
Guillaume Apollinaire, the French avant-garde poet, died in Paris in 1918. Weakened by wounds received fighting in the First

'Thirty-nine. It is a good age. One begins to appreciate things at their true value.' So reads the entry for October 19 in *An Almanac* by Norman Douglas. The American comedian Jack Benny would agree with him, 39 being the greatest age to which he would admit, right up to his death at age 80. Benjamin Franklin would also agree for he was 39 when he wrote his extraordinary letter of sexual advice, so explicitly worded that it was not released for publication by the US State Department until 1926, 181 years after it was written. His appreciation of the 'true value' of things is expressed in his preference for older women: 'As in the Dark all Cats are grey, the Pleasure of Corporal Enjoyment with an old Woman is at least equal and frequently superior; every Knack being by Practice capable of improvement . . . and . . . They are so grateful!!!'

Life Expectancy Life expectancy at birth for males in Afghanistan, Guinea, Laos, Mali, Mauritania, Niger, Senegal, and Somalia is only 39.

Accession at 39 **William the Conqueror** was crowned William I, King of England, on Christmas Day, 1066, just seventy-two days after his decisive victory at Hastings. He was a just and resourceful leader who brought political stability, centralized government, new architecture and a new language to the country.

Romance at 39 **Jackie Kennedy**, the world's most glamorous widow, shocked the world on October 20, 1968, when she married the much older, wealthy Greek shipping-tycoon, Aristotle Onassis. At 39, she was still a great beauty and universally adored as the tragic heroine of the presidential assassination. Overnight the adulation (which had been suffocating her) turned to dismay at what appeared to be her 'appreciation of the true value of things' in purely monetary terms.

Horatio Nelson was approaching his fortieth birthday when he began his affair with Emma Hamilton in September 1798. The romance started in earnest when he returned to Naples after destroying the French Fleet at the Battle of the Nile. She was a blacksmith's daughter who had risen to become an ambassador's wife and had met Nelson briefly five years earlier. When she first set eyes on him after his long absence she was so horrified by his appearance that she fainted into his one arm. This is not surprising: he had suffered from malaria and yellow fever, had lost his right eye, his right arm and most of his teeth and had been savagely wounded in the forehead. He must have had a remarkable personality to have charmed her in this condition, but charm her he did and the affair flourished to become one of the most celebrated romances in history.

Late Start at 39 **James Robertson Justice**, the Scottish character actor, did not make his first film until he was nearly 40. Earlier, he had been a naturalist and was an expert falconer who had taught Prince Charles the sport. He found fame as the bad-tempered and overpowering Sir Lancelot Spratt in the 'Doctor' films, stealing every scene in which he appeared.

Aristide Maillol, the French sculptor, was the same age before he abandoned tapestry-designing at the turn of the century and turned his attention to sculpture. He was driven to this by an eye disease brought on by the strain of meticulous tapestry work.

Achievement at 39 **Roald Amundsen**, the Norwegian explorer, was the first man to reach the South Pole, on December 14, 1911, beating Scott to it by using sledges with fifty-two dogs.

Aristophanes, the Athenian playwright, produced the first successful piece of feminist propaganda, in 411 BC. The play *Lysistrata* tells the story of a sex strike by the women of Greece. They refuse to grant any sexual favours until the men have stopped making war.

Donald Bailey, the British engineer, invented the 'Bailey Bridge' in 1941. A quickly transportable bridge for urgent river-crossings, it proved invaluable in the later stages of the Second World War.

William Beaumont, the US army surgeon, was able to make the first direct study of human digestion, using a patient with a gunshot wound that never healed, allowing direct observation of digestion as it occurred in the stomach.

Leonard Bernstein, the US composer and conductor, achieved his most popular success, in 1957, with *West Side Story*.

Coco Chanel produced the world's most famous perfume 'Chanel No. 5' in 1922, its success giving her the financial basis for her empire.

William I

Cleopatra

Georges Claude, the French engineer and chemist, invented the neon light in 1910. While studying inert gases he found that passing electrical currents through them produced light without using any filaments. The first use was for fancy display lights of strange shapes.

Vasco da Gama, the Portuguese navigator, returned from one of the most important voyages in the history of Europe, after opening up the sea-route to India via the Cape of Good Hope. He was 37 when he set sail and 39 when he returned, triumphant, to make Portugal a world power.

D. W. Griffith, the pioneer of American epic films, made his masterpiece *The Birth of a Nation* in 1915. He turned cinema into an art form, introducing such basic techniques as the close-up, the long-shot, the fade-in, the fade-out and cross-cutting.

Ben Nicholson, the English painter, famous for his austere geometric designs of circles and straight lines, produced his first totally abstract picture in 1933 while on a visit to Paris.

Jonas Salk, the US microbiologist, was the first to develop a vaccine for the prevention of polio. He tested it successfully on children in 1952 and by 1954, when he was 39, it was in full scale production, saving countless people from the misery of this crippling disease.

Scipio Africanus the Younger, the Roman general, wept as his orders for the destruction of Carthage were carried out and the great city was flattened in the spring of 146 BC. It was his great victory, ending the Third Punic War after a prolonged land and sea blockade of the Carthaginian capital, but the enormity of his destructive act filled him with dread.

Misfortune at 39

Franklin Delano Roosevelt, the thirty-second American President, was paralysed by polio in 1921 and never walked again. He was stricken suddenly while on holiday at Campobello Island, New Brunswick, but his wife insisted that he remain active in politics and her energy saw him through his early agonies and on to his election as President.

Life-span of 39

Lucrezia Borgia, Italian noblewoman and political pawn, who was three times married into prominent families as part of her scheming family's power struggle, and who had a child possibly by her brother Cesare, or more probably by her father, the Pope, died in the summer of 1519.

Lenny Bruce, the most notorious of the sick comedians of the 1960s, was repeatedly at odds with authority, accused of obscenity. First imprisoned in 1961 and repeatedly refused permission to perform, he was eventually convicted of a drugs offence in Los Angeles in 1963. Three years later he was found dead in his Hollywood home.

Frederic Chopin, the Polish composer, died in 1849 of tuberculosis, from which he had suffered for over ten years.

Cleopatra, Queen of Egypt, fled to her mausoleum and barricaded herself in with three attendants when Octavian's forces arrived. He soon captured her but, unlike Julius Caesar and Mark Antony, was not impressed by her seductive powers. On learning that she was to be paraded through the streets of Rome when he made his triumphant return, she killed herself, possibly with a venomous snake, but more probably by taking poison.

Che Guevara, the Argentinian-born revolutionary, restless after his Cuban success, travelled to Bolivia to stir up revolt. He failed miserably, was wounded, captured and killed without trial, his body burnt and his ashes scattered to the winds.

Wild Bill Hickok, American Marshal of Abilene, Kansas, was shot through the back of the head while playing poker. In his hand at the time were two pairs – aces and eights – the 'dead man's hand'. His murderer, Jack McCall, was hanged for his crime.

Stonewall Jackson, Confederate General in the US Civil War, died of pneumonia following surgery, after being shot by his own men. Tragedy struck at the moment of victory. He had ridden forward to organize the pursuit of the enemy and when he returned at dusk he was mistakenly fired upon by his troops and seriously wounded. He died eight days later, following an amputation.

William Joyce, known as Lord Haw-Haw, was a Second World War traitor who offered his services to Goebbels's Nazi propaganda ministry in the summer of 1939. Throughout the war he made pro-Nazi broadcasts from Germany in English. Arrested at the end of the war, he was found guilty of treason and hanged.

Martin Luther King, the American black civil rights leader, was shot in Memphis, Tennessee, in 1968, while standing on the balcony of Room 306 of the Lorraine Motel. The bullet entered his jaw, went through his neck and severed his spinal cord.

Sabu, the Indian actor who became an American citizen and a decorated war hero, died of a heart attack.

Dylan Thomas, the Welsh poet, master of the music of words, might have lived a long, creative life had his magical writings brought him riches as soon as fame. But fame without money turned him into an epic boozer who, at 2 am on November 4, 1953, struggled out of his New York sick-bed and returned later to utter his last words: 'I've had 18 straight whiskies, I think that's a record.' He was treated by a doctor the following day but fell into a coma and died five days later in hospital.

Malcolm X, American black Muslim leader, was a black racist shouting hatred of white people. In 1964 he visited Mecca and became converted to a multi-racial approach, as a result of which he was murdered by extremist black racists, who shot him down in a Harlem ballroom.

40

Forty is the year of the 'middle-aged', although nobody who has reached this age is in a hurry to acknowledge the fact.

Many people positively resent the suggestion that they are middle-aged at 40. There are three acid tests they can apply. Bob Hope offers a simple one: You are middle-aged when your age starts to show around your middle. Mark Twain claims that you are middle-aged when your friends start telling you how young you look. Dr Laurence Peter (of the Peter Principle) provides the best test of all: 'Middle Age is when you stop criticizing the older generation and start criticizing the younger one.'

This last point is echoed by Quentin Crewe: 'The children despise their parents until the age of forty, when they suddenly become just like them – thus preserving the system.'

A number of sour notes have been sounded on age 40 – Shaw: 'Every man over 40 is a scoundrel'; Schopenhauer: 'On passing his fortieth year, any man of the slightest power of mind ... will hardly fail to show some traces of misanthropy'; Anouilh: 'When you're forty, half of you belongs to the past'.

Anatomists are no kinder. The gradual shrinkage of the body that becomes conspicuous with the elderly begins at 40. The measurement called 'arm-span' (the distance from fingertip to fingertip with outstretched arms) starts to decline at this age.

In contrast to the sour notes, there are a few bravely cheerful statements, led by Sophie Tucker's famous song 'Life Begins at Forty', which assails the shallowness and coarseness of youth. A century earlier, Thackeray thought along the same lines: 'Forty times over let Michaelmas pass,/Grizzling hair the brain doth clear, – /Then you know a boy is an ass,/Then you know the worth of a lass,/Once you have come to forty year.'

At the beginning of the nineteenth century, the pleasure-loving Prince Regent, later to become George IV of England, also favoured this age. His description of the ideal wife was: 'Fat, fair and forty.'

Earlier still, however, in 1716, Mary Montagu feels it necessary to defend the woman who has reached this year: 'At the age of forty she is very far from being cold and insensible; her fire may be covered with ashes, but it is not extinguished.' Today, modern attitudes have raked aside the ashes and allowed the fire to blaze on. Australian author Colleen McCullough, writing in 1977, provides a typically uninhibited, modern comment: 'The lovely thing about being 40 is that you can appreciate 25-year-old men more.'

Forty-year-old men often have the same sort of lovely thing on their minds. The threat of being 40 sparks off a sudden need for proving that a young girl will respond sexually to them. Brief affairs are commonplace at this age, not so much out of disloyalty or irresponsibility, but rather as reassuring virility displays by the male to his own private ego.

When married couples become embroiled in this type of mid-life crisis, few of them realize that the mood of compulsive lechery is only a temporary phenomenon. It soon passes, as the dreaded 40-mark fades into history, and a new age-mood arrives. Those who have not allowed the steel of their marriage to snap, may find to their surprise that the emotional heat has instead tempered and strengthened it.

Life Expectancy In the UK if you have reached this age you can still expect to live another 32 years if you are male, 37 years if female. At the turn of the century, the expectancy was only another 27 years for a male and 29 for a female, but even that was far better than many Third World countries today. The age of 40 remains the life expectancy at birth for males in Malawi, Tanzania and South Yemen; and for females in Afghanistan, Angola, Bhutan, Ethiopia and Guinea-Bissau.

> Among the animals, 40 is the record age for any kind of snake. It is held at present by a *Boa constrictor* that lived from 1936 to 1977.

Accession at 40 Catherine I became Empress of Russia at this age in 1725. A Lithuanian peasant girl, she was captured and sold to Prince Menshikov, adviser to Tsar Peter I. She soon became the Tsar's lover, gave him a child and was crowned Empress-Consort of Russia in 1724. When Peter died the following year without naming an heir she was almost immediately proclaimed Empress in her own right. **Kenneth Kaunda** became President of Zambia at this age, in 1964. He started life as a teacher, opposed British rule, was imprisoned for civil disobedience, and became a national hero in the process. **Julius Nyerere** became President of Tanganyika at this age, in 1962. After attending a British University he set about transforming his native country from a British colony to an independent nation. When it was declared a republic he was made its first President and two years later, when it was joined by Zanzibar, renamed it Tanzania.

Romance at 40 Elizabeth Barrett Browning, the English poet, was secretly married to Robert Browning when she was 40. Dominated by her despotic father, she had lived a shy, retiring life. Then, in 1845, she received a letter from the poet Browning declaring his love for her. They had never met, but he had fallen in love with her through reading her verses and they were married the following year, on September 12, 1846.

Achievement at 40 Nancy Astor, the British politician famous for her remark 'I married beneath me – all women do', took her seat in the

House of Commons on December 1, 1919, the first woman ever to do so.

John Buchan was this age when he wrote his most popular work *The Thirty-nine Steps*. He wrote fifty books, all in his spare time while engaged in his main activities of politics, diplomacy and publishing.

Pearl Buck, the American author noted for her novels on China, won the Pulitzer Prize for her novel *The Good Earth* when she was 40, in 1932. She turned over most of her earnings (in excess of $7,000,000) to a foundation for illegitimate children of US servicemen in Asian countries.

Charles Cruft, the impresario of the dog world, was this age in 1891 when he staged the first of what was to become the world famous 'Cruft's Dog Show'.

Francis Drake returned from his circumnavigation of the globe in 1580, when he was 40, was knighted by Queen Elizabeth on the deck of his ship and made Mayor of Plymouth.

Michael Faraday, the English scientist, and one of the world's great experimentalists, discovered electro-magnetic induction in 1831, when he was 40.

James Joyce published his great work *Ulysses* in Paris, on February 2, 1922, his fortieth birthday.

Claude Lévi-Strauss, the French anthropologist, was 40 when he published his first major work, *The Elementary Structures of Kinship*.

Alberto Moravia, the Italian novelist, produced his best-known work, *The Woman of Rome*, at this age. A sympathetic study of a prostitute, it manages to avoid the twin pitfalls of salacity and sentimentality.

Paul Revere, the American patriot, was 40 when he made his famous 'midnight ride', immortalized in Longfellow's poem. He travelled from Boston to Lexington in two hours to warn the American revolutionary leaders that British troops were about to arrest them. This enabled the leaders to escape and the colonists to prepare for the first battles of the War of Independence, fought the following day, April 19, 1775.

Jacques Tati, the French film director and actor, made his first great comedy *Jour de Fête*, in 1949, when he was this age. A perfectionist, he made only five major film comedies during his long life, the others being *Monsieur Hulot's Holiday*, *Mon Oncle*, *Playtime* and *Traffic*.

Revelation at 40

Muhammad, the founder of the religion of Islam, was 40 in the year AD 610 when he had a vision of a majestic being, later identified as the angel Gabriel, and heard a voice saying to him, 'You are the messenger of God.' This was the first of many such contacts. The messages from God were written down and form the basis of the Koran.

Defection at 40

Guy Burgess, the British Foreign Office official who had been a communist agent since his student days at Cambridge, was 40 when he learnt that he had been discovered, in 1951. He fled to Russia before he could be intercepted.

Retirement at 40

Joseph P. Kennedy, the President's father, who had become a bank president at 25, a millionaire at 30, and then a master at the art of manipulating the stock market, retired at 40 having made enough money to leave each of his nine, highly competitive children a million-dollar trust fund. He then became chairman of the Securities and Exchange Commission and proceeded to outlaw the very same speculative practices that had made him his fortune.

Life-span of 40

Edward IV, one of the few English kings to have been a good businessman, also had a great fondness for the pleasures of the flesh and his excesses in bed and at table are said to have contributed to his early death, just a few days short of his forty-first birthday.

Franz Kafka, a tormented, sickly, father-dominated bachelor, left instructions that his novels, *The Trial* and *The Castle*, were to be destroyed after his death. Fortunately, his friend Max Brod disregarded his wishes when Kafka succumbed to tuberculosis at the early age of 40.

John Lennon, the most influential figure in modern popular music, was murdered outside his New York apartment in December 1980. He and his wife Yoko Ono were returning late one night to the massive Dakota Building where they had been living a reclusive existence for some years, when a deranged fan of Lennon stepped up and shot him five times at close range. He died soon afterwards, aged 40 years and 2 months.

Jack London, the American author, gained enormous popularity and wealth from his fifty books, lived a full, adventurous and exciting life, and yet sank into alcoholism and died at 40 from an overdose of morphine at his ranch in California.

Glenn Miller, the American big-band leader who during the Second World War was made an Air Force Major and leader of the US Air Force Band in Europe, flew from an airfield in England to Paris on December 16, 1944. The plane disappeared. No bodies or wreckage were ever found. His brother believes that he had cancer and disappeared deliberately to avoid causing suffering to his mother. He claims the US government has facts they will not release.

Jean Seberg, the St Joan of the cinema, was found dead in 1979, her body curled up in the back of her car. A massive overdose of barbiturates laced with alcohol was the official cause of death. When she was a teenager she had said 'I won't live past 40', and she proved herself correct. But there is a mystery connected with her death. According to one theory, her association with the Black Panthers had led to persecution by the FBI which shocked her into a miscarriage, and eventually drove her to kill herself.

41

Forty-one is the year of anxiety, for those who fear the ageing process. For them, being forty was bad enough, but they could still cheat with that age, calling it 'one end' of their 30s decade. Now, however, they are well and truly 'into their 40s', and this can cause panic in some people. With health products advertising that they are to 'fortify the over-forties', the 41-year-old may suddenly feel a sharp pang of 'elderliness' and respond to it irrationally. Viv Nicholson, the pools-winning author of *Spend, Spend, Spend*, was rushed to hospital suffering from a drugs overdose at this age. She had tried to kill herself, she is reported as saying, because she had suddenly become 'frightened of getting old'. She recovered soon after the incident and solved her problem by getting engaged to a 25-year-old man who excited her and made her feel young in his company.

Life Expectancy

In early-Victorian England, before the problem of infant mortality was dealt with, life expectancy at birth was only 41·2 years. The dramatic rise since then, of more than thirty years, has been due to advances in hygiene and medicine. Other countries have not all benefited to a similar degree.

Forty-one is the life expectancy figure at birth for males in Cameroon, Equatorial Guinea, Ghana, Mozambique, Rwanda and Sierra Leone; and for females in Laos and Tanzania.

> Among the animals, 41 is the age of the oldest goldfish with a recorded history.

Accession at 41

Kemal Ataturk, the 'Father of the Turks', became the first President of the Republic of Turkey at this age, in 1922, and proceeded to make sweeping reforms, drastically modernizing the legal system, education, religion and politics.

Late Start at 41

Henri Rousseau, 'Le Douanier', the father-figure of Naïve painting, gave up his job as a customs office clerk at 41 to devote his time to his art. Despite this late start, and a personality made up of peasant shrewdness, bland self-esteem and gullible simplicity, he went on to produce a series of haunting masterpieces.

Achievement at 41

Simone de Beauvoir, the French existentialist feminist and intimate of Jean-Paul Sartre, published her major work *The Second Sex*, in 1949. In it, she put forward views that were twenty to thirty years ahead of their time. She was also ahead of her time in her lifestyle, adopting an open relationship with Sartre in which each was free to indulge in sexual activities with other partners.
Karl Benz, the German mechanical engineer, designed and built the world's first practical motor car powered by an internal combustion engine, in 1885.

It was a three-wheeler driven by coal-gas and travelled at 15 mph.
Yul Brynner, the multi-lingual ex-trapeze-artist who shaved his head for the Broadway production of *The King and I*, was this age when he won his Oscar for the film version in 1956.
Christopher Columbus, who was born in 1451, was 41 years old when he discovered the Americas, landing on San Salvador Island in the Bahamas on October 12, 1492, during the first of his four voyages to Central America. He went ashore in his best scarlet doublet, with the royal standard of Spain in his hand, fell to his knees and kissed the ground. He then formally took possession of the island, reading out a proclamation and saying a prayer, watched somewhat apprehensively by a group of naked islanders who had emerged from the trees. He made his peace with them by distributing red caps and glass beads and they reciprocated with parrots and cotton threads.
Captain James Cook, the son of a Yorkshire farm labourer who became one of the greatest English explorers, was also 41 when he discovered the east coast of Australia in 1770, and named the region New South Wales. Like Columbus, he gave the natives he encountered gifts of beads, but with less success, the inhabitants hurling lances at the intruders.
Joan Crawford, the Hollywood star, won an Oscar at this age for her role in *Mildred Pierce*, a film which was a turning point in her career, making her the leading interpreter of the 'suffering heroine'. In real life, if we are to believe the biography written by her step-daughter Christina, it was those around her who were the true sufferers.
John Flanagan, the Irish-American athlete, became the oldest world record-breaker in sport when he set his last world record for hammer-throwing on July 24, 1909 at the age of 41 years and 196 days.
Akira Kurosawa, the master of the Japanese cinema, was 41 when his film *Rashomon* won the top prize at the 1951 Venice Film Festival. It made Japanese film-making famous on a world-wide scale for the first time.
Robert Peel was British Home Secretary in 1829 when he created the Metropolitan Police Force and established the first disciplined policing of London.
Plato was 41 when he established his Academy in Athens in 387 BC. So-called because it was near the grove of the mythical hero Academus, it became a centre for philosophy and mathematics and established him as one of the greatest philosophers.
Marco Polo, the Venetian traveller and explorer, was 41 when he completed his mammoth journey, lasting twenty-four years, to the Far East. Shortly after he returned home in 1295, he was imprisoned in Genoa and while there he dictated his story to a fellow-prisoner. The book caused a sensation, although many believed it to be a fantasy invented by him. On his deathbed he was asked to retract his 'fables', but replied that he had told barely half of the wonders he had seen.

Henry Royce, the British engineer and co-founder of Rolls-Royce, saw his first car driven out of the factory on March 31, 1904, just four days after his forty-first birthday. It was the epitome of 'silence, smoothness and reliability' and so impressed Charles Rolls that he contracted to sell all future Royce cars. Two years later, in 1906, the firm of Rolls-Royce was formally established.

Johann Strauss, the Austrian waltz king, composed *The Blue Danube* in 1867 at the age of 41.

Max Weber, the German sociologist, was this age when he published his most famous work, *Protestant Ethic and the Spirit of Capitalism*, in which he challenged the theories of Karl Marx.

Misfortune at 41 Camille Pissarro, the French Impressionist landscape painter, returned from London to his home near Paris in the summer of 1871 to find that the Germans had destroyed nearly his whole life's work - all but forty of his 1,500 paintings. He had travelled to London with Monet to avoid the Franco-Prussian War.

Crime at 41 Clifford Irving, the American author, committed the most entertaining crime of recent years, at this age. In 1971 he persuaded the giant New York publishing firm, McGraw-Hill, to pay an advance of more than $300,000 for an 'official' biography of millionaire recluse Howard Hughes. Hughes himself ruined the hoax by making his first public statement for fifteen years, speaking over the telephone to a group of selected newsmen and denying the Irving story. By the following March the swindle had collapsed and Irving and his wife Edith were both jailed.

Retirement at 41 Donald Bradman, one of the most famous cricketers of all time, and one of the most prolific scorers in the history of the game, retired in 1949. The first Australian cricketer to receive a knighthood, his 117 centuries in 338 first-class innings make him one of the most successful batsmen ever.

Life-span of 41 Anne of Cleves, the fourth wife of Henry VIII, was a great disappointment to him. He married her for political reasons, but finding her far less attractive than the portrait he had been shown before the marriage was arranged, he quickly had it annulled and sent her to live out her life in quiet seclusion at Richmond, where she died at 41 in 1557.

Howard Appledorf, the American nutritionist, suffered one of the most bizarre deaths of recent times. The 41-year-old professor was murdered in 1982 by health-food fanatics who objected to his defence of 'junk foods'. A group of them broke into his Florida apartment, forced his head into a canvas bag filled with ice-cubes and then sat down and ate a ritual meal while the cubes slowly melted and he drowned. The word 'murder' was left scrawled on the walls, not in blood, but in peanut butter.

Jane Austen, four of whose novels were published anonymously in her lifetime, died of Addison's disease in 1817. After her death, her authorship was made public by her brother Henry and a further two novels published - *Persuasion* and *Northanger Abbey*.

Brendan Behan, the Irish playwright who had been an alcoholic since the age of 8 and who described his art as 'building comedy from the rubble of despair', finally succumbed to the abuses he heaped on his body, in 1964.

King Canute, who had become King of England in 1016, King of Denmark in 1019, and King of Norway in 1028, died at this age in 1035. He once tried to stop the tide coming in, to demonstrate to his sycophantic followers that his powers were limited.

Wallace Carothers, the American chemist who invented nylon, the first completely man-made fibre, died in 1937, the year in which the first nylon stockings were made.

Mata Hari, one of the first 'exotic dancers' to show herself naked to the public eye, created a sensation in the capitals of Europe at the turn of the century. In 1917 she was accused of spying for the Germans, but the evidence was dubious. A tube of 'secret ink' in her possession when she was arrested turned out to be a contraceptive chemical. She made the jury laugh at her trial, but they convicted her all the same, and she was shot by a French firing squad on October 15. Her body was unclaimed and was therefore sent to a medical college for student dissection classes.

Brendan Behan

Forty-two is the year of disillusionment for some, satisfaction for others. In the film *10*, Dudley Moore plays a Hollywood songwriter who celebrates his forty-second birthday with the phrase: 'They said that life begins at forty but they lied.' He promptly abandons his attractive female partner and sets off in hot pursuit of a beautiful young girl who rates '10' on his private scale of sex appeal.

The character in this film has been patient. He has waited two years for the song 'Life Begins at Forty' to come true, but now, convinced that he is stagnating, he over-reacts with a melodramatic, romantic pursuit which, in real life, would probably cause chaos and further disillusionment. But whatever the cost, he is, at least, no longer stagnating, and for some 42-year-olds this may be more important than anything else, including quiet, humdrum success. It creates, as it were, a new set of 'growing pains', and gives an impression of dynamic force in life.

For others, such drastic measures are not needed. If their work is full of increasing challenges, they may hardly notice the mid-life period in their lives. The painter, Rudolf Lehmann, was so busy with his art during this phase of his life that, for him, it was a perfect time, and 42 the perfect year of that time: 'A boy may still detest age,/But as for me I know/A man has reached his best age/At 42 or so.'

Life Expectancy
The life expectancy at birth for males living in Bhutan and Nepal, and for females in Guinea, Mali, Mauritania, Niger, Senegal, South Yemen and Somalia is only 42.

> Among the animals, 42 is the age of the oldest known racehorse. 'Tango Duke', born in 1935, lived until 1978.

Elvis Presley   Edward VIII and niece

Romance at 42
Edward VIII abdicated at this age, on December 10, 1936, because of his romance with Mrs Simpson, whom he married six months later, in France.

Achievement at 42
Woody Allen, who presents himself as the small, plain, shy, but secretly imaginative, sex-obsessed man whose life is full of fantasies (some of which he realizes), won three Oscars on the same day, for the film *Annie Hall* (Best Director, Best Screenplay and Best Film awards), but failed to turn up for the ceremony. It was rumoured that he was playing the clarinet in a jazz group at the time.

John James Audubon, the French-born naturalist, published the first volume of his great work *The Birds of America* in 1827. The coloured illustrations have since become collector's items of great value.

John Davis, the English navigator, discovered the Falkland Islands on August 9, 1592, when he was seeking a passage through the Straits of Magellan. He failed and, after provisioning his ship with 14,000 dried penguins, returned to England.

Christian Dior, the French fashion designer, was this age when he introduced the 'New Look' in 1947, a major shift in style for female clothes.

Alexandre Dumas, called Dumas père, was 42 when he produced his most famous work, *The Three Musketeers*, in 1844. After he achieved fame he stepped up his output of books to as much as forty in a year, by using a factory of hack writers. He provided the plots and the purple passages and his assistants did the rest. This left him time to spend the vast amounts of money he earned, a task which he carried out with great flamboyance and incredible sexual vigour. He boasted that he had sired 500 illegitimate children, but this was a wild exaggeration.

Margot Fonteyn, the English ballerina, was already 42 when she started her famous partnership with Rudolph Nureyev who, in 1961, had just defected to the West.

Richard Gatling, the American inventor, was this age when he had the idea for a rapid-fire machine-gun. Called the Gatling-gun, it could fire 1,200 rounds per minute, and was soon adopted by the US Army, in the 1860s. It was a clumsy gun, with rotating barrels, and often jammed, but it offered a great increase in fire-power and caused many casualties.

John Huston, the American film director, was 42 when he won two Oscars for his direction and screenplay of *The Treasure of the Sierra Madre*, in which he also coaxed an Oscar-winning performance out of his father, Walter Huston.

Al Jolson, the American entertainer, was 42 when he starred in *The Jazz Singer* in 1927, the very first feature-length talking film.

David Livingstone, the Scottish missionary, discovered and named the Victoria Falls in 1855 during a major expedition into the African interior.

André Michelin was 42 in 1895 when he and his

brother entered a motor car in the Paris to Bordeaux race, using pneumatic tyres for the first time in a competitive event. Although they did not win, the new tyres aroused such interest that, within a decade, all motorists had adopted them. The French brothers went on to become one of the leading tyre manufacturers in the world.

Elisha Otis was also 42 when he demonstrated his safety elevator in New York on May 25, 1854. He did this by riding the platform high in the air and then ordering the rope to be cut, proving that, in such an emergency, the lift would not fall. His invention was adopted and within a few years the first passenger lift was in operation in the five-storey Broadway china shop of E. V. Haughwout and Co. It heralded the development of the skyscraper and the eventual total transformation of the city skyline.

Misfortune at 42

Eadweard Muybridge, one of the most important pioneers of moving-picture photography, suffered a serious setback when making his first experiments with fast-moving animals. Asked to settle the question of whether a trotting racehorse ever has all four legs off the ground at the same time, he started his investigations in 1872 at the Sacramento racecourse in California, but was forced to abandon them while being tried for the murder of his wife's lover. Although acquitted, he found it expedient to travel for a few years and could not resume his experiments until 1877.

James Whistler, the American artist, was nearly 43 when he was savagely attacked in an article by leading art critic John Ruskin, who accused him of 'flinging a pot of paint in the public's face'. The article appeared on July 2, 1877, and Whistler decided to sue him. He won his court case but received only a farthing in damages, and the costs ruined him.

Life-span of 42

Prince Albert, the consort of Queen Victoria, died at this early age from typhoid fever. He is supposed to have damaged his health irreparably in a fit of hysterical outrage at the Prince of Wales's improper adventures, and Victoria never forgave her son Edward for this. Albert was so loved by the Queen that she stayed in mourning throughout the remaining forty years of her life, and she is said to have slept with his nightshirt clutched in her arms and with a portrait of him on her pillow beside her.

Joachim Bonnier, the bearded Swedish racing driver, who looked like a Viking chieftain and who was respected in motor-racing circles around the world, was 42 when he was killed at Le Mans in 1972.

Robert Kennedy was 42 when he was murdered in Los Angeles on June 5, 1968, during his bid to secure the Democratic Presidential nomination. He had just delivered a victory speech in the ballroom of the Ambassador Hotel, after winning the California Democratic Primary and was on his way to speak to the press. The crush was so great that he made a detour through the hotel kitchens. There, a Jordanian called Sirhan Sirhan leapt forward and started firing wildly. Kennedy fell to the floor with two shots in his back and one in his head. Within a few hours he was dead and Sirhan, who had been caught easily, was accused of his murder. However, as with his brother John's death, all was not what it seemed. The shots entered his body from behind, but Sirhan was in front of him when he fired. One eyewitness claimed that he had seen a security guard fire at Kennedy and hit him three times. The truth was never discovered and another 'cover-up' has been suggested.

Guy de Maupassant, the French author, tried to kill himself in 1892 by cutting his own throat when in despair over the early death of his psychotic brother in an asylum. He was staying with his mother on the Riviera at the time and she reluctantly agreed to have him committed too. He was removed to an asylum in Paris where he died in 1893, a month short of his forty-third birthday.

R. J. Mitchell, who designed what is arguably the most beautiful propeller fighter plane ever built – the Spitfire – died at 42 in 1937. It first flew in March 1936 and was operational in June 1938. Nearly 19,000 were built and became the stars of the Battle of Britain in 1940.

Modest Mussorgsky, the Russian composer famous for his opera *Boris Godunov*, was an unstable character who drank himself to death at this early age, robbing the world of a highly original musical talent.

Peter the Great, Emperor of Russia, also died at this age, in 1725, as the result of an act of bravery. In the autumn of 1724, seeing some soldiers in danger of drowning from a ship aground on a sandbank, he plunged into the icy water to help them. Catching a chill he became seriously ill and within a few months was dead.

Madame de Pompadour, the official mistress of Louis XV of France, who became a great patroness of the arts, died of lung cancer in her apartment at Versailles in 1764.

Elvis Presley, the religious, hymn-singing boy who became the world's greatest rock star, died at his Memphis mansion in 1977 following a strenuous ball game with his friends. The cause of death was given as heart failure, but this had undoubtedly been brought on by the excessive use of drugs to control his increasing weight and to stimulate him for his stage performances.

Anthony Van Dyck, the Flemish artist, died in 1641 after an immensely successful career as *the* court painter of his period. Unlike so many great artists, he lived a life of wealth and luxury, surrounded by beautiful ladies. He became the favourite of Charles I of England who not only knighted him but also found him a suitable wife. This infuriated Van Dyck's mistress so much that she attempted to mutilate his painting hand. The cause of his early death, however, was not female jealousy, but a prolonged excess of both work and pleasure

43

Forty-three is traditionally the greatest age which a much older woman will accept as flattering. The year is pin-pointed by W. S. Gilbert in *Trial by Jury*, with the lines: 'And a very nice girl you'll find her!/She may pass very well for forty-three/In the dusk, with a light behind her.'

Life Expectancy The life expectancy at birth for males in Burundi, the Comoros, Sudan and Vietnam is still only 43.

Achievement at 43

Leo Hendrik Baekeland, the Belgian/American inventor who was one of the great pioneers of the plastics industry, invented Bakelite in 1907 when he was 43 and patented it two years later. It was the first heat-proof plastic in the world and an excellent electrical insulator.

Pierre-Augustin Beaumarchais was this age when he wrote *The Barber of Seville* in 1775, which has gained lasting fame as the basis for Rossini's opera.

Donald Campbell, the English speed record-breaker, was 43 when he broke the land speed record on July 17, 1964. He already held the water speed record of 260 mph, achieved in 1959, but his new goal was to beat John Cobb's land record of 394 mph set up in 1947. In his first attempt he crashed at 365 mph, but miraculously escaped almost unhurt. Another 'Bluebird' car was then built and with this he achieved his new record of 403·1 mph, at Lake Eyrie Salt Flats in Australia. He was to die three years later, attempting to break the world water speed record, when his boat disintegrated at a speed of over 300 mph.

Marie Curie, the Polish-born French physicist, won her second Nobel Prize when she was 43, in 1911. She and her husband had received a joint Nobel Prize already, in 1903, for their discovery of radium and polonium. Then, after his death in 1906, when she became the first woman professor at the Sorbonne in Paris, she continued the research and eventually succeeded in isolating metallic radium, for which she was given the 1911 Nobel Prize for Chemistry.

Sigmund Freud was 43 when he published his most famous work, in 1899, called *The Interpretation of Dreams*.

Wassily Kandinsky is usually credited with having painted the first totally abstract picture in the history of modern art – a water colour produced in 1910 when he was 43. Rival claims have been made for the 'first abstract', but Kandinsky certainly remains the major innovator of this important trend in modern art.

John Kennedy was this age when he became the youngest ever American President, in 1960. He was the first Catholic President and the first of Irish descent. A hero of the Second World War, he had driving energy, a keen intellect and a cool head. His emotional detachment combined with his progressive idealism was both dangerous and exhilarating. In private he is reputed to have been a sexual athlete and once said to an aide: 'This administration is going to do for sex what the last one did for golf', and set about proving it with a series of brief conquests of some of Hollywood's most beautiful stars.

Gregor Mendel, the Austrian monk who became the 'father of genetics' was not recognized as such until many years after his death. His research, establishing the basic principles of heredity, was originally presented at meetings of the Natural Science Society held in 1865, when he was 43. Published the following year, his findings made no impact on the scientific world and it was not until the early part of the twentieth century that his importance was appreciated.

Louis Pasteur, the French chemist who became the 'father of bacteriology', was the same age when he published *Études sur le Vin* in 1866, in which he presented his findings in the field of research that was later to become known as 'Pasteurization' – the application of moderate heat to destroy harmful micro-organisms.

Scott of the Antarctic was 43 when he reached the South Pole on January 17, 1912, only to find that the Norwegian Roald Amundsen had beaten him to it by one month. Scott's achievement became even more of a tragedy when, on the return journey, ill-health and a blizzard caused the death of both himself and his team. The explorers were forced to remain inside their tent, their provisions gradually running out. Scott recorded everything in his diary, including their last hours. It was later found and published, and a film of their sad adventure was made in 1948, with John Mills playing the role of Scott.

Conversion at 43

John Donne, the greatest of the metaphysical poets, took holy orders at 43 in 1615 and went on to become Dean of St Paul's in London and 'the most eloquent pulpit orator of his day'. This was a major change of direction for the poet who had previously been looked upon as a witty man-about-town and had enjoyed the worldly life of a gentleman adventurer, taking part in a military expedition against Cadiz. After these exploits he was imprisoned for arranging a clandestine marriage with the 16-year-old daughter of Sir George More and when he was released was forced to live on charity. James I offered to help him, but only if he took holy orders. His change of direction, therefore, seems to have been more of a 'shotgun conversion' than the result of a sudden, divine awakening.

Misfortune at 43

John Milton began to lose his sight in 1644 and when he was 43 in 1651 he became completely blind. Despite this misfortune, he went on to write *Paradise Lost* and *Paradise Regained*, employing paid assistants who, in addition to his nephews, his daughter and other friends, helped with the task of taking dictation, reading aloud and making corrections.

Crime at 43 **Roman Polanski**, the Polish film director, was arrested by Los Angeles police at this age for allegedly drugging and raping a 13-year-old girl at the house of film actor Jack Nicholson. He was supposedly taking pictures of the girl, a model and aspiring actress, for the French *Vogue* magazine. At first he denied the charges, then admitted 'partial guilt' and was allowed to leave for Tahiti where he was due to make a film called *Hurricane*. Instead of directing the film, and returning to America to stand trial, he left for Europe and was officially declared a 'fugitive from justice' by the American court.

Roman Polanski

Retirement at 43 **Max Schmeling**, the German World Heavyweight Boxing Champion, became increasingly unpopular in the 1930s because of his admiration of Adolf Hitler and the German race myth. Later, he turned against the Nazis who punished him by assigning him to the paratroops in 1941. He was wounded in the airborne invasion of Crete that year, but survived to return to the boxing ring after the war, winning three of five fights in Germany in the 1940s, before retiring at 43, in 1948.

Life-span of 43 **Marcus Brutus**, the Roman general who led the conspirators who murdered Julius Caesar and who was portrayed sympathetically by Shakespeare, was not only dignified and idealistic, but also obstinate, indecisive, extortionate and cruel. He killed himself after a military defeat in 42 BC.
Butch Cassidy, the American outlaw, was killed with his friend the Sundance Kid by Bolivian cavalry in 1909. He was a Utah cattle rustler ('Butch' was short for 'Cattle Butcher') who joined a gang and robbed trains and banks, but never the 'common people'. He claimed that he only wanted to hurt cattle barons and big bankers. His gang, the 'Wild Bunch', came under such pressure from the law, that he, the Sundance Kid and the Kid's girl-friend decided to go to South America and become legitimate ranchers, but they eventually returned to crime until their violent deaths in 1909.
Edward II, King of England from 1307 to 1327, was hated by his powerful barons for the way in which he fell under the spell of certain male companions and granted them excessive privileges. He was eventually deposed in favour of his son and met a brutal end while imprisoned at Berkeley Castle, traditionally murdered by the insertion of a red hot iron into his body in such a way that it left no external marks. At the end of the sixteenth century his death was dramatized by the playwright Christopher Marlowe.
Lucretius, the Roman poet, is supposed to have been driven insane by a love potion and to have written his important works during lucid periods before killing himself at 43. His great work *De Rerum Natura* sets out to attack religion and superstition for creating unnecessary fears about death. He argued that the soul is born and dies with the body and there is no afterlife – a remarkably modern view for one writing in the first century BC.
Romy Schneider, the beautiful international film actress, was found dead in her Paris apartment in 1982. First reports suggested suicide, but these were later denied.
Alexander Scriabin, the Russian composer whose musical aim became the achievement of ecstasy, to which end he used bells, incantations and blaring trumpets, was increasingly fascinated by the supernatural. In the later years of his short life he became an occultist and evolved a 'mystic chord' of ascending fourths. He died at his birthplace, Moscow, in 1915.
Bessie Smith, the American blues singer, died in 1937 after a car accident. Legend has it that she was refused entry to a white hospital and that this delay caused her death, but there appears to be some doubt about this. The truth will probably never be known. A large, physically violent, hard-drinking woman, she had a stormy and complex love life, in keeping with the powerful lyrics of her songs.
Jan Vermeer, the Dutch master, was a slow worker apparently having produced no more than forty, usually rather small, paintings by the time of his death at 43. He was the supreme exponent of light and shade in Dutch interior painting. When he died he left his wife and eleven children penniless. The local baker held two of his paintings because of unpaid bills and refused to return them to his widow, thereby making his bread the most valuable in the history of baking.
Natalie Wood, the American actress who married Robert Wagner in 1957, divorced him in 1963 and re-married him in 1972, was on their boat, off Catalina Island, late on the night of November 29, 1981, when she slipped, fell into the water and drowned, cutting short a brilliant Hollywood career.

Forty-four is the year of self-examination – the age when many people, for the first time, start to consider their own physical ageing. Author Irene Friese comments that, starting at this age, people 'become aware that they no longer look young, may no longer feel attractive, and can no longer rely on their body to function unflaggingly ... The feeling that time is running out creates a psychological pressure to make the most of one's remaining good years; to seek out desired experiences before it is too late.'

Life Expectancy The life expectancy at birth for males in Benin, Botswana, Cambodia, Ivory Coast, Saudi Arabia, Swaziland, Zaire and Zambia, and for females in Liberia and Malawi is only 44.

Achievement at 44 **James Barrie**, the Scottish playwright, saw his most famous play, *Peter Pan*, given its first performance in 1904, when he was 44. Enduringly popular, it has been revived on the London stage every Christmas since then, except in 1940.
Frank Baum, the American author, was a magazine editor until his enormous success with his children's book, *The Wizard of Oz*, in 1900, when he was 44, enabled him to devote all his time to writing. He went on to create a total of fourteen *Oz* books.
Arthur Bliss, the English composer, was this age when he reached a wide audience with his score for the H. G. Wells film, *Things to Come*, in 1935.
Albert Camus, the French author whose work represented a search for significant values in a meaningless world where the dogmas of both Christianity and communism were rejected, won the Nobel Prize for Literature at 44 in 1957.
Edward Elgar, the English composer, was 44 in 1901 when he composed the first of his *Pomp and Circumstance* marches, the middle section of which has since become famous as *Land of Hope and Glory*.
Ian Fleming, the English author, wrote his first James Bond book in 1952. He described the book, *Casino Royale*, as a 'story written in less than two months as a piece of manual labour to make me forget the horrors of marriage'. Despite this claim that it was a mere diversion in the last days of bachelorhood, he paid fanatical care to the technical details of his plot and started a new genre of thriller writing.
Charles Kingsley, the English clergyman, was also 44 when he wrote a very different kind of book, *The Water Babies*, in 1863. It is said to have been inspired by his thoughts on evolution, and he is one of the few clergymen who wholeheartedly accepted Darwin's theory when it first appeared.
Bernard Lovell was 44 when he completed the world's largest steerable radio telescope, at Jodrell Bank, in 1957.
Machiavelli, the Italian statesman, wrote his most famous book, *The Prince*, in 1513, when he was 44. A thinker of genius, he has often been misrepresented as an immoral cynic, the epitome of devious dealings, but this was merely part of a propaganda attack by the French against the Italians. In reality, *The Prince* was a savage satire against tyranny and was based largely on his observations of Cesare Borgia.
A. A. Milne was 44 when he wrote his immortal children's book, *Winnie-the-Pooh*, in 1926. Among English children's literature it is second only in popularity to *Alice in Wonderland*.

Winnie-the-Pooh by Ernest Shepard

Roger Moore, the English actor, was this age when he took over from Sean Connery as Ian Fleming's fictional spy, James Bond, in 1973.
Edward Teller, the Hungarian-born American physicist, developed the first Hydrogen Bomb in 1952 when he was 44. Often referred to as the 'Father of the H-bomb', he clashed with Oppenheimer over the further development of the Atomic Bomb. He won, and President Truman gave his blessing to Teller's further research. Teller's quoted aim was 'to defend our science, our culture, our American freedom, our civilization'. Not satisfied with his thermo-nuclear solution to this problem, he has since been busy working on the Neutron Bomb, a weapon with only one-tenth of the heat of the H-bomb, which kills by radiation and leaves buildings standing. Seen as an ideal battlefield weapon, a typical bomb of this kind would kill everyone within a mile's radius, but after only twenty-four hours the radioactivity would have dissipated and the zone could be occupied by troops.

Misfortune at 44 **Doris Day**, Hollywood's professional virgin, heroine of a string of light comedies in the 1950s and 1960s, discovered on becoming a widow in 1968 that her husband and her lawyer had secretly relieved her of her $20,000,000 fortune, leaving her penniless at 44 after twenty years as one of the most highly paid female stars of the cinema. She suffered a nervous breakdown, understandably, but soon bounced back as a television star and eventually, in 1974, was awarded $22,000,000 in damages from her former lawyer, giving her story the happy ending to which she had become accustomed.
Emperor Hirohito of Japan, ruler through the whole of the aggressive phase of twentieth-century Japanese history, including the conquest of Manchuria and much of China, and the great expansion into the Pacific region during the Second World War, faced defeat at the age of 44. After the United States

dropped the Atomic Bombs, he supported surrender in 1945 to avoid further bloodshed. The Americans stripped him of his 'Imperial Divinity', but allowed him to continue as a powerless ceremonial Head of State.

Friedrich Nietzsche, the German philosopher, was 44 in January 1889, when he suddenly flung his arms around the neck of a maltreated cab-horse in a Turin street and collapsed in tears, after which he was removed to a lunatic asylum for the final eleven years of his life.

Life-span of 44

Pieter Bruegel, the Flemish artist, portrayed with great vigour and inventiveness bustling scenes of peasant life in village landscapes. He was unusual in that he put the emphasis on the common people, and refused to become a slave to religious pomposity and the grandeur of the formal scenes that dominated so many major paintings. Although he died at this young age in 1569, he lived long enough to see his work become much loved by his contemporaries.

Anton Chekhov, the Russian playwright, died of tuberculosis in 1904, a few months after being taken to see the first night of his final play, *The Cherry Orchard*, in Moscow. Already a dying man, and coughing uncontrollably throughout the performance, he also had to suffer through a series of long speeches commemorating (inaccurately) the twenty-fifth anniversary of his writing début.

F. Scott Fitzgerald, the American 'Jazz Age' author who became a despair-ridden alcoholic, died at 44, reputedly while in bed with his columnist friend Sheilah Graham. According to her, however, his heart attack came when he was sitting in a chair reading the *Princeton Alumni Weekly*.

Saint Francis of Assisi, the Italian founder of the Franciscan Order, died at the same age in 1226, looking exactly as if he had just been taken down from a crucifixion. In the months before his death he suffered great pain and had developed the 'stigmata' to an alarming degree, his body spontaneously developing the same wounds in the hands, feet and side as had been reported for Christ on the cross. It was his final, extraordinary act in a life devoted to imitating that of Christ as closely as possible.

Tony Hancock, the English comedian whose skilful portrayal of impotent aggression and threadbare grandeur had made him the most popular comic performer in Britain in the 1950s, killed himself in a fit of depression and despair in an apartment in Sydney, Australia, in 1968. Privately he was a dissatisfied, alcoholic perfectionist with ambitions that led him further and further away from the roles he played so superbly, until he brought about his own tragic end.

Heinrich Himmler, controller of Hitler's SS, and the second most powerful man in Nazi Germany, killed himself with poison on May 23, 1945, following an unsuccessful attempt to escape disguised as a common soldier, at the end of the war.

Billie Holiday, the American jazz singer, suffered a constant struggle with heroin addiction in her final years, before dying at this age in 1959.

D. H. Lawrence, the miner's son who became a highly controversial author, suffered from tuberculosis even as a youth and it finally killed him at 44, in 1930. His most famous book, *Lady Chatterley's Lover*, was not publicly published in full until nearly thirty years after his death.

Mary, Queen of Scots was beheaded in 1587, at Fotheringay Castle. For years she had been a threat to Elizabeth I, constantly plotting and scheming against her, while held in what amounted to 'house arrest', but eventually in October 1586 she was tried and condemned to death.

Jackson Pollock, the alcoholic, hypersensitive, self-obsessed American abstract artist who made his name with drip paintings killed himself at this age by deliberately crashing his car, in 1956.

Robert Louis Stevenson, the Scottish author of *Treasure Island* and *Dr Jekyll and Mr Hyde*, was making a salad on the verandah of his house when a blood vessel burst in his brain and he died almost immediately, aged 44, in 1894. He had been a sickly child and continued to suffer from ill-health as an adult, but despite this became fanatically prolific as a writer, driving himself near to death on at least three occasions. Eventually, in 1888, he took his family off to live in the South Seas, searching for his own 'Treasure Island', where the prize would not be a chest full of money, but a chance for improved health. They settled on the island of Upolu in Samoa, where he built a great house on a mountain slope overlooking the sea. It was at this idyllic spot that he died a few years later, and the natives, who loved Tusitala – the teller of stories – carried his body to the summit of the mountain and buried it there.

Friedrich Nietzsche

Forty-five is the year for re-stabilizing one's life. The 'mid-life crisis' is usually over by this age, and the individual seeks a new stability, order and structure in the daily pattern of living. This may involve a sobering return to an earlier lifestyle, following dramatic upheavals, or it may constitute the establishment of an entirely new routine. Either way, there is a move towards a calm equilibrium – although it may take some years before this is fully achieved.

For some, 45 rather than 40 is the year of 'middle age'. Antonia Fraser, speaking on her fiftieth birthday: 'I'm now a grandmother's age though I decided middle age didn't begin till 45, which is later than most people would put it.' John Brophy, in his book *The Human Face* says: 'I am forty-five: middle-aged ... When I look in the mirror I see white hairs above my ears: a damnable injustice, for I know plenty of men of my own age without any ... But I cannot deny that it is the face of a man – how do they call it in magazine stories – no longer in the first flush of youth ... But ... Unless I bend my mind to it, I think of myself still as a youngster ... Against the background of most of my thoughts, forty-five is a venerable age, and remote from me. It is not I who have attained that undesirable beacon, but my face and my silly white hairs. If I could disown them I would.' Despite these protests, he has accepted that he has changed. He has literally faced his face, and recognized that, whatever his wishes, a new phase of life has arrived.

Brophy would be sad to learn some of the names that have been given to this age in the past. Gilles Lambert in *The Conquest of Age* calls 45 the start of 'the first period of active pre-senescence'; Dante calls it the 'end of youth' and the start of 'old age'; Aschoff disagrees – he says that old age does not start for another twenty years and that 45 is the start of what he calls the 'ripe age'.

Medically, 45 is the start of the period when, for males, coronary heart disease becomes the No. 1 killer in most Western countries and, for females, the risk of giving birth to a baby with Down's syndrome (mongolism) increases dramatically (to 1 in 65 births).

Yet, despite these thoughts, 45 is a year of great achievement for many people. For some, it heralds a phase of exciting maturity, building on earlier foundations; for others, it is the start of a period of 'second wind' for creativity.

Life Expectancy
The life expectancy at birth of males in Bangladesh, Liberia and Nigeria and of females in Benin, Burundi, Cameroon, Equatorial Guinea, Gabon, Ghana, India, Mozambique, Nepal, Rwanda, Sierra Leone and Sudan is 45.

Romance at 45
Charles Dickens experienced his mid-life crisis later than usual, held back perhaps by his public image as the symbol of wholesome family life. He had for some years, however, been growing tired of the quiet wife who had given him ten children, and at 45 he fell in love with a teenage actress called Ellen Ternan who was the same age as his eldest daughter. The emotional upheaval was, he wrote to a friend, '... so strong upon me that I can't write, and can't rest one minute'. His reaction to this unrest was to start what amounted to a new career, as a performer of his own writings, in a long series of public readings, the first of which took place a few months later. After this, he became publicly separated from his wife, much to the shock of Victorian society. Ellen Ternan left the stage in 1860 to become his mistress and remained so for the rest of his life.

Achievement at 45
Christiaan Barnard, the South African surgeon, was 45 when he performed the world's first heart transplant operation in December 1967. The patient, an incurably ill South African grocer called Louis Washkanksy, was given the heart of an accident victim and, although he lived for only eighteen more days, the operation heralded a new era in spare-part surgery.

Wernher von Braun, the German rocket specialist, was this age on January 31, 1958 when he put the first American satellite, *Explorer I*, into orbit. As a young man he had developed the V2 rocket for Nazi Germany, 3,600 of which had been launched with devastating effect before the end of the war. In 1945 he fled before the advancing Russians and surrendered to the American forces. Taken to the United States, his talents were put to better use working for the American space programme.

William Burroughs, one of the heroes of the beat generation, published his most famous book, *The Naked Lunch*, at this age, in 1959, following a cure for drug addiction. The opening sentence reads: 'I awoke from The Sickness at the age of forty-five, calm and sane, and in reasonably good health except for a weakened liver and the look of borrowed flesh ...' He goes on: 'The Sickness is drug addiction and I was an addict for fifteen years ... I have used junk in many forms: morphine, heroin, delaudid, eukodal, pantopon, diocodid, diosane, opium, demerol, dolophine, palfium. I have smoked junk, eaten it, sniffed it, injected it in vein-skin-muscle, inserted it in rectal suppositories.' During his long period of addiction he accidentally shot and killed his wife. The book itself, viewed as a masterpiece by certain critics, is a nightmare fantasy of degradation, composed in the manner of a surrealist word-collage.

Christopher Cockerell, the English engineer, was 45 when he patented his invention, the hovercraft, on December 12, 1955. The first important 'flight' came several years later when the Saunders-Roe SR-N1 was launched at Cowes on May 30, 1959. That summer it made the first hovercraft crossing of the English Channel.

Lawrence Durrell, the English author, published

the first part, *Justine*, of his most famous work, *The Alexandria Quartet*, when he was 45.

Galileo, the Italian mathematician, physicist and astronomer, built the first telescope that could be used for serious astronomical studies, in 1609. With its magnification of × 32, it was possible to start a series of detailed observations of the heavens, leading to many major discoveries.

Carl Jung, the Swiss psychiatrist, after suffering through a haunting mid-life crisis, emerged to produce a major work, *Psychological Types*, in which he introduced the idea of 'introverts' and 'extroverts', in the year 1921, when he was 45.

Martin Klaproth, the German chemist, was this age when he discovered uranium in 1789, isolating it in a sample of pitchblende from Saxony, but little realizing the immense significance his find would have for the future.

Richard von Krafft-Ebing, the German physician, was 45 when he published his pioneering work, *Psychopathia Sexualis*, in 1886. Dealing with a wide variety of sexual deviations, it was described as 'the first classic of sexual science'.

Peter Medawar, the British zoologist, was awarded the Nobel Prize for Medicine at this age, in 1960, acknowledging his pioneering work in the field of acquired immunological tolerance – research which led to many advanced surgical techniques involving organ transplants.

Harry Selfridge, the American businessman, was this age when he introduced the department store concept to England, opening what was then the largest store in Europe, in London's Oxford Street, in 1909. He had started out at Marshall Field and Co., in Chicago, the first department store in the world.

Josiah Wedgwood, the English pottery designer and manufacturer, was 45 when his famous 'Jasper Ware' – in pale blue or green decorated with Greek figures in white – was introduced, in 1775.

Misfortune at 45

Napoleon was this age when he met his Waterloo, defeated by Wellington in a final, decisive battle on June 18, 1815.

Life-span of 45

Roald Amundsen, the Norwegian explorer who beat Scott to the South Pole and who later flew over the North Pole in a dirigible with the Italian explorer, Umberto Nobile, was killed in an air search for Nobile, who had become stranded on ice during a later exploration.

Giovanni Belzoni, the notorious Italian archaeologist, was, in reality, one of the greatest Egyptian tomb-robbers of all time, and possibly the only man in history to be mugged for an obelisk. French agents attacked him at gunpoint and made off with the huge obelisk which he had just stolen himself from an island in the Nile. He died at 45 on a trip to Timbuktu.

Martin Bormann, the Nazi leader who was Hitler's deputy, disappeared in 1945 and was tried in his absence at Nuremberg and convicted. It is likely that he died at the same time as Hitler, as he was with the Führer to the end, but there is no proof of this.

Montgomery Clift, the American actor who portrayed introspective, troubled heroes, suffered a serious motor accident in 1957 which scarred his face and led to alleged drinking and drug-taking. His behaviour towards the end of his short life became strange and it was feared that he may have suffered from some kind of brain damage. He died from a heart attack in 1966.

King Farouk of Egypt established himself in Italy as a European playboy after being sent into exile. It was there that he died, in 1965, after a night of gluttony at his favourite Roman restaurant. His famous last words are reputed to be: 'Soon there will only be five kings left: the Kings of England, Diamonds, Hearts, Spades and Clubs.'

Rocky Marciano, the bull-like Italian-American heavyweight boxing champion who beat everyone there was to beat and so retired with a fortune, died in an aircraft crash in 1969, on the eve of his forty-sixth birthday.

Rasputin, the Russian 'holy-man', was assassinated at the age of 45 on December 30, 1916. He was a monk who taught that repentance was the basis of religion and that, in order to repent, it was first necessary to sin. He assisted his female followers in this by debauching them as often as possible. The Empress allowed him to treat her son for haemophilia, and when he appeared to effect a cure he gained entry to court circles and set about debauching ladies of higher status. This was his undoing, for a group of powerful, jealous men were driven to contrive his removal. They invited him to a midnight feast where they fed him poisoned cakes and wine, then, when he was dazed and drowsy, they sexually abused him, shot him four times, castrated him, beat him, trussed him up and threw him, still alive, into the icy cold Neva River, where he drowned.

Rasputin

46

Forty-six is the year of defiance. Although the mid-life crisis is over for most by this age, there is still a group of defiant anti-agers who are determined to display themselves as pseudo-juvenile. If they are grandparents, they go to great lengths to conceal the fact. They read endless books and magazine articles on 'how to stay young'. In extreme cases they go to health farms, boil themselves in saunas, and pursue the dozens of fashionable fads available to the youth-dreamers. They may even abandon the family saloon, buy a beach-buggy and take up hang-gliding or skiing for the first time in their lives, to mimic the sportiness of youth.

For some, all this sudden activity works wonders; for others it merely looks ridiculous; for still others it may be downright dangerous. Jogging, like smoking, can be damaging to your health.

People used to be surprised that such outbursts of activity took place in the mid-40s, especially in men, until someone invented a term for it: they called it the 'male menopause' and this immediately excused it as a kind of masculine 'sickness'. Although this served a useful purpose in making the male antics more forgivable, it was a complete misuse of the term 'menopause'. There *is* a male equivalent of the female menopause, but this occurs much later (see age 53). What is happening in the mid-40s is merely a male rebellion against middle age – it is a mental rather than a hormonal phenomenon. The letters to agony columns blithely ignore this fact. A typical one starts off: 'They say men don't go through the menopause, but I am convinced they go through some kind of change of life. Last year, at 46, my husband suddenly started doing crazy things he had never done before ... He started body-building – though he has never lifted more than his daily pint. And he drove me and our daughters mad with his overdose of get-up-and-go ...' Hardly the picture of hot flushes and general malaise usually associated with a menopause. The phenomenon clearly needs a new name. It is real enough to deserve one, even though it applies to only one segment of the otherwise gracefully ageing community.

Life Expectancy The life expectancy at birth for males in Bolivia, Congo and Lesotho, and for females in Bangladesh, Cambodia, Comoros, Saudi Arabia and Vietnam, is 46.

If you lived in seventeenth-century London, you had only a 10 per cent chance of reaching this age, according to John Graunt's Life Table.

Among the animals, 46 is the record age for a seal and for a monkey. The seal in question was a Grey Seal, and the monkey was a White-throated Capuchin. The Capuchin, a South American monkey, lived in the Evansville Zoo in Indiana, USA and died on April 12, 1976, at the age of 46 years and 11 months.

Accession at 46 George V came to the British throne at this age on June 22, 1911. A quiet, old-fashioned, country gentleman in personality, he reigned quietly for a quarter of a century.

Achievement at 46 Samuel Beckett, the austerely pessimistic Irish playwright saw his play *Waiting for Godot* score a major success on its first presentation, at the *Théâtre de Babylon* in Paris in January 1953, when he was this age.

Joan Collins, the British actress, successfully defied her forty-six years when she appeared in sexually explicit scenes in the 1978 film *The Stud*, twenty-seven years after her first film role, in *Lady Godiva Rides Again*.

Juan Fangio, the Argentinian motor-racing driver, won his fifth and last World Championship in 1957. This made him the oldest champion in motor racing, at 46 years and 55 days.

General Franco, the Spanish dictator, won his complete and unconditional victory in the Spanish Civil War on April 1, 1939. He had been the youngest captain in the Spanish Army and then its youngest General, before becoming the Head of State for the new Nationalist regime in 1936. It then took him another three years to gain complete control of the country, which he finally achieved when he was 46.

George Hegel, the German idealist philosopher, was 46 in 1817 when he published his major work, the *Encyclopaedia of the Philosophical Sciences in Outline*. 'In religion', he said, 'the truth is veiled in imagery, but in philosophy the veil is torn aside ...'

Hippolyte Mège-Mouriés, the French chemist, invented margarine in 1863. He had been commissioned by Napoleon III to find an alternative to butter during a period of acute shortage.

James Murray, the Scottish lexicographer, was this age when he saw the first volume of his monumental work, *The New English Dictionary*, published in January 1884. Known today as *The Oxford English Dictionary*, it took forty-four years to complete, but he died before he reached Z, his last word being *turndown*.

Gregory Peck, the American film actor who specialized in portraying moral and physical courage in the face of danger, made *To Kill a Mockingbird* for which he was awarded an Oscar.

William Shockley, the American physicist, was this age when he shared the Nobel Prize for his role in inventing the transistor.

Shirley Temple, the Hollywood child star, was appointed US Ambassador to Ghana, in 1974, having abandoned acting for politics.

The Duke of Wellington was this age when he scored his great victory over Napoleon at Waterloo, on June 18, 1815. Respected more than loved by his men, he was a good soldier who later proved to be a bad politician - so bad, in fact, that as Prime Minister his house was stoned by an angry mob on two occasions.

Misfortune at 46 **Count Cagliostro**, the eighteenth-century Italian charlatan, magician and adventurer, was this age when he was arrested in Rome, in 1789, after his wife had denounced him to the Inquisition. He was condemned to death, but his sentence was commuted to life imprisonment. His long career of deceit had included selling love philtres and elixirs of eternal youth, summoning up the dead, foretelling the future, and transmuting base metals into gold.

Crime at 46 **Peter Cook**, the Cambridge rapist, was this age when he committed his first assault, in October 1974. Working as a local van-driver, he began as a burglar, then decided on the spur of the moment to rape his victim, whom he had already tied and gagged. Afterwards he asked her if she enjoyed it. During the next ten months he raped eight Cambridge women, causing a reign of terror in the city, and becoming more violent with each attack. He wore a curious black leather mask with a zip-fastener over the mouth and the word RAPIST painted across the forehead in white.

Life-span of 46 **Charles Baudelaire**, the French poet, died in Paris in 1867 after a life of notoriety and scandal. He contracted VD as a teenager, became addicted to hashish and opium, was convicted of offences against public morality, had some of his poems banned, was accused of blasphemy and obscenity, and was declared a bankrupt after extravagantly spending his inheritance. On a trip to Belgium to escape his creditors in 1866 he became paralysed, was taken back to Paris and died there from VD, in his mother's arms, the following year.

Pierre Curie, the French physicist who made major discoveries in the field of radioactivity and, with his wife Marie, discovered two new elements, radium and polonium, was awarded the Nobel Prize in 1903, but was tragically killed only three years later, being knocked down by a dray in a Paris street.

Graham Hill, the popular British racing driver who had competed in a total of 176 Grand Prix events, had been world champion twice, in 1962 and 1968, and had crashed at 150 mph and survived, was finally killed at this early age, only a few months after his retirement, in a light-aircraft crash near his home.

Hans Holbein, the sixteenth-century German painter and one of the world's finest portraitists, died of the plague in 1543.

John F. Kennedy was this age when he was assassinated in Dallas on November 22, 1963, at 1 pm. As his car passed the Texas School Book Depository three shots rang out and he was hit in the throat, head and back. His wife threw herself across his body to shield him, but it was too late and despite a wild rush to Parkland Memorial Hospital, he was dead within thirty minutes. Mystery surrounds the fact that the gun used by the supposed assassin was a non-automatic which took 2·3 seconds to re-load, while film analysis of the incident reveals that only 1·5 seconds elapsed between two of the shots. Rumours of a sinister conspiracy have persisted ever since.

Lawrence of Arabia (T. E. Lawrence) was killed in a motor-cycle accident in 1935, three months after being discharged from the Royal Air Force to face an early retirement at 46. He was filled with a sense of emptiness after the dramas of his earlier life and wrote to a friend that 'there is something broken in the works ... my will, I think.' The accident which killed him solved his problem.

Robert Schumann, the German composer who was a child prodigy, suffered from acute nervous complaints. In February 1854, he complained of painful sound illusions, during which he heard angels dictating a musical theme to him. After a few days, he asked to be taken to a lunatic asylum and then attempted to drown himself. He was removed soon after to a private institution where he remained until his death at 46 in 1856.

Pancho Villa, the Mexican bandit and revolutionary, was living quietly in retirement on a farm in Durango when he was shot down while on a shopping trip to a nearby town, in 1923. His killers were government agents making sure that his retirement was permanent.

Oscar Wilde, the extravagant, generous, eccentrically dressed, verbally outrageous, homosexual Irish playwright who was the rage of London society in the 1880s for his supreme conversational wit, died a broken man in France, on November 30, 1900, and his famous last words were: 'It really would be more than the English could stand if another century began and I were still alive. I am dying as I lived – beyond my means.'

Oscar Wilde

47

Forty-seven was the age of retirement for an ancient warrior – the year when it was considered that the rigours of hand-to-hand battle were too much for the human body. Montaigne, in his essay, *Of Age*, written in 1580, remarks that Servius Tullius, the popular Roman ruler of the sixth century BC, renowned for the thoughtfulness with which he cared for his people, 'dispensed the Knights, who were seven and forty years of age from all voluntary services of war.' The fact that in modern times the military retire their officers at a greater age does not mean that they are treated more harshly or that they have a better physique in middle age, but simply that they are now wielding pens and batons rather than heavy swords and axes.

Life Expectancy Life expectancy at birth for males in Haiti, India, Indonesia, Kenya and Papua New Guinea, and for females in Botswana, Indonesia, Ivory Coast, Papua New Guinea, Swaziland and Zambia, is 47.

Accession at 47 Lenin assumed power in Russia in November 1917. Almost overnight he swapped a fugitive's hideout for the leadership of the Revolutionary Government of the largest country in the world. 'It makes my head spin,' he confessed.

Achievement at 47 Antonioni, the Italian film director, scored his first international triumph in 1960 with his film *L'Avventura*.
Marshall Field, the American businessman, was 47 when he introduced the idea of high-volume sales and low prices, with a huge variety of merchandise available in one building, and opened the world's first department store (and the largest anywhere for twenty-five years), in Chicago in 1881.
Henry Fielding, the English author of *Tom Jones*, was still actively writing in the year of his death, 1754. Although only 47, he was already crippled by gout, with dropsy and asthma to add to his misery. Despite these misfortunes, he managed to write *The Journal of a Voyage to Lisbon*, in which he described his tortuously slow journey to Portugal to seek the sun. It depicts vividly the horrors of eighteenth-century medicine and is a remarkable achievement for a heavily bandaged, wheelchair-bound man, reflecting his indomitable courage and cheerfulness. Despite the sun, however, he died within two months of landing.
Alexander Fleming was 47 when he discovered penicillin in September 1928. By a lucky accident he left some culture plates of *Staphylococcus aureus* exposed to the air for a few days and, just as he was about to discard the contaminated plates, he noticed that the bacteria had failed to grow where a speck of mould had settled. The mould was *Penicillium* and he called the bacterium-killer 'penicillin'. It was first used on humans in 1941 and was of tremendous value to the Allied wounded of the Second World War.
Samuel Morse developed the Morse Code at this age in 1838, having devoted most of the 1830s to perfecting his invention – the single-wire electric telegraph. In the same year he obtained support for the first telegraph line in the US, from Baltimore to Boston. Within a matter of months he had transmitted his first message down the line: 'What God has wrought!'
Marcel Proust was 47 when he suddenly became world famous after winning the Prix Goncourt for his work *A l'ombre des jeunes filles en fleur*, in June 1919. It was the second part of his immense novel-cycle *Remembrance of Things Past*, considered by many to be the greatest French novel of the twentieth century.
Theodore Roosevelt, the twenty-sixth President of the United States, was this age when he received the 1906 Nobel Peace Prize for his mediation which brought to an end the Russo-Japanese War. A forceful politician who had been a vigorous corruption-fighter for years before taking office, as well as a rancher and a soldier, he was a national hero in the man-of-action mould who, despite the nature of his Nobel award, thrived on conflict.
Lord Rutherford, the New Zealand physicist, the founder of modern atomic theory, was 47 when he realized the age-old alchemist's dream of achieving the transmutation of matter. He published, in June 1919, the results of his experiments in which he succeeded in changing the element nitrogen into an isotope of oxygen, by bombarding it with alpha-particles. It was he who recognized that if an atom could be split, artificially, a huge amount of energy would be released – the idea that heralded the Atomic Age.
Lewis Waterman, the American inventor, was this age when he devised the first successful fountain pen, in 1884.

Misfortune at 47 Rudolf Hess, Hitler's deputy, was this age when he decided to bring the Second World War to an end in the spring of 1941. On May 10 he flew solo to Scotland with a set of peace proposals in which he offered the British government a compromise, leaving Germany in his hands in exchange for allowing the British Empire to stay untroubled by German aggression. His unauthorized offer was ignored and he was promptly interned as a prisoner of war.
John, King of England from 1199 to 1216, was forced to accept the Magna Carta when he was 47, curbing his royal powers. Dated June 12, 1215, it gave the barons much stronger authority. John continued to fight against them until his death from dysentery sixteen months later.

Retirement at 47 Beatrix Potter, the English author of 'nursery classics' such as *The Tale* *of Peter Rabbit*, retired from writing at this age in 1913 and spent the last thirty years of her life happily farming in the Lake District. She had been an acutely shy child, overwhelmed by domineering parents, had never been to school and had no friends. She smuggled pets back to her London nursery from her country holidays, including mice, bats, frogs, snails, rabbits and a hedgehog. She devoted all her energies to studying them and drawing them and gradually turned her obsession into children's books.

Life-span of 47

Attila the Hun, the 'Scourge of God', King of the Huns from 433 to 453, who devastated the Balkans and later attacked Gaul and Italy, died on the night following his marriage to the beautiful Ildico, from a sudden heart attack.

Simon Bolivar, the South American revolutionary known as 'The Liberator', whose victories changed the face of South America, ended a bitterly disillusioned man, his dream of a South American federation shattered by internal squabbling and hatreds. He resigned from power and died shortly afterwards, in 1830, at this comparatively young age.

Judy Garland, after an emotionally intense life involving five marriages, endless pills and psychiatrists, and an attempted suicide, finally died of an overdose of sleeping tablets in 1969. As her *Times* obituary put it: 'The burden of stardom was too much for her.' Her death at 47 was reported as 'accidental', but it followed a series of disastrous cabaret performances in which she forgot the lines of her songs and was booed by her audiences. The great irony of her death is that the vibrance of her electrifying personality could easily have overshadowed any physical ageing she might have suffered, had she not become a slave to anxieties, medicines and alcohol.

Joseph Goebbels, the arch-propagandist of the Third Reich, a club-footed cripple capable of stirring up mass hysteria in Hitler's followers, was himself loyal to the Führer to the very end, remaining with him in the Berlin bunker and poisoning himself and his family of six children after Hitler's own suicide, in May 1945.

Jim Jones, the founder of the 'People's Temple', a self-styled Messiah who insisted on complete obedience from his flock of religious fanatics, died at 47 as part of the mass suicide he ordered in November 1978. A year earlier, he and 1,000 of his faithful followers had migrated from California to a new base camp in a jungle clearing in Guyana. There they had practised 'White Nights', rehearsals for the mass suicide of all the men, women and children in the camp by means of cyanide-laced fruit-juice. This camp, called Jonestown, was visited by American Congressman Leo Ryan on November 19, 1978, to investigate, with the help of six journalists, the death of a friend's son following his attempt to leave the 'Temple'. They were shown around, but as they were leaving shots were fired. The Congressman and three journalists were killed. That night Jones ordered the real 'White Night' and 913 of the 1,000 died – men, screaming women, and children begging for mercy. Jones himself was shot dead, but it is not known by whom. He declared 'We are committing an act of revolutionary suicide, protesting the conditions of an inhuman world.'

Joseph McCarthy, the notorious American Senator who conducted a reign of terror in the 1950s, sniffing out communist sympathizers, died at 47 of hepatitis. 'McCarthyism' became a modern name for political persecution. On February 9, 1950, he announced, 'I have a list of 205 names ... of members of the Communist Party ... working and shaping policy in the State Department.' His hysterical attacks continued and in 1954 there were thirty-six days of televised hearings, but with Stalin's death and a thawing of the Cold War, his influence soon dwindled.

Horatio Nelson died at this age, shot in the spine at the Battle of Trafalgar, the scene of his greatest victory. He died on his flagship as it attacked the centre of the French line. His last words, on hearing that the French were defeated, were: 'Thank God, I have done my duty.'

Edith Piaf, the French singer who had been born on a pavement in a Paris slum, died after a life of heavy drinking, drugs, several car accidents and a turbulent sex life. Deserted by her parents when still a baby, she was reared in her grandmother's brothel. Skinny and only 4 feet 10 inches tall as an adult, she could reduce an audience to tears with her emotional singing and made a fortune from her performances. She gave it all away to her lovers, of whom there were hundreds, and she died with nothing, but thousands attended her funeral – the last great audience for the 'Little Sparrow' of Paris.

Pyrrhus, the King of Hellenistic Epirus, whose costly military successes gave rise to the expression 'a Pyrrhic Victory' – meaning a hollow triumph – died at this age in 272 BC.

Edith Piaf

48

Forty-eight was considered a ripe year by the ancients. When Marcus Porcius Cato (Cato the Younger) heard that his friends had been defeated in battle by Julius Caesar in 46 BC, he decided to kill himself. After dinner, he retired to his room and read Plato's discussion of immortality. Then he stabbed himself. His companions, hearing his fall, rushed in and dressed the wounds. He objected, asking them, 'Do I now live the age, wherein I may justly be reproved to leave my life too soon?' An essayist remarks: 'Yet he was but eight-and-forty years old. He thought that age very ripe, yea and well advanced, considering how few men come unto it.' When he had recovered his strength sufficiently, Cato removed the bandages, extracted his intestines, and died.

Today, when men enjoy a much longer average life-span, such a death seems premature. Forty-eight may be called a ripe year, but not in the sense of a ripe old age. It is ripe in the sense of full maturity. With mid-life traumas almost a thing of the past, there is time for the full fruition of an individual's character – a mature ripening that provides us with some of the greatest creative achievements known to mankind. Whatever Cato may have thought, 48 has proved a great year for human triumphs, especially in the sciences.

Life Expectancy
The life expectancy at birth for males in Burma, Guatemala, Uganda and Zaire and for females in Lesotho is 48.

Romance at 48
Emily Dickinson, one of America's foremost poets, was a shy, reclusive woman who rarely left her home and remained single all her life. She never published any of her work and it was not until after her death that a hoard of 1,800 poems and countless letters was found. At the age of 48 she enjoyed a 'December romance' with Otis Lord, a 64-year-old family friend whose wife had just died, but even then her passion appears to have been expressed largely through her writings rather than in a physical consummation of the relationship.

Achievement at 48
John Bardeen, the American physicist, joined the Bell Telephone Laboratories at the end of the Second World War, where William Shockley and Walter Brattain were carrying out experiments with semi-conductors. Together the three men produced a device called a 'transfer resistor', later re-named 'transistor'. For this invention they were awarded the Nobel Prize for Physics in 1956, when Bardeen was 48. He shared another Nobel Prize for Physics in 1972, becoming the first man to gain two such awards for the same field of study.

John Dunlop, a Scottish veterinary surgeon who settled in Belfast, decided to try to improve the ride of his 10-year-old son's tricycle on the rough Irish roads. He fitted it out with air-filled rubber tubes inside the canvas and rubber treads and thereby invented the first pneumatic tyre, which he patented in 1888 when he was 48.

King Camp Gillette, an American travelling salesman, hit on the idea of producing a thin, disposable razor blade that would require repeated replacement and would therefore be in constant demand. He was 48 in 1903 when he sold his first 'Gillette Razors', but during the whole year managed to dispose of only 51, plus 168 blades. By the end of the following year, however, the new fashion had caught on and his company had produced 90,000 razors and 12,400,000 blades. It was he alone who made beards less fashionable in the early part of the twentieth century.

Edmund Halley was this age when he published his great work on comets in 1705. He predicted that a great comet would appear in 1758 and, although he did not live to see it, he was correct. 'Halley's Comet' has been seen every seventy-six years since then, and records indicate that it has, in fact, been seen every seventy-six years since the time of Christ. It was last seen in 1910 and will appear again in the winter of 1985-6, becoming visible to the naked eye by November 1985, and disappearing again in spring 1986.

William James, the American psychologist and the brother of author Henry James, published his monumental work, *The Principles of Psychology*, in 1890, at this age, after working on it for twelve years. Although suffering from ill-health for most of his life, he was a pioneer of major importance to his subject and was responsible for creating America's first experimental laboratory for psychological research. It was he, more than anyone, who established psychology as a laboratory science, rather than as a genteel philosophical sideline. Strangely, despite this, it was also he who defined psychology as 'a nasty little subject'.

Johannes Kepler, the German astronomer, published his third principle of planetary motion in 1619, at the age of 48, completing his major contribution on the elliptical orbits of the earth and the planets around the sun. Kepler's Laws have made him one of the founders of modern astronomy.

Henry Wadsworth Longfellow, the American poet, was 48 when he published his most successful work *The Song of Hiawatha*, in 1855. Although immensely popular in the last century, his reputation has since declined sharply.

Konrad Lorenz, the Austrian naturalist, published his most famous book, *King Solomon's Ring*, in 1952. It brought him world-wide fame, much to his horror, as it was intended as a light-hearted children's book and he feared it might damage his scientific reputation. Instead, it reinforced it, because it lacked the ponderous heaviness of his more academic writings, and projected more accurately the magical appeal of Lorenz's personality in the presence of animals. Throughout his life he has always been a genius as an observer of animal life, but something of a disaster as a philosopher – a field in which sadly he has always wished to impress.

Thomas Macaulay, the British historian who was an infant prodigy, was 48 when his immensely successful *History of England* was first published, volumes 1 and 2 appearing in 1849. Volumes 3 and 4 were not produced until six years later and the final volume came out in 1861 after his death.

Pericles, the Athenian statesman, was responsible for the rebuilding and beautifying of Athens, including the construction of the Acropolis, which was begun in 447 BC when he was 48, when work started on the world's most magnificent temple, the Parthenon.

Auguste Piccard, with his twin brother **Jean,** was this age when he developed a new type of airtight gondola cabin for balloons and with it, in 1932, reached a height of 54,150 feet, as part of his pioneer exploration of the stratosphere.

Joseph Stalin was 48 when he instituted the first of his 'Five Year Plans' in which he set about a radical industrialization of Russia, in 1928. He also forcibly introduced the intensely hated collectivization of agriculture and ruled as a ruthless dictator, backing his wishes with a reign of terror by secret police.

Harold Wilson became Prime Minister of Great Britain at 48 in October 1964, lasting until 1970; then resumed the office from 1974 until 1976.

Harold Wilson

Misfortune at 48

Charles Boycott, the English land agent, brought misfortune upon himself at this age, in September 1880, giving the English language a new verb, 'to boycott'. He had refused to lower the rents of his tenants in times of hardship and was punished by being ostracized by the entire community (in Co. Mayo, Ireland). People refused even to sell him provisions and the necessities of life – a pressure tactic to which his name has become permanently attached.

Crime at 48

Dr Crippen, the American-born English murderer, was 48 when he poisoned his wife, the actress Belle Elmore, with hyoscine, in 1910. He was a dentist who had become caught in a love triangle, hating his wife and loving his young female assistant, Ethel Le Neve. After killing his wife, he cut up her body and buried it under the cellar floor in his house. Then, with Ethel dressed as a boy, and the couple posing as father and son, he sailed off to Quebec. The captain became suspicious because father and 'son' were too amorous and cabled back to the UK with his observations. Crippen and Ethel were arrested when the ship docked in Canada and sent back for trial. He refused to allow her to take any of the blame and left her his entire estate (£268) when he was hanged on November 23. This was the first case in history of a criminal being caught through a radio message.

Life-span of 48

Prince Aly Khan, the playboy son of the Aga Khan who married film star Rita Hayworth, died at this young age in 1960. A delicate child, he did not attend school, but grew stronger as a young adult and became a gentleman jockey, riding at least 100 winners. He also became active in running extensive racehorse studs. Later he joined the French Foreign Legion and eventually became the head of the Pakistan delegation to the United Nations.

Enrico Caruso, the Italian tenor, died of pleurisy in 1921. He was the very first singer to enjoy worldwide fame, thanks to the invention of the gramophone, for which he made 154 recordings.

Charles I, King of England, was beheaded in 1649. His belief in the 'divine right of kings' precipitated the English Civil War and led to his execution by his subjects. He faced the axe bravely and became known as 'The Martyr King'.

Ethelred the Unready, King of England, died in 1016. Best remembered for his mocking name, he was a vacillating and ineffectual monarch who allowed his country to be overrun by the Danes. His name 'Unraed' meant 'evil counsel'.

Friedrich Krupp, the German industrialist who had tripled the Krupp fortune in seven years, killed himself in 1902. Although married and with two daughters, he had become a homosexual and, when photographs exposing this were published in the German newspapers, he felt so disgraced that he committed suicide.

Anna Pavlova, the Russian ballerina, died of pneumonia in 1931, at just the point when her immense skills were beginning to falter with advancing age.

Cecil Rhodes, the South African financier and 'Diamond King' who was the 'father' of Rhodesia, died of heart disease in 1902. He built his fortune on the principle of 'philanthropy plus five per cent'. A strong believer in Anglo-Saxon superiority, he dreamt of painting the map of Africa red from the Cape of Good Hope to Cairo.

Forty-nine is the age of the mind, according to Aristotle. In his *Rhetoric* of 322 BC he comments: 'The body is at its best between the ages of thirty and thirty-five: the mind is at its best at about the age of forty-nine.' He seems to be judging the brain not so much on inventiveness and creativity, which reach their peaks at earlier ages, but on maturity of thought and concept.

Life Expectancy Life expectancy at birth for males in Zimbabwe and females in Guatemala and Nigeria is 49. To find a similar figure for Britain, it is necessary to go back to the period 1900 to 1902, when it was 49·2 years.

> Among the animals, 49 is the record age for a goose, one of the most long-lived of all birds.

Achievement at 49 **Lucille Ball**, the ex-Ziegfeld Girl with red hair and a quacky voice, was 49 when she graduated from being a film and television comedienne to become a powerful Hollywood tycoon in 1960. It was then that she assumed the role of President of Desilu Productions and became boss of her old Hollywood studio – RKO.

William Booth, the English evangelist, founded the Salvation Army when he was this age, in 1878. He introduced military organization, uniforms, and discipline to the world of evangelizing and called himself General Booth, placing himself in supreme command of his new army. Although autocratic, restless, explosive and emotional, with savage prejudices against both science and philosophy, and complete ignorance of theology, he nevertheless instituted valuable social aid programmes in the down-and-out areas of big city slums. He himself grew up in poverty and spoke of his 'blighted childhood'. Among the activities he condemned totally as social evils were football, cricket, horse-racing and card-playing.

John Bunyan, the English tinker who became a preacher and then an author, published his famous work *The Pilgrim's Progress* in 1678 when he was 49.

Howard Carter, the English Egyptologist, was this age in 1922 when he opened the tomb of Tutankhamen, the greatest find in the history of archaeology. It was on November 4 that the first signs of the tomb were discovered. Three days later he reached the sealed entrance and for the next ten years he supervised the removal of the contents to the Cairo Museum.

Jean Dubuffet, the French artist and creator of *L'Art Brut* (Raw Art), outraged the art world with his brutally crude images, applied in a superficially childlike manner, called his *Corps de Dames* series, in 1950.

Hugh Gaitskell, the Labour politician, succeeded Attlee as leader of the Labour Party in 1955, but was always in opposition and never had the chance to express fully his undoubted powers as a socialist supremo.

Kenneth Grahame, the British author, was forced by ill-health to retire from his job at the Bank of England in 1907 and published his children's book *The Wind in the Willows* the following year when he was 49. The book began as bedtime stories for his son and continued in letters to the boy when he went away to school. His animal characters – Mole, Rat, Badger and Toad – were dramatized by A. A. Milne in 1930, in the play *Toad of Toad Hall*.

Inigo Jones, the English architect, was this age when he completed his greatest work, the Banqueting House at Whitehall, in 1622. He was the founder of the classical school of English architecture.

Burt Lancaster, the ex-circus acrobat who became a professional swashbuckler and eventually matured into a major screen actor, won an Oscar in 1962, at 49, for his role in the film *Elmer Gantry*.

Trygve Lie, the Norwegian politician and lawyer, became the first Secretary General of the United Nations Organization, in 1946.

Karl Marx, the German political philosopher who was born a Jew and baptized a Christian, published his great work *Das Kapital* in 1867. Intellectually arrogant, he avoided mass audiences and spent most of his adult life in London where he haunted the British Museum Library and lived in poverty.

David Niven in *Separate Tables*   Puccini

David Niven, the British film actor, was 49 when he won an Oscar for his dramatic part in Rattigan's *Separate Tables* in 1958. A dapper ex-soldier from a military family, he drifted towards Hollywood in the 1930s and became one of its most enduring and best-loved stars.

George Stephenson, the English inventor responsible for the first practical railway service, was only a small boy when he first went to work in a colliery. While there he began studying engineering in his spare time and by July 25, 1814, had constructed his first colliery locomotive, travelling at 4 mph. On September 27, 1825, he himself drove the engine of his first public railway – between Stockton and Darlington, drawing thirty-eight carriages at 12–16 mph. In 1829 he built his famous *Rocket* and his crowning glory was at 49, with the opening, on September 15, 1830, of the first practical passenger railway line in

the world. This ran between Manchester and Liverpool and the train travelled at an average speed of 29 mph and with top speeds exceeding 35 mph.
Bram Stoker, the Irish novelist who had suffered an invalid childhood, unable to stand or walk until he was 17, but who then became an outstanding athlete and football player at Dublin University, was 49 when he published his most famous story, *Dracula*, in 1897.
George Washington, the first American President, was this age when he defeated the British, forcing the surrender of the British commander, Lord Cornwallis, on October 19, 1781, at Yorktown, Virginia. He was a practical, persistent, inflexibly just man of high integrity.

Misfortune at 49

Puccini, the Italian operatic composer, famous for his romantic masterpieces such as *La Bohème*, *Tosca*, *Madame Butterfly* and *Turandot*, himself suffered a love-life as traumatic as that of the characters in his productions. After the death of his mother, he fled from his home town of Lucca with a married woman and caused a scandal. Later they were able to marry, but there was then an even greater scandal in 1908, when Puccini was 49. His wife became jealous of a young servant girl and drove her from their home, threatening to kill her. The girl poisoned herself and, when examined, was proved to be a virgin. Puccini's wife was found guilty of causing the girl's death and he had to buy off the servant's parents. From then on the relationship between him and his wife became a mere formality.

Life-span of 49

Queen Anne, the dull, dowdy and devout monarch of Great Britain and Ireland from 1702, died at this early age in 1714. She had been pregnant eighteen times, but only five babies were born alive and none of these survived to adulthood. Although she was a rather foolish woman, she ruled at a time of great intellectual activity, distinguished by such authors as Defoe, Pope and Swift.
Saint Thomas Aquinas, the foremost philosopher of the Roman Catholic Church, who attempted the impossible task of reconciling reason and faith, died at the Cistercian Abbey at Fossanova after falling ill on a mission for the Pope in 1274.
Stanley Baker, the Welsh film actor famous for his tough, tight-lipped man-of-action roles in many British films, died from lung cancer in 1976, one month after being knighted by the Queen.
Edith Cavell, the English nurse who became a popular heroine of the First World War for assisting Allied soldiers to escape from German-occupied Belgium, was executed at the age of 49 – shot by the Germans on October 12, 1915.
Robert Clive, Clive of India, who laid the foundations for Britain to rule that country, killed himself in his London house in 1774. After having been the Governor and Commander-in-Chief in Bengal, with full powers, his health declined and his melancholic temperament got the upper hand. He had already, in his younger days, proved moody and quarrelsome and had once fought a duel. On one other occasion he had attempted suicide, but this time he succeeded.
Davy Crockett, the American frontiersman and politician, whose hunting background included the pursuit of bears and raccoons and occasionally Indians, was elected to Congress on the strength of his homely, yarn-spinning speeches. After his eventual political defeat he joined the American forces and was killed at the Alamo by the Mexican army on March 6, 1836.
Isadora Duncan, the eccentric, dramatic dancer from San Francisco, unappreciated in her own country, left for Europe as a young woman and achieved great success both in England and in Russia. Eventually tragedy struck when her two children by different lovers were killed together in a motor accident. After this she never fully recovered and died young herself, at 49, accidentally strangled by her long scarf which became entangled in the wheel of a car she was driving, in 1927.
Marty Feldman, the bug-eyed Jewish comedian from the East End of London, died of a heart attack during a comeback attempt in Mexico. After many successful comedy roles, he had turned his hand disastrously to film direction. Then, in 1982, he wisely went back to a comedy role in a Monty Python film called *Yellowbeard*, but collapsed and died after the end-of-shooting party. One week before, he had said: 'I am too old to die young, and too old to grow up, so I must go on running all my life.'
Conrad Gesner, the Swiss naturalist who produced some of the most impressive early natural histories, including his massive *Historia Animalium*, which covered all known kinds of animals, died of the plague in 1565.
Alexander Hamilton, the first Secretary of the Treasury in the US, under George Washington, was shot dead in a duel with a political opponent, Aaron Burr, in New Jersey in June 1804.
George Lincoln Rockwell, the founder of the American Nazi Party, was shot dead by an ex-member of his party in 1967. The son of a vaudeville performer, he had been a navy commander in the Second World War. His slogan was 'White Power' and his avowed aims were the deportation of all negroes to Africa; the sterilization of all Jews and the confiscation of all their property; the hanging of all traitors including Truman, Eisenhower and Chief Justice Earl Warren and his own election as President.
Andreas Vesalius, the Flemish anatomist whose dissections of the human body revolutionized medical knowledge, and made him the founding father of modern anatomy, was condemned to death by the Inquisition, for heresy. The sentence was reduced to a holy pilgrimage, which had the same effect, since he was shipwrecked returning from it and died on the island of Zante.

50

Fifty is a mellow year. It is an age when old friendships are treated more warmly; when the harsher relationships with colleagues, and possibly relatives, begin to soften, as the intense competitiveness of the 40s starts to wane.

There is a feeling that, at 50, you have completed the formation of your personality. The final entry in George Orwell's working notebook read: 'At fifty, everyone has the face he deserves.' And in *More in Anger* Marya Mannes comments: 'By the age of fifty you have made yourself what you are, and if it is good, it is better than your youth.'

According to old proverbs, 50 is the age by which one is supposed to be rich, but wealth at this stage of life can bring a certain sadness – sadness that it did not come earlier, during the years of youthful love and vigour. Struthers Burt, in *Fifty Years Spent*: 'Fifty years spent and what do they bring me?/Now I can buy the meadow and the hill:/ Where is the heart of the boy to sing thee?/ Where is the life for thy living to fill?'

T. S. Eliot views 50 as the start of a trying time for a celebrity, when 'you are always being asked to do things, and yet you are not decrepit enough to turn them down.'

Sexually, 50 is looked upon as bad news. A German proverb says bluntly 'Desire ends at 50,' and Napoleon remarked: 'At fifty one can no longer love.' Several authors agree. Thomas Hardy: 'Love is lame at fifty'; Gore Vidal: 'For certain people, after 50, litigation takes the place of sex.' Mark Twain sees the decline as exclusively masculine: 'After 50 his [the male's] performance is of poor quality ... whereas his great-grandmother is as good as new ... Her candlestick is as firm as ever, whereas his candle is increasingly softened and weakened by the weather of age ... '

A slight advantage in this decline in male lust emerged from a conversation between British Defence Secretary John Nott and the Head of Italy's Security Services, at a dinner in May 1982. The Italian commented: 'I know that today is your fiftieth birthday. May I congratulate you? It is a fine age for a man. When a woman says "Yes" he is flattered, and when she says "No" he is relieved.'

The biological facts are that male testosterone levels *are* beginning to fall more rapidly at this age and the frequency of male orgasm has fallen to about 50 per cent of what it was at 30. At 50 about 7 per cent of men suffer from 'erection impotence'. Apart from sexual matters, there is also a decline in the quality of the sense organs. Reading glasses become suddenly much more commonly needed and there is a decline in the sense of taste – which is why gourmets of this age prefer more spicing of their food. Fifty is also the age when body height begins to diminish, as part of the shrinkage of old age.

As always, however, there are many exceptions to these general rules and, for the lucky ones, the decade they are entering may prove to be one of the most enjoyable of their lives, with the frantic cut-and-thrust of youth far behind them and the befuddling of old age far in front.

Life Expectancy The life expectancy at birth for females in Congo and Haiti is 50.

> Among the animals, the oldest known age recorded for a chimpanzee is 50 years and 6 months, for a male called Jimmy at the Seneca Zoo, Rochester, New York. Fifty is also the age for the oldest known crustacean – a lobster.

Accession at 50 **Claudius I,** the Roman Emperor, came to power unexpectedly at 50, following the murder of his infamous predecessor, Caligula, on January 24, AD 41.
Papa Doc Duvalier was elected President of Haiti in September 1957 and started his reign of terror. Before gaining power he had been an important medical officer on the island, combating malaria, but once in charge, his personality changed and he became a brutal dictator.

Achievement at 50 **James Boswell,** the Scottish author, published his famous *Life of Samuel Johnson*, on May 16, 1791, when he was 50, generally regarded as one of the supreme biographical achievements.
Charles Bronson, the American film actor, was one of fifteen children of a Lithuanian coal-miner. His film career had been coasting quietly for twenty years when suddenly, and astonishingly, at the end of the 1960s he found himself, at the age of 50, becoming an overnight sex symbol and the recipient of the 1971 Golden Globe award as the world's most popular actor.
Charles Darwin was 50 when he published *The Origin of Species* on November 24, 1859. It had taken him thirteen months and ten days to write it. Every copy was sold on publication day – a modest total of 1,250. Fifteen years later it had sold 16,000 copies in England – a large figure for the period. It destroyed at a stroke the concept of 'special creation' and heralded the beginning of the end for many of the more naïve religious beliefs.
Henry Ford, the American motor-car magnate, was 50 when he introduced the novel idea of an 'assembly line' into his car manufacturing plant, as part of his dream to 'put America on wheels' by producing cheap cars that the ordinary person could afford. This was in 1913 and led to the mass production of his famous Model T Ford.
William Harvey, the seventeenth-century English physician, discovered the circulation of the blood and published his findings at the age of 50 in his 1628 treatise, a slender volume entitled *On the Motion of the Heart and Blood in Animals.*
Henry Kissinger, who emigrated to the USA in 1938 and became the American travelling-peacemaker and cease-fire negotiator extraordinary, was

50 in 1973 when he was made Secretary of State and was also awarded the Nobel Peace Prize. A physically less than handsome man who has always attracted women, he has been quoted as saying 'Power is the ultimate aphrodisiac'.

Wilhelm von Röntgen, the German physicist, was this age in 1895 when he discovered X-rays, for which he was later awarded the Nobel Prize, in 1901.

Gerhard Schmatz, the West German mountaineer, has a special claim to fame, being the oldest person ever to climb to the top of Mount Everest. He scaled the 29,028-foot mountain at the age of 50 on October 1, 1979.

Willie 'The Shoe' Shoemaker, the world's most successful jockey ever, had ridden over 8,000 winners by May 27, 1981, when he was 50.

Misfortune at 50
Rembrandt, one of the world's greatest masters, went bankrupt at 50. Despite the fact that he was the most prolific and the most popular portrait painter in Amsterdam, with commissions pouring in, and with a marriage to a wealthy heiress, he was so extravagant that he was declared a bankrupt on July 25, 1656, when he was just ten days past his fiftieth birthday. He had amassed a great collection of art, antiques and jewellery, but it all had to go. His house and his treasured possessions were sold at a series of three compulsory auctions. His rich wife was already dead and had put severe restrictions on the use of her estate. Twenty years after her death Rembrandt sold the tomb from above her grave to obtain some ready cash. One of the greatest portrait painters in the whole history of art, Rembrandt was a bulbous-nosed, coarse-featured genius who was also one of the epic spendthrifts of all time.

Retirement at 50
W. G. Grace, the bearded giant of cricket, the best-known figure in the history of the game, retired finally at this age, from test cricket. Altogether he had made 54,904 runs, 126 centuries, had taken 2,876 wickets and held 877 catches.

Stanley Matthews, the soccer genius who was knighted for his quicksilver skills, played his last professional match five days after his fiftieth birthday in February 1965. His team, Stoke, beat Fulham 3-1.

Life-span of 50
Alfred the Great lived up to his name, a skilful military leader and a wise lawgiver. But as Saxon king of Wessex in the ninth century, he was plagued by a recurrent illness and died in the year 899.

Madame du Barry, the last mistress of the French King Louis XV, was sent to the guillotine at 50, in 1793.

Captain James Cook, the English explorer, was killed at the same age, but by blunter instruments. He was clubbed to death on a beach in Hawaii, by local natives, during a fracas over a stolen boat.

Sir Humphry Davy, the great English chemist who had saved so many lives beneath the surface with his miner's safety lamp, was unable to prolong his own life past the age of 50.

Errol Flynn, the Australian actor, whose colourful screen roles were matched by his extravagant private life, collapsed and died of a heart attack in Vancouver in 1959.

Lady Hamilton, Nelson's mistress, died at 50, fat and alcoholic, after losing all her money on drink, gambling and party-giving, ending with nine months in a debtor's prison. Apart from her involvement with Nelson, her life had not been a happy one. At 16 she had bartered her body in exchange for the release of a young cousin from a British warship. The captain made her pregnant and she had to farm out the child. She then took to performing naked dances before a bed on which impotent men lay for a quack's cure. Later, she married the elderly Sir William Hamilton, to spite his callous nephew.

Leslie Howard, the English actor, was shot down in a plane, vanishing like Glenn Miller in a mystery flight over wartime waters. His aircraft left Lisbon in 1943, but never arrived in London.

Alan Ladd, the avenging angel of a string of classic action-movies of the 1940s, took to alcohol when his career collapsed in middle age. Some months before his death he 'accidentally' shot himself, but recovered. Then, in 1964, aged 50, he finally succumbed to an overdose of sedatives mixed with alcohol.

Steve McQueen, like Alan Ladd, a blue-eyed, athletic, dynamic star of action-films, also died at this age, but in his case the cause was chest cancer against which he fought desperately right to the end. Trying extreme cures, he suffered a heart attack following surgery in Juarez, Mexico, in November 1980.

Ross McWhirter, the co-editor, with his twin Norris, of *The Guinness Book of Records*, was murdered in 1975 by political terrorists. He had established a reward fund for payments to informers who led to the arrest of terrorist bombers, and it cost him his life.

Tsar Nicholas II was shot by the Bolsheviks, along with his entire family, then bayoneted, burned and thrown into a mine-shaft, in July 1918. A timid autocrat, he believed that he was solely responsible to God, but the deity failed to reciprocate.

Mike Todd, American hustler extraordinary, was killed in an air crash in 1958, a year after marrying Elizabeth Taylor. A razzmatazz promoter of great imagination, he introduced 65 mm Todd-AO to answer the threat of television, and scored a smash hit with his film *Around the World in 80 Days*.

Virgil, the Roman poet, died at 50, in 19 BC at Brundisium. The *Aeneid*, his greatest work, is an epic poem telling of the wanderings of survivors of the sack of Troy, until they settled in Italy and became the ancestors of Rome.

Fifty-one is the year of the menopause. A study carried out in Europe in the late 1970s revealed that, although it is common for women to experience the 'change of life', climacteric, or menopause anywhere between their mid-40s and their mid-50s, the average age for it to occur is 51·7 years. This is three months later than in the late 1960s, but when women who have been using oral contraceptives are eliminated from the figures, then there is no difference. It would seem that the action of the 'pill' somehow delays the onset of the menopause.

For some women, this is the age when they feel that their sex lives are over – if they cannot reproduce, what is the point of sexual activity? They may even think of the menopause as a kind of 'illness', accompanied as it often is by the well-known 'hot flushes', caused by the lowered levels of oestrogen circulating in the body. But the problem is more psychological than physiological. Few of the body changes of menopause need have any adverse effect on behaviour. If there is a period of anxiety or depression, it comes from an attitude of mind rather than physical state. All that really happens is that women stop shedding eggs, and for some this comes as a great relief. It means that they can at last indulge in 'wild sexual abandon' without the slightest fear of pregnancy. Masters and Johnson in *Human Sexual Response*, discussing the sexual implications of the menopause, comment: 'Many a woman develops renewed interest in her husband and in the physical maintenance of her own person, and has described a "second honeymoon" during her early fifties.'

Life Expectancy
Life expectancy at birth for males in Iraq, Libya, Morocco, Pakistan and Nicaragua, and for females in Bolivia, Burma, Kenya, Pakistan and Uganda is 51.

Achievement at 51
Field-Marshal Alexander was this age when he achieved a Second World War triumph in North Africa, defeating the Germans in Tunisia and taking 200,000 prisoners after a series of brilliantly conducted manœuvres. He accepted the German surrender in Tunis in May 1943.

Dirk Bogarde, the English film actor whose performances grew steadily better with age, reached his peak at 51 in Visconti's 1971 production of *Death in Venice*.

Samuel Cunard was 51 in 1839 when he established the Cunard Line, to provide a regular transatlantic steamship service for the first time. Within a year the inaugural trip was undertaken, using the paddle-steamer *Britannia*, with Cunard himself on board, making the crossing in fourteen days and eight hours. This was also the start of a regular postal service between England and America, at first fortnightly, and then, after seven successful years, weekly.

William Faulkner, the American author of the Deep South, was 51 when he won the Nobel Prize for Literature in 1949. Although he hated public formalities, he interrupted a marathon drinking bout long enough to sober up for his acceptance speech. The effort proved fruitful, for the award resulted in a greatly increased international recognition of his talents.

Ernst Heinkel, the German aircraft designer, made history at 51 when his Heinkel He 178 achieved the first successful flight by a jet plane, on August 27, 1939. Despite this breakthrough, it was with conventional propeller aircraft that he made his name as Hitler's most important aircraft manufacturer during the Second World War, designing bombers with a characteristic and much feared drone, heard over London night after night.

Justinian I, the Byzantine Emperor, known as 'the Emperor who never sleeps' because of his tireless energy and unusual ability to go without rest or food for long periods, was this age when he completed his great code of Roman law, the *Codex Justinianus*, in the year AD 534. It was a work of lasting importance that provided the basic pattern for many later codes of law.

William Pitt the Elder, the manic-depressive eighteenth-century politician, was, in his more manic moods, one of the greatest orators ever heard in British politics. His malady, not understood at the time, was referred to as 'gout in the head', for which he took the waters at Bath. It was his forceful personality, when in full voice, that was responsible for leading Britain into the role of Imperial Power, culminating with his 'year of victories' in 1759, when he was 51. Even his rival Walpole wrote with reluctant admiration 'Our bells are worn out threadbare with the ringing for victories.'

Samuel Plimsoll, the British MP and social reformer, achieved a triumph on behalf of all sailors when he was 51, at the culmination of a long campaign to secure safer conditions on board ships. He had earlier published *Our Seamen*, a powerful attack on 'Coffin Ships' which were unseaworthy, overloaded vessels, usually heavily insured, in which unscrupulous owners risked the lives of their crews. For several years Plimsoll tried his hardest to persuade Parliament to take steps to stop these abuses, but without success. Exasperated, he launched an attack on shipowners on July 22, 1875, so vicious that it caused a scene in the House of Commons. He was reprimanded by Disraeli and apologized to the House for his behaviour, but he had won the day and by the following year the Merchant Shipping Act was at last passed and the famous 'Plimsoll Line', preventing the overloading of ships, became a permanent feature of all merchant vessels.

Misfortune at 51
Gustave Courbet, the French realist painter, was arrested on June 7, 1871, and brought before a military court, accused of being

the instigator of the destruction of a column commemorating the Grand Army of Napoleon Bonaparte, in the Place de Vendôme, Paris. Although he was innocent, he was a suitable scapegoat, being an atheist, a socialist and an anti-militarist, so he was sentenced to six months' imprisonment, and fined. In prison he became seriously ill but after he was freed, as if to add insult to injury, his case was reopened and he was sued for the cost of rebuilding the column he had not knocked down. His entire personal property was seized, including all his paintings, and he was fined a further 500,000 gold francs. Unable to pay, he fled to Switzerland, taking refuge in an old inn. He died there, a broken man.

Ovid, the Roman poet, was banished to the Black Sea when he was 51, towards the end of AD 8, and never allowed to return. The reason is a mystery, but he was probably involved in some scandal concerning the Emperor's daughter, Julia.

Crime at 51

Kim Philby, the British intelligence officer who acted as a Soviet spy, had his crimes uncovered in 1963 and was forced to flee to Russia, disappearing from Beirut in January of that year.

Retirement at 51

Galina Ulanova, the Russian ballerina, a leading dancer of the Bolshoi Ballet Company, retired at this advanced age, in 1963, to become ballet mistress at the Bolshoi Theatre in Moscow.

Life-span of 51

Honoré de Balzac, the great French novelist, who died in 1850, is not the only famous author to have drunk himself to death, but is probably the only one to have done so with coffee. His extraordinary work-pattern consisted of starting to write at midnight, dressed in a long white robe, and continuing for sixteen hours, interrupting the scratching of a series of quill pens only with pauses for endless cups of coffee. During a single writing session he consumed several dozen cups of strong coffee and, when one recalls that he published over 100 novels and a number of plays, it is hardly surprising that his stomach finally succumbed to the relentless onslaught of caffeine, and brought his life to an early close at 51. When he was not writing he somehow found time to indulge in countless sexual affairs with an enormous variety of women, from whores to aristocrats. He summed up his attitude to women with the comment: 'A woman is a well-served table that one sees with different eyes before and after the meal.'

Napoleon Bonaparte, the Corsican-born Emperor of France, also died at this age, in 1821, of a gastric ulcer aggravated by defeat, and exile on the island of St Helena.

Roger Casement, one of the Irish martyrs in the revolt against British rule, was tried for treason in 1916 and hanged. In Germany he had tried to form an 'Irish Brigade' to invade Ireland and free it from British control, but he was caught when returning to Ireland by submarine. There were rumours that his 'black diaries', revealing a variety of homosexual activities, had prejudiced his trial.

Jan van Eyck, the founder of the Flemish School of painting, revolutionized the technique of oil painting, because turpentine became available. He used multi-glazing techniques to heighten his colours and became a master of light and colour, and precision of detail. He died at 51 in 1441, at Bruges.

Gordon of Khartoum   Dirk Bogarde in *Death in Venice*

Gordon of Khartoum, the British general, became a national hero for his exploits in China and for his ill-fated defence of Khartoum against the Sudanese rebels led by the *Mahdi*, where he was killed as the city was overrun, following a siege, on January 26, 1885.

Calamity Jane, the popular heroine of the American West, famed as a beautiful crusader for justice, was in reality a frontier prostitute who wore men's clothing, carried a gun and went on wild drinking sprees. She died in poverty at 51, in 1903, and was buried next to her old friend, Wild Bill Hickok, in Deadwood cemetery.

Molière, the French dramatist and actor, collapsed during the fourth performance of his last play, *Le Malade Imaginaire* (The Imaginary Invalid). It was a story about a hypochondriac who is terrified of dying, but there was nothing imaginary about Molière's condition. He was carried back to his house, but did not survive the night. His death at 51, in 1673, seems to have been caused by acute stress from overwork.

Robert Newton, the actor with the wild, rolling eyes and the thunderous voice, best remembered for his extravagant portrayal of Long John Silver in *Treasure Island*, became an alcoholic in his later years, but continued acting to the end, in 1956.

Marcel Proust, the French author, was still working on his great masterpiece, the novel-cycle *Remembrance of Things Past*, when he died of pneumonia, in 1922.

Rainer Maria Rilke, the Austrian poet, was 51 when he died of poisoning after being pricked by a thorn, when picking roses for his last love, a beautiful Egyptian woman called Nimet Bey, in 1926.

52

Fifty-two is the year of the 'oldest swinger in town'. Instead of accepting that he is moving well into his mellowing 50s, he rebels against the idea, buys an expensive 2 + 2 hard-topped sports car (the young men they are meant for rarely being able to afford them), wears deeply opened shirts, a few gold ornaments, and perhaps some fancy facial hair. When age-researchers showed a photograph of this type of 52-year-old male to people and asked them to assess his true age, it was put at between 28 and 30. Shedding over twenty years in this way is not achieved lightly. It requires a great deal of discipline and effort. All forms of drugs are avoided, including alcohol, coffee, tea and tobacco. Mineral water is acceptable – tap water never. Health foods are essential – honey, yoghurt, raw nuts, fresh fruits and the like. Jogging and sports such as tennis, swimming and squash are obligatory on a daily basis. Meditation and postural exercises are usually thrown in for good measure. It is a hard life being an ageing stud.

The only famous quotation associated with this age is that of Lady Astor: 'I refuse to admit that I am more than 52 even if that does make my sons illegitimate.'

Life Expectancy Life expectancy at birth for males in Algeria, Honduras and Jordan, and for females in Jordan is 52.

Romance at 52 Colette, the French novelist, at last found happiness at 52 with Maurice Goudeket, a man seventeen years her junior. The author of seventy-two books, largely concerned with love, she had had a complex love-life of her own, scattered with unsuccessful marriages and lesbian affairs. Then, in what she termed 'the autumn of her womanhood', she finally managed to enjoy a deeply loving companionship with Goudeket, which renewed her creative energy and made it possible for her to remain fully active well into her old age. Their happiness was interrupted only by her young husband's imprisonment by the Gestapo in the Second World War.

Achievement at 52 David Attenborough, the English naturalist, was 52 in January 1979 when he presented *Life on Earth* on BBC television, the most spectacular and most impressive natural history series ever screened.

Francis Bacon, the English painter of tortured, agonized humanity, was this age when London's Tate Gallery gave him a major retrospective exhibition, with more than half of his surviving works on display. An intensely self-critical artist, he has destroyed more of his paintings than now exist, including some of his finest works. It was estimated some years ago that he has obliterated over 700 works by painting new pictures on top of them.

Paul Delvaux, the Belgian artist, became Professor of Painting in Brussels in 1950, at this age. What makes this appointment unusual is that Delvaux was one of the leading members of the surrealists, a group fervently opposed to all academies and art institutions.

William Jenney, the American civil engineer and architect, was 52 in 1882 when he invented the skyscraper and changed the direction of modern city planning. He designed the Home Insurance Company Building in Chicago which is generally considered to be the world's first 'tall building'. A ten-storey construction, it incorporated an internal skeleton of iron and steel, instead of being supported by load-bearing walls.

John Maynard Keynes, the English economist, was this age when he virtually destroyed classical economics with the presentation of his 1935 magnum opus *The General Theory of Employment*. This led to what has been called the Keynesian revolution, in which the idea of a self-regulating economy was replaced by one in which governments consciously adjusted and modified the economy, employing a variety of financial techniques such as varying the interest rates. He argued that, if left alone, the economy would only find equilibrium at a high level of unemployment and would remain there indefinitely, because total spending by consumers, investors and the government was too low to employ all the economy's resources.

Colette

Hiroo Onoda achieved a remarkable feat of survival in the Japanese Army. He was sent to the Philippine island of Lubang as a 23-year-old Lieutenant in 1944, where he failed to learn of the end of the Second World War on August 6, 1945. Living alone in the jungle, he continued to fight the war for another twenty-nine years, eating a variety of wild foods and occasionally shooting a domestic animal to boost his diet. He sniped at islanders from time to time and when police tried to catch him they were met with a hail of bullets. He believed that all stories of the war's end were enemy tricks to deceive him into surrender, until one day in 1974 when he met a Japanese tourist on a camping holiday, and learned the truth. Even then, he refused to lay down his arms unless ordered to do so by his old Commanding Officer. The man in question was now a bookseller and had to be flown out to the Philippines to order Onoda to give himself up, which he did on his fifty-second birthday, returning to Japan as a national hero, where he was able to visit his own tombstone.

Wilfred Rhodes, the English cricketer, was the oldest man ever to play in test cricket. He took part in a test match for England against the West Indies on April 12, 1930, when he was 52 years and 165 days old.

Misfortune at 52

George Wallace, the American politician, was paralysed in an assassination attempt when he was 52, in 1972. On May 15, he was shot in the car-park of a shopping centre at Laurel, and has since been confined to a wheelchair. In his earlier days, as Governor of Alabama, he became famous for the vigour of his segregationalist views, epitomized in his 1963 statement, 'I draw the line in the dust and toss the gauntlet before the feet of tyranny, and I say segregation now, segregation tomorrow, segregation for ever.'

Crime at 52

Jack Ruby, the American saloon-keeper, was this age when he shot down Lee Harvey Oswald on November 24, 1963. He slipped into the basement of the Dallas jail and mingled with television crews awaiting the transfer of Kennedy's assassin, stepped forward and shot him once with a ·38 revolver, in full view of the television audience. He himself died in hospital while receiving treatment for cancer and it has been claimed that he was paid to silence Oswald and was then himself killed, to conceal a conspiracy to murder Kennedy. Ruby's background, to say the least, was murky. As a teenager he was a message-runner for Al Capone. In Dallas, he operated sleazy night-clubs, acting as his own bouncer and pistol-whipping drunks for fun. He was arrested nine times but never convicted, until his last momentous crime in 1963.

Life-span of 52

Mark Antony, the Roman leader who became Cleopatra's lover, committed suicide at this age. Following their defeat at Actium in 31 BC, and again outside Alexandria the following year, they had little option but to kill themselves. He died first and then, after failing in her negotiations with their enemy Octavian, she too took her life. They were buried together at Alexandria.
Thomas à Becket, a proud and ostentatious priest who persuaded his king, Henry II, that he was a trusted servant of the crown, and in this way managed to become his archbishop, quickly turned against the king's policies and fell from favour. Henry spoke harshly about Becket in the presence of some of his knights. Four of them took his spoken wish - to be rid of the archbishop - literally and, on December 29, 1170, they entered Canterbury Cathedral where,

at twilight, they cut him down with their swords. He was canonized two years later and Henry II did penance at his tomb.
Lou Costello, the American comedian, died of a heart attack at 52, in 1959. As the short, tubby member of the Abbott and Costello comedy duo, he earned a huge fortune from a string of over thirty films in the 1940s and 1950s, but eventually fell out with his 'straight man' partner after insisting that, as the 'funny man' partner he should receive 60 per cent of their joint earnings.
Christian Dior, the French fashion designer, died at 52, in 1957, after a life dogged by ill-health.
William Hazlitt, the English essayist and critic who was a precocious child, died in 1830, exhausted by overwork and harassed by financial problems.
Harry Houdini, the world's greatest escapologist, who defied death on many occasions, including being shackled with irons and locked into a roped-up, weighted box, which was then dropped from a boat into the water, eventually died at 52, in 1926, suffering from peritonitis following a stomach injury. He hated mind-readers and spirit mediums and wrote books attacking those who peddled the supernatural. He arranged with his wife that he would try as hard as possible to communicate with her after his death but, as he would have predicted, the experiment was a failure.
Gamal Abdel Nasser, the forceful President of Egypt who dreamt of a United Arab Republic - a world force of combined Arab states led by himself - died at 52 in 1970, his spirit broken by the massive humiliation he suffered in the Six-Day War with Israel in 1967. He offered to resign, but his people wanted him to stay. The defeat, however, seems to have destroyed his health.
Erwin Rommel, the much admired German Field-Marshal, known as the Desert Fox, was 'visited' on October 14, 1944, on Hitler's orders. The Führer suspected him of plotting an assassination and a coup. Although he was not involved in the plot, when he was given the choice of a trial or suicide, he chose to poison himself. His last words to his wife were, 'In 15 minutes I will be dead.' He then joined the two generals who had been sent for him and killed himself in their car. After his death Hitler ordered national mourning.
William Shakespeare, who had written thirty-two plays in twenty years, died on April 23, 1616. He had been so successful that he had been able to buy the largest and finest house in his native town of Stratford-upon-Avon (for the sum of £60). According to the local vicar, John Ward, the Bard had been entertaining his old friend Ben Jonson and 'had a merry meeting' but 'it seems drank too hard, for Shakespeare died of a fever there contracted'. It would be wrong to think that he died solely as the result of a drinking party, however, for he had been ill for several months and had already made his will, as if expecting the end.

53

Fifty-three is the year of the 'male menopause'. It has been estimated that about 15 per cent of men go through a genuine physical climacteric, with a rapid, sharp decline in testosterone levels. For them, there is a recognizable time when sexual vigour decreases and it may be accompanied by a variety of symptoms of general malaise, such as morning lethargy, insomnia, dizzy spells, heart palpitations and even the traditional (female) hot flushes. For most males this specific age-event does not happen; their decline in sexual vigour is so gradual that it is barely noticed. They may have experienced something popularly called a 'male menopause' back in their 40s, but that had more to do with refusing to accept middle age than with any genuine physiological fall-off.

Life Expectancy
Life expectancy at birth in Egypt and Turkey for males and in Turkey and Zimbabwe for females, is 53.

Romance at 53
Peter Paul Rubens, the maestro of vast, fleshy nudes, who produced 3,000 works of art and was immensely rich and successful during his lifetime, was this age when he fell in love with and married his 16-year-old model, Helena, after the death of his first wife. Helena *was* the 'Rubens nude', fair and plump, pink and white, dazzling and buxom. He ravished her in paint, depicting her being captured, raped, rescued, or simply lost in playful thought, unselfconsciously naked and voluptuous. It was a great love affair conducted on canvas, while in real life she gave him five children and a second, happy home life. They married on December 6, 1630, and his continued success was such that, when he died nine and a half years later, his estate was so huge that it took five years to make an inventory of it.

Achievement at 53
Lawrence Bell, the American aircraft designer, was this age in 1947 when his experimental X-1 rocket-propelled aeroplane was the first to break the sound barrier in level flight, heralding the era of supersonic flight. The plane, *Glamorous Glennis*, was piloted by Chuck Yeager, over California on October 14, and reached a speed of 670 mph.
Robert Goddard, the American physicist, was also 53 when, in 1935, he became the first person to shoot a liquid-fuelled rocket faster than the speed of sound. From his early days he had been convinced that rockets could take people into space, but was derided as 'moon mad'. The United States ignored his ideas, but Germany made use of them in the Second World War, to create its devastating V-2 rockets. Goddard died in 1945, not knowing that within a few decades he would be looked upon as the 'father of space flight'. In 1969, while the US astronauts were circling the moon preparing for their historic moon-landing, the *New York Times* printed a formal retraction of its 1920 editorial in which it heaped ridicule on Goddard's claim that one day rockets would fly through a vacuum to the moon.

Edward Heath, the British politician, became Prime Minister in June 1970, a few days before his fifty-fourth birthday.
Maurice Ravel, the French composer, was this age when he produced his most famous work, *Bolero*, in 1928.
Adam Smith, the Scottish economist, was 53 when he published his major work, *The Wealth of Nations*, which became the source of most future writings on the subject of political economy. It took him nine years to dictate and re-write it before it finally appeared in 1776.
Margaret Thatcher, an ex-research chemist and lawyer who entered politics in 1959, broke the tradition of centuries when she became the first female British Prime Minister, in May 1979, when she was 53.
Kurt Waldheim, the Austrian diplomat, was this age when he became Secretary General of the United Nations in 1972.

Crime at 53
Thomas Blood, the British adventurer known as Captain Blood, committed an audacious crime at 53. On May 9, 1671, disguised as a clergyman, he overpowered the keeper of the Crown Jewels and escaped with the crown under his arm, while an accomplice took the orb. They were caught and imprisoned. Blood refused to make a confession to anyone other than the king. Charles II visited him in prison and was told by Blood that if he was executed his death would be avenged. Pondering this thought, the monarch wisely pardoned him. Later, he was imprisoned again, this time for alleged conspiracy, and when released, his health broken, he died in fourteen days.

Retirement at 53
Robert Baden-Powell, the founder and mastermind of the Boy Scout and Girl Guide movements, retired from the British Army in 1910, to devote himself full-time to the scouting movement for the rest of his life.

Life-span of 53
Isambard Kingdom Brunel, the British engineer who designed the greatest of the early transatlantic steamers, the *Great Western*, the *Great Britain* and the *Great Eastern*, each being the largest in the world at the time of its launching, died in 1859.
Dr Peter Burrows, the England football team's doctor and a health fanatic, died at 53 as a result of jogging exercises, proving what many indolent critics had suspected for a long time, namely that, like smoking, jogging can seriously damage your health.
Maria Callas, the Greek-American operatic soprano, famous both for her magnificent voice and her intense relationship with Aristotle Onassis, died in 1977.

René Descartes, the French mathematician and philosopher, died in 1650 after catching a chill that later developed into pneumonia. He had been invited to Sweden by Queen Christina, an avid collector of learned men, and with great reluctance went to Stockholm to tutor her in philosophy. She chose 5 am as the time for her tutorials and Descartes found the icy climate where 'men's thoughts freeze in winter' too much for him. Within four months of his arrival he was dead.

Robert Donat, the English actor, best remembered for his title role in *Goodbye Mr Chips*, for which he won an Oscar, was a lifelong sufferer both from asthma and from self-doubt. During the making of his last film, *The Inn of the Sixth Happiness*, in 1958 he was so ill that he had to be supplied with oxygen between takes. He died at 53 before the film was released.

Princess Grace of Monaco   René Descartes

Hermann Goering, the bulky Nazi leader designated by Hitler to be his successor, took poison at this age when his request to be shot instead of hanged was refused by the Nuremberg Tribunal. He died in his cell on October 16, 1946, the night his execution had been ordered. He left a note explaining that the poison capsule had been hidden all the while in a container of ointment.

Princess Grace of Monaco, the poised, blue-eyed, milky-skinned grand-daughter of an Irish bricklayer and daughter of an ex-cover-girl, suffered a stroke at 53 while driving a car on a dangerous road near Monte Carlo. Her teenage daughter Stephanie tried to take control of the wheel but failed and the car crashed down a sheer drop of 120 feet. Princess Grace ... in hospital the following day. As actress Grace ... she had enjoyed an Oscar-winning career in ...wood in the 1950s, but abandoned the film ...s to marry Prince Rainier in 1956. From then until her death in 1982, she gave an immaculate, regal performance as the glamorous figurehead of her tiny Mediterranean principality.

Veronica Lake, the peekaboo blonde star of Hollywood in the 1940s, played the ultimate slinky, husky-voiced temptress in immensely popular thrillers, but her career collapsed in the 1950s and she

disappeared for years, surfacing only once in a while for bankruptcy proceedings or for arrests on charges of public drunkenness. She was re-discovered in the 1960s working as a barmaid in a downtown New York hotel and made a brief comeback, but an excess of alcohol had permanently damaged her liver and she died at this comparatively early age in 1973, of acute hepatitis.

Vivien Leigh, the unusually beautiful English actress, died of tuberculosis in 1967. On the night of her death the exterior lights of London's West End theatres were extinguished for one hour as a sign of mourning. She was one of the few actresses ever to win more than one Oscar – her first being for her role in *Gone With the Wind*, and her second for *A Streetcar Named Desire*.

Margaret Leighton, the sensitive, elegantly vulnerable English actress, equally at home on stage or screen, died of multiple sclerosis at 53 in 1976.

Lenin, the Soviet founding father, was shot by a would-be assassin in 1918. Four years later, when he fell seriously ill, doctors extracted one of the two bullets that had struck him and he recovered rapidly. A month later he relapsed, suffering a stroke with partial paralysis and loss of speech. After further illness, he eventually died of another stroke on January 21, 1924, at 53. His embalmed body is on view in Red Square, Moscow.

Henry Morgan, the notorious Welsh pirate with a reputation for savagery and torture, carried on a private war with the Spanish in Central America in the seventeenth century. He led as many as 36 ships and 2,000 men on murderous raids, but when he returned to England he was knighted by Charles II and sent back to the West Indies as Governor of Jamaica, where he enjoyed his final years harassing his old pirate-companions, and eventually dying a wealthy man in 1688, aged 53.

Pier Paolo Pasolini, one of the most controversial figures in the film world, died in an appropriately bizarre and violent manner while working on one of the most horrible films ever made – the Marquis de Sade's *120 Days of Sodom*. He was bludgeoned to death in 1975 by a teenage boy who then proceeded to drive the film director's own Alfa Romeo over his prostrate body. This happened on a patch of waste land just outside Rome when Pasolini allegedly made homosexual advances towards the boy. It was as if a scene from one of his films had come to life and destroyed him.

Tchaikovsky, the Russian composer, was a hypersensitive boy whose mother's death when he was 14 turned him into a mentally delicate adult, writing over-emotional music. Plagued by his powerful homosexual feelings and dogged by a disastrous marriage, he experienced both a nervous breakdown and suicidal feelings. He once stood in the icy waters of the Moscow River – up to his armpits – in an attempt to die of pneumonia, but his death, in 1893, was caused by cholera.

Fifty-four is the year when 'old age' becomes visible in the distance. Previously out of sight, it now becomes clearly seen as a small speck on the horizon. The young imagine they have for ever, but suddenly in their mid-50s it dawns on people that they only have a limited number of healthy, active years left. Some find this depressing, but others react by using their time more greedily and doing the things they used to feel could be left until 'later on' or 'next year perhaps'.

Life Expectancy
Life expectancy at birth for males in Ecuador, Syria and Tunisia and for females in Iraq, Libya, Morocco and Nicaragua is 54.

Accession at 54
Oliver Cromwell became Lord Protector in 1653 when he was this age. The only dictator in British history, he used his few years in power to bring in a number of religious reforms and also to sweep away certain corrupt and barbaric practices. But his influence was short-lived and soon after his death the people were cheering the return of the monarchy. When this happened Cromwell's body was exhumed from its place of honour in Westminster Abbey and, somewhat belatedly, hanged.

George I, the first Hanoverian King of Great Britain and Ireland, came to the throne in 1727. Unable to speak the language of the country he was supposed to be ruling, and openly supporting German mistresses at court while imprisoning his own wife for alleged infidelity, he was a less than popular monarch.

Achievement at 54
William Caxton, the first English printer, was 54 in 1476 when he returned from Bruges and set up his press at Westminster. His first dated book, *The Dictes and Sayings of Philosophers*, appeared on November 18 the following year and he went on to print practically all the English literature available to him at the time, including Chaucer's *Canterbury Tales*.

John Cockcroft, the British physicist, was this age when he shared the 1951 Nobel Prize for Physics with Ernest Walton, for pioneering the use of particle acceleration in studying the atomic nucleus.

William Caxton's printer's mark

James Eads, the American engineer, saw the climax of his greatest masterpiece – the opening of the Eads Bridge at St Louis – when he was 54, in 1874. It was the largest bridge of any type built up to that time and was recognized throughout the world as a landmark in engineering achievement. It made pioneering use of structural steel, its foundations were planted at record depths, and his design solved problems of span and clearance which had defeated a team of twenty-seven leading engineers.

Joseph Guillotin, the French doctor, was 54 on April 25, 1792, when his humane killer was used for the first time. For several years he had been pressing for the introduction of an efficient decapitating device to render the execution of criminals painless, and the German-built machines were soon busy filling baskets with condemned heads. Apparatus similar to the guillotine had been used earlier in Germany and Italy, and also in Scotland where it was known as 'The Maiden', last employed in 1685 to behead the Earl of Argyll.

Benito Juarez became President of Mexico in January 1861, at 54. A Zapotec Indian, he started life as a servant, married the boss's daughter, became a lawyer and then rose through the political ranks to achieve the highest office ever held by a local Indian.

Alfred Kinsey, an American zoologist respected for his gall-wasp studies, suddenly found himself internationally notorious at 54 when he published his first 'Kinsey Report' in 1948, on *Sexual Behavior in the Human Male*. Its admirable objectivity heralded a healthy new phase of sexual openness, helping to sweep away much of the vicious puritanism of earlier times.

The First Duke of Marlborough, John Churchill, one of England's greatest generals, was this age when he scored his triumph at the Battle of Blenheim, inflicting the first major defeat the French Army had suffered in over fifty years. 52,000 English and Austrian troops faced 60,000 French and Bavarians. At the cost of 12,000 casualties, Marlborough and his allies captured 13,000 enemy troops and wounded or killed about 18,000 more. The following year, 1705, a grateful English nation presented him with Blenheim Palace at Woodstock.

Joshua Slocum, an American sailor, was this age in 1898 when he successfully completed the first solo circumnavigation of the globe. He set off on the 46,000-mile trip in his 36-foot sloop *Spray* in 1895, from Newport, Rhode Island, and returned home three years and two months later. In 1909 he set out once more but was never seen again.

Misfortune at 54
Hadrian, the Roman Emperor famous for constructing a great defensive wall in the north of England, was a homosexual whose languid favourite Antinious fell into the Nile when they were on a trip up the river together in AD 130. He drowned and the 54-year-old Hadrian was so desolated that he wept publicly for the youth. In

response to the Emperor's grief, the dead boy's memory was honoured with cults of Antinious and statues of him became a common sight.

Life-span of 54

Albert Anastasia, the American mafioso gangster, was 54 when he had his head blown off while it was wrapped in a towel at the barber's shop of the New York Sheraton Hotel, in 1957. Until his death at the hands of rival mobsters, he had survived for thirty years as the Lord High Executioner of gangland, trading in prostitution and vice and helping to set up a gang of contract killers called Murder Incorporated.

Michael Ayrton, the English painter and sculptor, died at this age in 1975. The youngest member of the BBC *Brains Trust*, an art critic, novelist, stage designer, documentary film-maker, essayist and mythologist, his brain was too busy for its own good and this reduced the quality of his painting and sculpture, although his visual presentations of the myths of Daedalus and Icarus will be long remembered.

Laventry Beria, the notorious chief of the Soviet secret police under Stalin, was shot at this age in 1953. When Stalin died, Beria's determination to make his security organization the supreme power in Russia led to his early arrest and his execution for treason.

James Boswell, the Scottish biographer of Dr Johnson, died at 54 of venereal disease, from which he had suffered most of his adult life.

Guy Burgess, the British Foreign Office official who spied for the Russians, died at this age in 1965 after defecting to Moscow in 1951.

Charles II, King of England in the seventeenth century, the greatest libertine to have occupied the English throne, was a self-indulgent man of pleasure, amoral and charming, much liked by his people. On February 1, 1685, he suffered an apoplectic fit and five days later was dead. Rumours that he was poisoned seem to be groundless.

John Davis, the Elizabethan English navigator who discovered the Falkland Islands, was murdered by Japanese pirates in 1605, when he was on a voyage to the East Indies. The pirates, who had been shipwrecked and were on an overcrowded junk, attacked Davis's *Tiger* in an attempt to capture the better ship. They killed twenty of his men who had boarded the junk and then assaulted the *Tiger* itself, hacking and slashing Davis to death.

Paul Gauguin, the French painter, left Europe in March 1895, never to return. He had already spent two years in the South Seas, painting on the island of Tahiti, and decided to go back to the Pacific for a further spell. After three years he attempted suicide, but was to live for another five years, his bed seldom empty of a teenage native girl. Despite the fact that he was by now in the advanced stages of syphilis, with open sores on his body, he reported that 'My bed has been invaded every night by young hussies running wild.' In 1901 he moved to the Marquesas,

where he eventually succumbed to a heart attack on May 8, 1903. At the time of his death he was under threat of three months' imprisonment because he had been taking the side of the natives against the authorities. His final message to a friend was 'all these worries are killing me' – hardly the romantic image cherished by city commuters of 'doing a Gauguin'.

Bill Haley

Bill Haley, the American band-leader, died in 1981, at this age. He was an overweight guitarist and vocalist who suddenly became world famous as the 'Father of Rock 'n' Roll'. He invented this new style by blending white Country-and-Western with black Rhythm-and-Blues, and had the first ever hit record of Rock 'n' Roll, in 1953 with *Crazy Man Crazy*. In 1954 his recording of *Rock Around the Clock* became a world-wide success and led to widespread rioting in cinemas, with thousands of seats being ripped up in a frenzy of excitement caused by this new form of music, in which Haley and his band, the Comets, performed crude acrobatics while playing. Haley failed to sustain his early lead and was quickly eclipsed by other Rock 'n' Roll performers.

Gertrude Lawrence, one of the leading stars of theatre review in the 1920s, died of cancer of the liver, in 1952. She was Noël Coward's favourite actress and he wrote his play *Private Lives* especially for her. Despite her enormous success on the stage, however, she failed to make the transition to the cinema screen with any success.

Montezuma, the last Aztec Emperor of Mexico, died at 54 in 1520, under mysterious circumstances. According to Spanish accounts he was stoned to death by his own people, with whom he had become unpopular, because he had not met the Spaniards with violence. But his people believed he had been murdered by the Europeans.

Peter Sellers, a founder-member of the Goons, who became one of the world's most brilliant voice-mimics, suffered a major setback when he had a serious heart attack in 1964. After a while, however, he continued with his demanding international film career, and went on working right up to his death from a second, similar attack, in July 1980.

55

Fifty-five is the beginning of 'old age' according to Hippocrates, and Thomas Parnell, writing in 1722, says: 'If truth in spite of manners must be told,/Why really fifty-five is something old.' Montaigne, in 1580, sees 'five and fifty' as the earliest age 'to send men to their place of sojourning' (i.e. to retirement).

These early attitudes seem out of step today, when many people are still functioning at full pressure and are in good physical shape. Modern diet and health care have added a good ten years to these older assessments. Sir Arthur Keith, the famous anatomist, commented in 1925: 'The human body becomes deformed from over-work just as much from laziness ... most of us could be upright and supple at 55, if ... exercises ... were practised daily.' He was writing about certain dance exercises that were popular at the time, but the fact remains that today far fewer people are prematurely aged by grotesque overwork or over-indulgence than was the case in earlier centuries. So it would be more accurate to say that 55 is 'yesterday's old age' (or, sadly, today's old age in some Third World countries).

Despite these improvements in many countries it has to be reported that 55 is the start of the peak period for suicides. In the USA this peak phase runs from 55 to 64, at 21·6 suicides per 100,000 per annum. We sometimes think of suicide as being the act of the 'desperate young', but statistics show that it is those in the older age group who are more likely to kill themselves. America is not alone in this. In all countries studied, 55 was the starting point for this peak in suicide rate. Illness, immobility, loneliness and depression were the most common triggers. The typical suicide is single, divorced or a widow(er), childless, urban and a drinker or drug-user. There are more males than females involved in successful suicide. Only 10 per cent of females who attempt suicide succeed at it – for most women it is more of a cry for help than a real wish to end it all.

One anatomical change that takes place at 55 is an appreciable loss in body weight for males. Following a weight increase during middle age, there is now a weight decline, as part of the general shrinkage of ageing. For females this change comes ten years later.

Life Expectancy
Life expectancy at birth for males in Peru and for females in Algeria and Honduras is 55.

> Among the animals, 55 is the greatest age ever recorded for any kind of amphibian – a Giant Salamander.

Achievement at 55
Johann Gutenberg, the German printer and goldsmith, was 55 when he completed the famous Gutenberg Bible in 1455 and heralded a new era of human documentation. Although printing from movable type had been known in the Orient for centuries, it was Gutenberg's independent invention of this printing process that lit the fuse for the great explosion of book production in the western world. He himself saw no reward for his invention, being involved in law suits over loans that had not been repaid and having his printing equipment confiscated in lieu of payment. He would have been surprised by the price paid for one of his books in 1978 – $2,400,000 for one of the twenty-one complete copies of the Gutenberg Bible still surviving. It was bought by Texas University at a sale in London.

Ho Chi-Minh became President of North Vietnam in 1945, at this age, and remained so until his death in 1969. He was that great rarity – a communist leader with a sense of humour – a charming, debonair man who chain-smoked Lucky Strike cigarettes and was once a staff worker at the Carlton Hotel in London and an assistant to the French chef Escoffier. He was also a poet who spoke English, French, Vietnamese and Mandarin, but his main impact was as the prime mover of the post-war anti-colonial campaign in Asia.

William Jackson Hooker, the English botanist, was 55 when he became the first Director of Kew Gardens in 1841. Under his leadership Kew became the world's leading botanical institution.

Mao Tse-tung, the Chinese communist leader, was an assistant librarian who passed his time reading the works of Karl Marx, with devastating effect for a quarter of the human race. In 1949, at this age, he became the first head of state for the newly established People's Republic of China, after driving out Chiang Kai-shek and the Nationalist armies who took refuge on the island of Formosa (now Taiwan). At a stroke, Mao swept away the traditions and conflicts of old China and introduced a revolutionary, centralized communist state of an extreme kind.

Montgomery of Alamein was 55 when he pursued the German forces under Rommel to their surrender in Tunisia in May 1943 – the most triumphant phase of his long army career.

Edward Teller, the Hungarian-born American physicist, was this age when he achieved the dubious success of developing the Neutron Bomb and testing it in the Nevada Desert, in 1963.

Misfortune at 55
Earl Haig, Commander-in-Chief of the British Forces in the First World War, was this age in 1916 when he sent huge numbers of his troops to their deaths, with little or nothing to show for their loss. In his unsuccessful offensive on the Somme River there were 450,000 British casualties, and there were also vast losses the following year in the Third Battle of Ypres, when he again failed to reach his objective. Although honoured in his day he has since been viewed as an unimaginative failure.

Life-span of 55
Louisa May Alcott, the American author best known for *Little Women*, suffered lengthy periods of ill-health as a result of contracting typhoid when acting as a nurse in the American Civil War. She died at 55 in 1888.

Elizabeth Barrett Browning, the English poet who had been an invalid since the age of 15, when she became seriously ill after a spinal injury, eventually died in her husband's arms, 'smiling, happily, and with a face like a girl's', at the age of 55 in 1861, following a severe chill.

Julius Caesar was murdered at 55 in the year 44 BC, stabbed to death by a group of conspirators. He was killed in the Senate house at 11 am on March 15, suffering twenty-three wounds, receiving with particular horror the blow from Marcus Brutus, a man he had trusted and loved. One of the world's greatest generals, his name was immortalized in several ways including calling the month of his birth 'July' (from 'Julius') and giving names such as 'kaiser', 'tsar' and 'gaysar' (all based on 'Caesar') to the great leaders of the German, Slavonic and Islamic worlds.

John Christie, the English murderer, was hanged at 55 in 1953. Harmless-looking, bespectacled, balding, neatly dressed and sexually repressed, he was in reality a sadistic killer. The bodies of his wife and three other young females were found under the floorboards of his West London home. More were buried in the garden. He sought what he considered to be the perfect sexual consummation, gassing his victims to sleep before raping them and then killing them to conceal his crimes.

Christopher Columbus returned from his final voyage crippled by arthritis and hardly able to move. Sick in body and mind he was bitter and frustrated that he could not persuade the King of Spain to entrust him at this stage of his life with the governorship of the Indies. When offered instead a profitable tenancy in Castile, Columbus rejected the suggestion with great indignation. His condition worsened rapidly, with gout adding to his misery and he finally died, aged 55, at Valladolid on May 20, 1506.

Claude Debussy, the French composer whose work was a musical counterpart to the paintings of the Impressionists, died at this age in 1918.

Emily Dickinson, the American poet, was 55 when she died in 1886, virtually unpublished. Examining her belongings after her funeral, her sister Lavinia discovered literally hundreds of poems, neatly tucked away, and she arranged for their posthumous publication.

Rudolph Diesel, the German engineer who invented the 'Diesel Engine', was 55 when he disappeared at sea, presumed drowned in the English Channel in 1913.

Henry VIII, King of England from 1509 until his death in 1547, ended his life a sad, embittered, grossly overweight man with a violent streak. Prematurely aged, he had become moody, unpredictable and paranoid. His physical problems began after a fall in his last tournament in 1536, which left him with recurring headaches. Later he developed a chronic ulcer on his leg, preventing him from exercising. It was this that caused the increase in his size and heralded a final phase of life made miserable by ill-health.

Henry VIII

Yukio Mishima, the Japanese author, committed *hara-kiri* at this age in 1970. This ancient form of ritual suicide, consisting of self-disembowelment by sword (*hara* means belly and *kiri* means cut), was traditionally completed by the victim's decapitation, to spare him prolonged pain, and Mishima's death followed the same pattern. He had built up a private army called 'The Shield Society' to preserve the martial arts and to fight communism. His honourable suicide was a form of protest at weakness in Japan's military policy. He and his followers seized control of the military headquarters in Tokyo and he gave a ten-minute speech from the balcony to a crowd of 1,000 servicemen, attacking the restrictions on Japanese rearmament. He then publicly tore out his entrails and a loyal follower lopped off his head.

Friedrich Nietzsche, the German philosopher, died of syphilis at Weimar, in 1900. He had developed a superman philosophy in which the virile, powerful individual who was anti-dogma and anti-tradition, was given the dominant role, with Christianity attacked as a slave morality and democracy the cult of mediocrity. This attitude meant that, although he was opposed to nationalism and racism, his ideas were later adopted by the Nazis.

Pliny the Elder, the Roman author of *Historia Naturalis*, died at 55 in the year AD 79, because of his insatiable curiosity. As commander of the fleet in the Bay of Naples he went ashore to investigate a strange cloud-formation over Pompeii. It was the result of Mt Vesuvius erupting, and he was soon overcome by the volcanic fumes and perished along with most of the inhabitants of Pompeii and Herculaneum.

Cardinal Thomas Wolsey, the son of an Ipswich butcher, who had risen to power as the chief minister, adviser and champion of Henry VIII, died, probably of acute stress, at this age in 1530. He had been the king's great favourite, able to build himself the magnificent palace at Hampton Court, but when Henry broke with Rome, Wolsey lost favour and in 1530 was accused of treason, arrested in York and taken to London. On the way there he stopped at Leicester, sick and distressed, and died a few days later. Graspingly ambitious and rapacious in his years of power and success, he seems to have been physically incapable of facing disaster.

56

Fifty-six is a crisis year for modern warriors. Retirement from the British armed forces is today fixed at 55 (eight years later than for ancient warriors – see age 47) and, after the first excitement of civilian freedoms, there is usually a period of acute unease about the future. At 56 the modern ex-serviceman, physically fitter than his age-peers, feels himself ready to tackle a bigger job, but invariably finds he is facing a smaller one, or no job at all. The services, having retired their men ten years earlier than civilians and having made them ten times as healthy, create for them a crisis period of readjustment.

Life Expectancy The life expectancy at birth for males in Cape Verde, Dominica, El Salvador, the Philippines and Tuvalu, and for females in Egypt and Tunisia, is 56.

Achievement at 56 Alexandre Gustave Eiffel, the French engineer, was this age when he completed his 'Eiffel Tower' in Paris in 1889, and earned himself the nickname 'magician of iron'.
William Gilbert was 56 in the year 1600 when he published his great work on magnetism, a pioneering achievement that led to his being called the 'father of electrical studies'.
George Goethals was this age in 1914 when his masterminding of the construction of the Panama Canal came to fruition with the opening of the canal to commercial traffic. The obstacles he had overcome were tremendous. In addition to the technical difficulties posed by the massive locks, there were also the dramas involved in caring for some 30,000 employees, many of whom were troubled by disease.
Rex Harrison, the English actor, was this age in 1964 when he won an Oscar for his performance as Professor Higgins in *My Fair Lady*.
Charles Macintosh, the Scottish chemist, was 56 in 1823 when he invented the waterproof garment that has ever since been named after him.
André Malraux, the French novelist, was appointed Minister of Cultural Affairs by de Gaulle in 1958, at the age of 56. He held the post for ten years.
Nikolai Rimsky-Korsakov, the Russian composer, was 56 in 1900 when he wrote his most popular work, *The Flight of the Bumble-bee*.
Sulla, the Roman military leader, entered Rome triumphant in 82 BC, when he was 56, and immediately put to death between 6,000 and 7,000 prisoners of war, in the Circus. Scenes of cruelty followed all over Italy as he massacred his opponents. Despite these tyrannical acts, he gave full power back to the Senate and retired from his dictatorship before his death in 78 BC.

Misfortune at 56 Jack Hawkins, the stoical British actor, master of the stiff upper-lip, tragically lost his splendidly rich voice in 1966, at this age, when cancer of the larynx meant the removal of his vocal cords. True to type, he persisted with his career, against all the odds, using other actors to dub his voice. He kept this up from 1967 to 1973, through seventeen films, but then, in a desperate attempt to speak again, he had a 'voice box' implanted and died of haemorrhaging following the operation.

Retirement at 56 Sarah Siddons, often claimed to have been the finest tragic actress in the history of the stage, retired on June 29, 1812, one week before her fifty-seventh birthday. Her farewell performance was as Lady Macbeth and the ecstatic audience would not allow the play to proceed beyond the sleep-walking scene, in which she gave an exhibition of perfection. Hazlitt wrote of her that 'passion emanated from her breast as from a shrine. She was tragedy personified.'

Life-span of 56 Ludwig van Beethoven, who had been totally deaf for eight years, died in 1827, after contracting pneumonia. Bedridden for some months, he finally succumbed to cirrhosis of the liver. His final words were 'I shall hear in heaven!'
The Earl of Carnarvon, the rich patron who sponsored Howard Carter's discovery of Tutankhamen's Tomb in Egypt, died at 56, less than five months after the opening of the tomb. He had been present on November 25, 1922, when the first stone was removed from the tomb's wall and, staring into the gloom, he saw 'strange animals, statues and gold' – the greatest treasure ever found in the history of excavation. He had returned to England for Christmas, but was soon back in the Valley of the Kings, visiting the site almost daily. Then, in March 1923, he was bitten by a mosquito, the bite turned septic and he was taken to Cairo for treatment. Pneumonia developed and he died at five minutes to two on the morning of April 5. At that precise moment all the lights in Cairo went out and stayed out for some time, with no technical explanation of any kind. Also at that precise moment, the Earl's dog, back in England, howled inconsolably and died. These extraordinary coincidences led to the widespread belief that the disturbed mummy had 'taken its revenge'.
Francis Drake, the greatest of the Elizabethan sailors, who had circumnavigated the globe and destroyed the Spanish Armada, died in 1596 of dysentery and exhaustion in Central America, where he was busy plundering Spanish possessions. He was buried at sea in a lead coffin, near Porto Bello, Panama.
Albrecht Dürer, the supreme genius of German art, died at 56 in 1528, after a short illness. He was buried in the cemetery of St John's Church in Nuremberg, where his lasting creations were honoured with the epitaph, 'Whatever was mortal in Albrecht Dürer is covered by this stone.'

Douglas Fairbanks, the screen's greatest swash-buckler, exuberant, dashing and agile, was one of the founder-members of United Artists, the film company he masterminded with his wife Mary Pickford, Charles Chaplin and D.W. Griffith to distribute their own films. He retired in 1936 and died in his sleep at 56 in 1939, from a heart attack.

Ian Fleming, the English author who invented James Bond, died of a heart attack, in 1964. His first ambition as an author had been to collaborate with Edith Sitwell on a study of the Swiss philosopher Paracelsus, but this project was abandoned and he turned instead to writing a string of Bond thrillers employing 'sex, snobbery and sadism' as his basic ingredients. He himself said of James Bond: 'I can't say I much like the chap.'

George Formby, the Wigan-born English comedian who raised gormlessness to the level of an art form, also died of a heart attack at this age, in 1961. Next to Gracie Fields he was the biggest box-office star in England in the 1930s and 1940s. He had been born blind, but recovered his sight when he had a coughing fit, and went on to star in music-hall from an early age and then moved into films where he had a long series of slapstick comedy successes.

George VI, King of Great Britain and Ireland from 1936 to 1952, died at this age of lung cancer, a few months after an operation to remove his left lung.

Betty Grable, the star of Hollywood musicals in the 1940s and 1950s, also died of lung cancer, in 1973.

Dag Hammarskjöld, the Swedish political economist who was Secretary General of the United Nations from 1953 to 1961, was killed in a plane crash near Ndola in Zambia. There were rumours that his death was not an accident. Two of the bodies recovered from the crash, in which all on board died, were riddled with bullets, and many disbelieved the official theory that a box of ammunition had exploded, causing these injuries. Former American President Harry S. Truman was quoted as saying: 'Dag Hammarskjöld was on the point of getting something done when they killed him. Notice that I said, "when they killed him".'

Adolf Hitler killed himself at 56, in his Berlin bunker, only hours after marrying his mistress, Eva Braun, who also took her life, on April 30, 1945, as the Allies closed in on Berlin and his capture appeared imminent.

Captain William Kidd, the Scottish privateer, was hanged at 56 in 1701. Commissioned by King William III to subdue piracy, it was claimed that he had turned pirate himself and he was branded the 'Scourge of the Indies'. He was arrested and sent for hanging at Wapping. The rope broke at the first attempt, but he was despatched at the second and his tarred body was hung up in chains to deter others.

Gypsy Rose Lee, the respectable American stripper who disliked nudity and who spiced her striptease acts with quotations from Aldous Huxley and Spinoza, died at this age in Los Angeles in 1970.

Abraham Lincoln, American President from 1861 to 1865, was this age when he was murdered in April 1865, shot in the head by John Wilkes Booth at Ford's Theatre in Washington.

Hans van Meegeren, the greatest of all art forgers, was an embittered Dutch painter whose original work had been dismissed with contempt by the famous art critic Abraham Bredius. In retaliation, he set out to fool Bredius by producing a fake Vermeer, a task which took him five years to complete. He sold the painting and to his delight Bredius authenticated it and boasted in art magazines about his wonderful find. Van Meegeren then produced five more Vermeers and became extremely rich, buying fifty properties, including a nightclub. He explained his wealth by saying he had won the state lottery. Unfortunately he sold one of his fakes to Hermann Goering and after the war was accused of being a collaborator. To prove his 'innocence' he painted a Vermeer in court, but he was jailed for a year none the less. Being confined in prison was more than he could bear and his health quickly declined. Within six weeks he was dead, at the age of 56.

Cloudesley Shovell, the English Admiral who was Commander-in-Chief of the British Fleet from 1705, was murdered in 1707, when his flagship, *Association*, sank with all 800 hands off the Scilly Isles in cloudy weather. He himself was thrown ashore still alive, at Porthellick Cove. A woman, seeing the great ring on his finger, killed him and stole it. Thirty years later, on her deathbed, she confessed and gave the ring to her priest.

An illustration by Dürer to the *Treatise on Measurement*

Fifty-seven is the oldest age for motherhood. The world age record for giving birth, according to *The Guinness Book of Records*, is held by Mrs Ruth Kistler of Portland, Oregon, USA, who gave birth to a daughter, Susan, at Glendale, near Los Angeles, on October 18, 1956, at the age of 57 years and 129 days. There are many other claimants to this record, some of them up in their 70s, but most such instances turn out to be attempts to conceal an illegitimate child, with the grandmother standing in for the real mother. Mrs Kistler is the oldest maternity case that is fully authenticated.

Life Expectancy Life expectancy at birth for males in Brazil, the Dominican Republic, Iran and Thailand, and for females in Iran, is 57.

Accession at 57 George IV became King of Great Britain and Ireland in 1820 at this age. As Prince Regent for his sick father George III he had effectively been Sovereign for some years, although he was far from ideal for the role. Indolent, devious, profligate and self-admittedly 'too fond of women and wine', he had been held in contempt by his father. But he was nevertheless a great patron of the arts and was responsible for the most exotically amusing building in England – his special private folly, the Brighton Pavilion.

Achievement at 57 Frederick Ashton, the English choreographer, was 57 when he succeeded Ninette de Valois as the Director of the Royal Ballet. William Beebe, the American explorer, was 57 in 1934 when he descended in his bathysphere to a record depth of 923 metres in Bermuda waters, with Otis Barton as his companion.
Leonid Brezhnev became Secretary of the Central Committee of the Communist Party in Russia in October 1964, when Khrushchev fell from power. He remained the virtual ruler of the Soviet Union for the next eighteen years, until his death in 1982.
Handel's *Messiah* was given its first performance on April 23, 1742, when he was 57. It was written to provide a work for Dublin charities, at the request of the Duke of Devonshire, the Lord Lieutenant of Ireland.
Henry Irving, the leading actor of Victorian England, was this age when he was honoured with the first knighthood ever bestowed on a member of the acting profession. Queen Victoria knighted him in July 1895.
Immanuel Kant, the Prussian philosopher, published his *Critique of Pure Reason* in 1781, the result of a decade of thinking and meditation. It brought about a revolution in philosophical ideas.
Gerardus Mercator, the Flemish map-maker, was 57 when he introduced the technique of using curved lines on maps to designate degrees of longitude and

Handel

latitude. First used by him in 1569, it later came to be called a 'Mercator's Projection'.
Eadweard Muybridge, the pioneer photographer, was this age when he published his first major work on *Animal Locomotion* in 1887, which showed animals in all phases of movement, in long sequences that permitted scientific study of the actions involved. The reason why the book enjoyed wide sales was not, however, connected with science, but with the fact that a large section of the volume was illustrated with pictures of nude women in a wide variety of poses.
Alexander Parkes, the English chemist, was 57 in 1870 when his invention of celluloid was named and marketed commercially by John Hyatt in the United States. Although its major weakness of rapid inflammability has seen its replacement in most spheres of use, the international sport of table-tennis is still entirely dependent on it.
Georges Pompidou became President of France in April 1969 when he was this age, and remained so until his death in 1974.

Misfortune at 57 Fulgencio Batista, the Cuban dictator, fled from the rebel forces of Fidel Castro on New Year's Day, 1959, when he was two weeks short of his fifty-eighth birthday. He had held power in Cuba for a quarter of a century. At first he had been a strong, efficient ruler, although financially corrupt. Later he became a more brutal tyrant and an embezzler, before he was finally overthrown and driven into exile, first to the Dominican Republic, then Madeira, and finally Portugal.

Lord Cardigan had just passed his fifty-seventh birthday in October 1854 when he led the Charge of the Light Brigade in the Battle of Balaclava, taking his men through a valley surrounded by Russian guns. Viewed by unthinking patriots as an act of great courage and daring, it was in reality one of the most infamous acts of folly ever perpetrated by a military leader. Although Cardigan himself escaped almost unscathed, 505 of his 700 men died, making him one of the most disastrous commanders in British military history.

Geronimo, the Apache Indian leader, finally surrendered to US troops on September 3, 1886, when he was this age. He had for many years resisted the intrusions of both the Spaniards and the Americans, and became particularly bitter when he lost his mother, wife and children at the hands of Mexicans. After this he developed into a cunning, skilful leader organizing revenge raids in Mexico and elsewhere. It was the Mexicans who gave him the name of 'Jerome' which was changed into Geronimo. His reign of terror was eventually brought to an end and he was confined on a reservation where, at last, he settled down to quiet farming and became converted to Christianity.

Life-span of 57
John Logie Baird, the Scottish pioneer of television, who had suffered from poor health all his life, died at this age, still experimenting with new television techniques, in 1946.

Humphrey Bogart, today the most popular of the Hollywood stars of the Golden Era of the cinema, died from throat cancer at 57, early in 1957. He owed much of his unique appeal to the war-wound he received when serving in the US Navy in the First World War. During the shelling of his ship, *Leviathan*, his face was damaged, causing partial paralysis of his upper lip. It was this that gave him his characteristic snarling lisp and fitted him so well to his many gangster roles in the 1930s. His increasingly crumpled face and battered persona, combined with a sense of dogged authority, led to more sympathetic roles in the 1940s and 1950s.

Randolph Churchill, the British politician and the only son of Winston Churchill, died in 1968 after spending many years working on a massive biography of his formidable father. Completely overshadowed, his own political career something of a failure, he became a near-alcoholic in later years, irascible and arrogant, and a threat to the peace and calm of any social occasion.

André-Gustave Citroën, the French industrialist, died at 57 in 1935. He had introduced Henry Ford's mass production methods into Europe and had provided France and other European countries with small, cheap cars. Before long he had established one of the largest car manufacturing companies in France, but went bankrupt during the depression

years and died soon after losing control of his empire.

Sergei Diaghilev, the Russian ballet impresario, suffered a diabetic coma when on holiday in Venice in 1929 and died at the age of 57, having achieved outstanding success with his Paris-based Ballets Russes.

Edward Gibbon, the greatest British historian of the eighteenth century, most famous for his *Decline and Fall of the Roman Empire*, became excessively fat in his later years and suffered increasing ill-health until he died in 1794.

Horace, the Latin poet, died at 57 in the year 8 BC. He was most famous for his *Odes* which Tennyson referred to as jewels: 'That on the stretch'd forefinger of all Time,/Sparkled forever.'

Simon de Montfort, the Earl of Leicester, and one of the outstanding personalities of his day, was an early advocate of limiting the powers of the monarchy. He was slain at the Battle of Evesham on August 4, 1265.

Lord Northcliffe, the newspaper magnate, who founded the *Daily Mail* in 1894, pioneering popular journalism, died of infective endocarditis in 1922. By 1929 his paper had gained the largest circulation of any in the world.

Humphrey Bogart in *Casablanca*

Niccolo Paganini, the Italian virtuoso violinist, died in 1840 after a tempestuous career that had made him into a popular idol.

Robert Taylor, the last of the matinée idols, died of lung cancer in 1969, after one of the most quietly professional careers of any of the top stars in Hollywood. Handsome, earnest and rather stiffly heroic, he survived his woodenness on screen because he was a no-nonsense performer who was never temperamental and was therefore loved by studios for his utter reliability, a quality so lacking in his more flamboyant colleagues.

Fifty-eight is the oldest age for swimming the English Channel. This feat was achieved by American Doc Counsilman in 1979, when he was 58 years and 260 days old. The former US Olympic coach made the crossing from England to France in just over thirteen hours, a remarkable endurance test for a man approaching his sixtieth year.

Life Expectancy

Life expectancy at birth for males in North Korea and St Vincent, and for females in the Dominican Republic, Ecuador, Peru and Syria is 58.

Accession at 58

Charlemagne was 58 when, during a Christmas mass in St Peter's, the people of Rome acclaimed him Emperor, whereupon the Pope crowned him. He was the supreme power in western Europe at this date (AD 800) and one of the poets at his court referred to him as the 'King Father of Europe'.

Achievement at 58

Hastings Banda, the African politician, was this age when he became the first Prime Minister of Malawi, in 1963. During the Second World War he had been in England, working as a doctor in Liverpool, but he then returned to Africa to lead his country's independence movement. Later, in 1966, he became Life President of Malawi.

Miguel de Cervantes, the most important figure in Spanish literature, was 58 when his famous work *Don Quixote* was published in Madrid in 1605. It was an immediate success and became the most popular book of the period.

Fyodor Dostoyevsky, the Russian author, was this age when he created his masterpiece *The Brothers Karamazov*, shortly before his death at the beginning of 1881. Freud described it as 'the most magnificent novel ever written', and Dostoyevsky was generally recognized as one of the world's greatest writers, but Lenin was not among his fans, irritably declaring, with the insensitivity typical of a political leader: 'I have no time for such trash.'

John Milton

John Milton was 58 when the first edition of his epic poem *Paradise Lost* appeared in 1667, establishing his reputation as one of the foremost poets in the history of English literature.

Francisco Pizarro, the Spanish Conquistador, was this age when he conquered the Incas of Peru in November 1533, with a force of only 180 men and 37 horses. Within just a few years, he and other Spaniards had exterminated the whole Inca nobility and destroyed almost every trace of Inca culture.

Wilhelm Steinitz was the oldest World Chess Champion in the recorded history of the game, being 58 when he lost to Lasker in 1894.

Misfortune at 58

Robert E. Lee, the 'Rebel General' and Commander of the Confederate Armies during the American Civil War, was 58 when he surrendered at the Appomattox Court House on April 9, 1865, marking the virtual end of hostilities between North and South.

Charles Dickens

Life-span of 58

James Bailey, the circus impresario credited with the enormous success of the Barnum and Bailey Circus, died in 1906.

Bertolt Brecht, the German playwright who was awarded the dubious honour of a Stalin Peace Prize, died of a heart attack in 1956, in East Berlin.

William Congreve, the English Restoration dramatist, who shaped the English comedy of manners with his satirical portrayal of fashionable society and its affectations, died at 58 in 1729 when he was involved in a carriage accident in London.

Gustave Courbet, the leading Realist painter in France, who rebelled against the vacuous romanticism of his day, died in 1877, in Switzerland, where he had fled from the French authorities, unable to pay a huge fine imposed on him for an act of vandalism he had not committed.

Charles Dickens died at this age in 1870 from 'an effusion on the brain'. His health had suffered badly from the intense strain of a punishing American tour, giving performances based on his books (present-day authors be warned!), but despite his weakening condition he refused to listen to the counsel of his friends and persisted in giving more and more public performances, earning huge sums of money in the process, but not living to enjoy them.

Gustave Flaubert, the French novelist who had suffered from epileptic fits all his life, the dread of which had made him deeply pessimistic in his outlook, died of an 'apoplectic stroke' in 1880, leaving an unfinished page of manuscript on his table.

Heinrich Heine, the German Jewish poet, who became the most famous love poet of the nineteenth century through the musical setting of his words by composers such as Schumann and Schubert, died in agony after nearly eight years of torment from venereal disease, which confined him to his 'mattress-grave', paralysed and partially blind.

Daniel 'Chappie' James, Commander-in-Chief of the North American Air Defense Command, the highest-ranking black officer in the history of the American forces, died of a heart attack a few weeks after accepting early retirement, in 1978.

John Knox, the Scottish religious reformer who was largely responsible for shaping the Church of Scotland, died of a paralytic stroke due to stress and overwork, in 1572.

William McKinley, the twenty-fifth President of America, was assassinated at this age, shot by an anarchist, Leon Czolgosz, on September 6, 1901, and dying eight days later. The attack came when he was holding a reception in the Music Hall of the Pan-American Exposition at Buffalo, New York.

Bronislaw Malinowski, the pioneer anthropologist, generally recognized as the founder of the science of Social Anthropology, died of a heart attack at this age in 1942.

Thomas More, who was Chancellor of England from 1529 to 1532, died at 58 due to the incompatibility between high office and an unshakeable integrity. He incurred the wrath of King Henry VIII, first because he refused to attend the wedding of the King and Anne Boleyn and second because he refused to accept Henry as Head of the Church of England. He was arrested, imprisoned and condemned to a traitor's death by being 'drawn, hanged and quartered', but the King changed this to the more honourable death by beheading, which was carried out in July 1535. At his execution he insisted on blindfolding himself and told the onlookers to witness that he was dying 'in the faith and for the faith of the Catholic Church, the King's good servant and God's first'.

Ivor Novello, the Welsh actor, playwright and composer, matinée idol of the 1920s and 1930s, whose great variety of activities included such contrasting creations as the song *Keep the Home Fires Burning*

and the dialogue for the Hollywood film *Tarzan of the Apes*, died in 1951, suffering a coronary thrombosis a few hours after appearing on stage at the Palace Theatre, London, in *King's Rhapsody*.

Thomas More   John Knox

Adriano Olivetti, the Italian industrialist, died at 58, in 1960, after suffering a stroke while on board the Milan–Lausanne Express. He was an enlightened employer, deeply concerned for the welfare of his workers, whose advanced ideas on design made his typewriters and other products world famous.

Pompey the Great, the Roman General, was 58 when he fell prey to Egyptian duplicity. Fleeing from Julius Caesar in 48 BC, he went to Egypt to seek the assistance of Ptolemy. The Egyptian king went to the coast to welcome him and sent out a small boat to collect him from his ship. As Pompey stepped ashore he was struck down and his head removed from his body, the Egyptians having decided not to anger Caesar by helping his enemy.

Dennis Price, the English comedy actor, adorned many films from the 1940s to the 1970s with his dry-voiced urbanity and cynical charm. He specialized in playing cads and bounders whose upper-class credentials were usually counterfeit, an amusing irony since he was the Oxford-educated son of a General and his real name was Dennistoun Franklyn John Rose-Price. On television he was the definitive 'Jeeves'. He suffered an emotional crisis in mid-life and made an unsuccessful suicide bid in the 1950s, but continued acting until 1973 when he died of heart failure.

Vikdun Quisling, the Norwegian traitor, was this age when he was shot by a firing squad at Akershus Fortress, Oslo, in October 1945. He suffered the unenviable immortality of giving his surname to the English language as a modern synonym for traitor. In the 1930s he had formed a Norwegian equivalent of Hitler's Nazi Party and during the Second World War proclaimed himself leader of Norway at the time of the German Occupation, after having assisted the Germans to plan their invasion of his country. He was arrested in May 1945, tried in August, found guilty in September and shot on October 24.

Fifty-nine is the year of the final encore for 'being middle-aged'. The mid-life condition which began in the late 30s has become more and more stretched in recent times, in advanced countries, until it has spread right into the late 50s. But as this decade comes to a close and the early retirement age of 60 looms, it reaches its upper limit. Still able to say they are 'in their 50s', the well-preserved 59-year-olds make the most of this year, often indulging in some special project, journey or adventure before facing up to life in their 60s.

Life Expectancy Life expectancy at birth for males in Chile, Guyana and Mongolia and for females in Dominica, El Salvador, St Vincent, South Africa and Tuvalu, is 59.

> Among the animals, the greatest age ever recorded for any kind of ape is 59 years – in the case of an orang utan called Guas who died at the Philadelphia Zoo in 1977.

Achievement at 59 The Venerable Bede completed his great work *The Ecclesiastical History of the English People* in the year 731. It gained him the reputation of the greatest scholar in the English church.

Nikolai Bulganin was 59 in 1955 when he became Premier of the Soviet Union and straight man for Khrushchev in their famous double-act of 'Mr K and Mr B', making state visits throughout the world. But he was ousted within a few years and quickly sank to obscurity.

Daniel Defoe, the English author, issued pamphlets that were considered scandalous by the piously pompous. As a result he was accused of libel against the church and was sentenced to pay a fine, stand in the pillory and go to prison. He stood in the pillory for three days, but people draped him with flowers instead of throwing missiles at him. When he left prison his business-life was ruined and he decided to become a hack writer to earn money. He also turned his hand to producing novels, and his isolation in Newgate jail inspired him to create his most famous work, *Robinson Crusoe*, at the age of 59. Based partly on the marooning of Scottish seaman Alexander Selkirk for four years on a Pacific island, the book was a runaway success and the impoverished author was quick to react: *Robinson Crusoe* first appeared on April 25, 1719, and by August of the same year he had already published a sequel.

Allen Dulles, the American spymaster, was appointed Director of the CIA at this age, in 1953. Under his guidance the agency went through a period of growth and was soon busying itself with the toppling of foreign governments. After a fiasco at the Bay of Pigs in Cuba, he resigned, in 1961.

Ella Fitzgerald, one of the top-selling recording artists of all time, was 59 when she completed her mammoth recording project – a nineteen-album series called *Song Books* in which she preserved for posterity nearly 250 of the most outstanding popular songs of the twentieth century. She began the task in 1956 and finished it in 1967.

Friedrich Froebel, the German schoolteacher, was this age when he opened the first kindergarten in the world, in 1841. He himself had been only 9 months old when his mother died and he was denied schooling and contact with other children until he attended courses at university. As a result he was determined to promote the early social contact of young children. His invention of the kindergarten quickly spread to other countries to become a common element in the educational sequence.

Otto Hahn, the German chemist, was 59 when he achieved a momentous success that was later to haunt him. It was just before the Second World War when he made the discovery that uranium atoms could be split if bombarded with neutrons in an Atomic Particle Accelerator. His discovery of the process of nuclear fission heralded the era of nuclear bombs and nuclear energy. In his later years he campaigned against the further development and testing of nuclear weapons.

John McAdam, the Scottish inventor of modern road surfaces, was 59 when he was appointed Surveyor General of the Bristol roads. He was then able, in 1815, to put into practice some ideas he had been developing in theory. Old road surfaces were appalling and he argued that the simple secret was to raise them up and construct them with proper drainage. His ideas quickly spread and today much of the world enjoys the fruits of his inventive mind.

Jonathan Swift was 59 when his most famous work, *Gulliver's Travels*, appeared in November 1726. He began his intellectual life as an idle, undisciplined, tavern-haunting Dublin student, matured to write devastating satire attacking humbug and corruption in both church and state, aged with a complex, frustrated and enigmatic mind in increasingly painful

A scene from Defoe's *Robinson Crusoe*

turmoil, and ended his life with several years of mindless imbecility. Of his 230 prose works, 300 poems and 500 letters, nothing was published under his name during his lifetime and he received hardly any payment. In a letter to his friend Pope he once wrote: 'Principally I hate and detest that animal called man; although I heartily love John, Thomas, etc.'

Misfortune at 59

The Shah of Iran, Mohammed Reza Pahlavi, was driven out of his country in January 1979, at this age, by the Ayatollah Khomeini. He and his family went first to Morocco, then Mexico, then the United States and finally Egypt, where he soon died a broken man. The irony of his removal was that the brutalities of his torture squad were soon far exceeded by the mass executions of the new religious leaders, who seemed intent on dragging Iran back into the Dark Ages as swiftly as possible.

Life-span of 59

Constance Bennett, the Hollywood actress who specialized in lazily elegant sophistication and the bored poise of the worldly-wise in a series of escapist comedies in the depressed 1930s, died in 1965 of a cerebral haemorrhage shortly after completing her only film of the 1960s, titled *Madam X*.

John Brown, the American militant who fought to abolish slavery and the persecution of the blacks, was tried for murder, slave insurrection and treason against the state, following an unsuccessful raid on the Federal arsenal. He was convicted in 1859 and hanged at 59, an act that aggravated the animosities which led to the start of the Civil War.

Oliver Cromwell, Lord Protector of the Commonwealth, died of a 'bastard tertian ague' in 1658, victim of a malignant fever that was sweeping England at the time, his condition already weakened by the strain of war and government. After his death and the restoration of the monarchy, his embalmed remains were dug out of his tomb in Westminster Abbey and hung up at Tyburn, where criminals were executed. His body was then buried beneath the gallows and his head was stuck on a pole on top of Westminster Hall, where it remained until the end of Charles II's reign.

Clark Gable, the American actor, died of a heart attack in Hollywood in 1960, shortly after completing *The Misfits*, with Marilyn Monroe, in which he insisted on doing his own stunts. It was the last film for both of them and it was rumoured that it was the strain of her unprofessional conduct during filming that hastened his end.

Sonja Henie, the Norwegian ice-skater who won Olympic gold medals in 1928, 1932 and 1936, and went on to become a top Hollywood star in ice musicals in the late 1930s and the 1940s, died at 59 of leukaemia, on board a plane heading for her home

Clark Gable in *The Misfits*

city of Oslo, in 1969. It was she alone who transformed ice-skating from an esoteric sport for specialists into a popular entertainment for millions. There is a museum in her honour in Oslo, containing her impressive collection of modern art and her vast array of trophies, all of which she presented to the nation.

Charles Lamb, the English essayist and critic, died at 59 from complications to a wound received from a drunken fall, in 1834. Reduced to misery caring for his mad sister, he had become an alcoholic in his later years.

Peter Lorre, the Hungarian-born actor who played to perfection his roles as whining, wheedling, baby-faced psychopaths, at once sad and sadistic, died of a heart seizure in 1964.

Archibald McIndoe, the pioneer plastic surgeon who founded the Guinea Pig Club of the 600 men he operated on during the Second World War, died shortly before his sixtieth birthday, in 1960. He remodelled smashed faces and reshaped damaged limbs, and also gave his patients something of his defiant spirit and the will to conquer their disfigurements.

Alfred Sisley, the French painter who was the purest of all the Impressionists, lived a life of unrelieved poverty. His spirit was eventually broken and he left Paris to live in the village of Fontainebleau where he survived in solitude and died forgotten in 1899.

John L. Sullivan, the first professional World Heavyweight Boxing Champion, died at this age in 1918. He gave the language a new word on February 7, 1882, when, in a bare-knuckle fight, he knocked his opponent, Paddy Ryan, unconscious and his trainer coined the expression 'a knockout'.

Virginia Woolf, the brilliantly imaginative but intensely neurotic English author, was easily driven into moods of suicidal depression if put under any undue strain or pressure. Too fragile for ordinary schooling, she grew up in her father's library, and later surrounded herself with her 'Bloomsbury Circle', making her house near the British Museum a literary nerve-centre. With her manic-depressive temperament, she suffered a number of nervous breakdowns and eventually, when 59, committed suicide by drowning herself, because she feared a complete mental collapse.

60

Sixty is the year of the home-lover. It was Lord Lytton who said that 'At sixty a man learns how to value home.' It is a time when people venture forth less in exploration and turn their attentions more intensely to the care of their houses and gardens. It is at this age that real adventures are usually replaced by imaginary ones. As E.W. Howe puts it: 'After a man passes sixty, his mischief is mainly in the head.'

Although sixty is the start of the period referred to as 'quiescent pre-senescence', it is still a time when many people feel young inside. Herbert Beerbohm Tree, the famous actor-manager, said, 'I was born old and get younger every day. At present I am sixty years young,' and Coco Chanel on reaching 60 remarked, 'Cut off my head and I am thirteen.' Ignoring the external wrinkles, the heart still beats romantically. Emerson: 'Spring still makes Spring in the mind,/When sixty years are told;/Love wakes anew this throbbing heart,/And we are never old.'

Sadly, whatever goes on in the mind, or the heart of the 60-year-old, there is less taking place elsewhere. At this age, 28 per cent of men suffer from erection impotence. For the majority who are still capable of making love, Masters and Johnson recommend 'that the man of 60 will find greater sexual contentment if, on two out of three occasions, he reserves his ejaculation altogether. In that way, sexual tension will accumulate to a climax worthy of his expectations.' As regards women, a study at Duke University found that 60 was 'the average age for ceasing sexual activity'. (For men the average was eight years later.) This does not mean that women *want* to cease then or that they are biologically forced to do so but that cultural attitudes influence them in an unnecessarily negative way.

In some countries and in some contexts, 60 is the official retirement age, but it is more common today for this moment to be delayed for another five years.

Life Expectancy
Life expectancy at birth for males in Grenada, Mauritius, Paraguay and Samoa, and for females in Cape Verde and the Philippines is 60.

Romance at 60
George Eliot, the Victorian novelist whose real name was Mary Ann Evans, married for the first time when she was this age. Her bridegroom, John Cross, twenty years her junior, was her investment banker on whom she became dependent for sympathy and advice when her lover, writer George Lewes, died in 1878, after living with her for a quarter of a century. Her novels, full of perceptive characterization, include *The Mill on the Floss* and *Middlemarch*.

George Sand, the French novelist whose real name was Amandine Aurore Lucile Dupin, baronne Dudevant, started a love affair with the painter Charles Marchal when she was this age. He was twenty-one years her junior and she called him her 'fat baby'.

She had previously enjoyed a period of calm serenity for fifteen years as the lover of engraver Alexandre Damien, and before that had experienced a series of unhappy and sometimes dramatic affairs with younger men. Short, swarthy, brusque and cigar-smoking, she relied on her passionate intellect for her appeal. Her lovers included frail, blond, sensitive writer Jules Sandeau, seven years younger; writer Prosper Mérimée; Michel de Bourges, bald, ugly and married, but who made her tremble with desire; poet and dramatist Alfred de Musset, to whom she sent her shorn hair; and the aristocratic, tubercular, opium-smoking composer Frédéric Chopin, six years younger than she was. Her private life proved more interesting than anything she could invent in her novels, as is proved by her twenty-volume *Histoire de ma vie*.

Achievement at 60
Nicholas-François Appert, the French confectioner, was this age in 1810 when he was awarded a major prize of 12,000 francs by the French government for his invention of airtight containers for preserving food.

Alexei Kosygin became Prime Minister of the Soviet Union at this age, in October 1964, after the fall from power of Khrushchev. Although he criticized his predecessor's 'subjective' approach, he followed it closely after he assumed control.

Izaak Walton wrote *The Compleat Angler* in 1653 when he was this age. A shop owner in the City of London, he started by writing biographies, then retired to rural Clerkenwell where he set to work on his great angling book – a hymn to the joys of fishing.

Misfortune at 60
John Paul II, the Polish Pope, was shot and seriously wounded at 5.17 pm on May 13, 1981, five days before his sixty-first birthday. The assassination attempt was made in St Peter's Square by a Turkish gunman, Mehmet Ali Agca, from South East Anatolia. It was claimed that the attack had been inspired by the Russian KGB following the pontiff's anti-Russian interference with Polish politics. As Soviet troops waited on the Polish border, the Vatican had sent a letter to Moscow threatening that, if Russia invaded, the Pope 'would lay down the Crown of St Peter and return to Poland to stand shoulder to shoulder with his people'. The invasion never came but the gunman did.

The Marquis de Sade, a well-born French nobleman with a taste for sexual cruelty, was repeatedly in trouble with the authorities for his gross depravities. He ended up in the Bastille where he found a cure – the writing of pornography. From that moment on his sexual debaucheries ceased in reality and were entirely confined to the written page. The great irony of his life, however, is that when he was found criminally insane at the age of 60, in March 1801, it was not for his obscenely vicious novels, but for the writing of a pamphlet lampooning Napoleon and Josephine. As a result he was confined for the rest of his life to the Charenton asylum.

Life-span of 60

Thomas Barnardo, the British philanthropist, died of heart failure in 1905. He was a missionary who stayed at home to do his good works. Originally he had been due to travel to China to join a mission there, but he was so horrified by the plight of the destitute children of the East End slums in London that he remained there to help them. Altogether he opened ninety homes for waifs and orphans, with the slogan, 'No destitute child ever refused admission.' By his death he had rescued and trained 59,384 children and given assistance to 250,000 others. Dr Barnardo's Homes became a major contribution to Victorian welfare.

Geoffrey Chaucer

Geoffrey Chaucer, the father of English literature, was this age when he died in 1400, with only 22 of his 120 intended *Canterbury Tales* completed. Even so, it is a masterpiece and the first major work written in vernacular English.

John Constable, the Suffolk landscape painter, died at this age in 1837. His work was not popular during his lifetime and he became deeply depressed after his wife's death, writing in his late 50s that 'my life and occupation are useless'. But he kept on to the end, still devoted to his landscapes, and was working on 'his best' on the day he died. His style became a major influence among the Victorian artists that followed him.

Calvin Coolidge, the thirtieth President of the United States, died when he was 60, early in 1933. A conservative man, he was epitomized by his slogan 'Keep Cool with Coolidge'.

Gary Cooper, the softly-spoken, laconic, sad-eyed 'Strong, Silent American Hero' of many great Hollywood films, who in real life seduced almost every one of his leading ladies, won three Oscars during his long career: the first in 1941 for *Sergeant York*; the second in 1952 for *High Noon*; and the third in 1960 for his services to the cinema. The last award was accepted for him by a tearful James Stewart who had just learned that Cooper had spinal cancer. He died of the disease the following year, six days after his sixtieth birthday.

Peter Finch, the Anglo-Australian actor, died of a heart attack in 1977 before the screening of his best performance, in the film *Network*, for which he received a posthumous Oscar.

George V, King of Great Britain from 1910 to 1936, was this age when he died, after being in poor health since a serious chest illness in 1928.

Erving Goffman, the most original anthropologist of his generation, who turned his back on traditional studies among remote tribes to look closely at people in his own, North American culture, and who wrote a series of outstanding books including *Presentation of Self in Everyday Life* and *Behaviour in Public Places*, died at 60 in 1982 following major surgery.

Herodotus, the first and greatest objective, scientific historian of Ancient Greece, died at this age in 425 BC.

Paul Klee, the Swiss-German genius of modern art, who could say more in a small watercolour than most artists could achieve in a vast canvas, died at 60 in 1940 of a heart attack that was the terminal stage of a rare disease called scleroderma in which the mucous membranes gradually dry up. His tombstone is inscribed with a curious passage from one of his diaries: 'I cannot be grasped in this world. For I am as much at home with the dead as with those beings who are not yet born.'

David Livingstone, the Scottish missionary and explorer, died at this age, made prematurely elderly by his hardships in Africa. His servants found him dead, kneeling by his bedside as if in prayer. In order to embalm his body, they removed his heart and viscera and buried them in African soil. Then, in a difficult journey of nine months, they carried his body to the coast. From there it was shipped home and buried in Westminster Abbey with an impressive Victorian funeral.

Leon Trotsky, the Russian co-founder of the Bolshevik Party and the 'Father of the Red Army', was an internationalist who believed in 'World Revolution'. He fell foul of Stalin after Lenin's death and was exiled. Condemned to death in his absence in 1937, he was tracked down by one of Stalin's agents and murdered, at the age of 60, in Mexico, in 1940. Trotsky was feeding his pet rabbits when a nervous young assistant asked him to read an article he had written. Studying the document at his desk, Trotsky was hit by a mountaineer's ice-axe which the young man had hidden in his raincoat. The three-inch skull hole took twenty-six hours to kill Trotsky. The murderer, a young Spaniard whose domineering mother was the mistress of one of Stalin's secret police generals, was severely beaten, arrested and jailed for murder. He served twenty years, was released in 1960, and is alive and well in Czechoslovakia.

61

Sixty-one is the start of the 'spring-time of old age' – the name given to the 60s by a French professor of medicine at Lyons. He accepts that, once over 60 and into this new decade, one must recognize that the body is no longer bursting with youthful vigour, but at the same time he feels that a clear distinction must be made between this early phase of 'old age' and the later stages. With appropriate caution, the 60s can still be a lively time, both mentally and physically.

Life Expectancy Life expectancy at birth for males in Lebanon and the Seychelles and for females in Brazil is 61.

Achievement at 61 David Ben-Gurion, the Zionist statesman revered as the 'Father of the Nation' by Israelis, saw his dream realized when he was 61, in May 1948, when the State of Israel was established and he became its first Prime Minister.
Joseph Glidden, the American farmer and inventor, was this age when he patented barbed wire, in 1874. It was his answer to the cattle barons who drove their herds indiscriminately across the lands of the American West, and it became immensely popular among the arable farmers who wished to fence off areas for crops. Its use intensified the range warfare between the cattlemen and the farmers and, before long, it was to find another, more sinister role in the First World War, when huge entanglements of the savage wire were employed on the battlefields to restrict human movement.
Katharine Hepburn was 61 in 1968 when she won her third Oscar for a starring role, making her one of the most honoured actresses in the history of the cinema. Her first award was for *Morning Glory* in 1932, her second for *Guess Who's Coming to Dinner* in 1967, and her third for *The Lion in Winter* in 1968. Coming from a distinguished New England family, she always shunned the usual Hollywood ballyhoo, refusing interviews, never giving autographs and generally keeping to herself, proving in the process that, if you are good enough, promotion and publicity are superfluous.
Antonio Moniz, the Portuguese neurologist, was this age in 1936 when he performed the first brain operation for the treatment of a mental disorder. It was a prefrontal leucotomy and led to him being hailed as the 'father of psychosurgery'. He was encouraged in his radical procedure by a strange case in America, where a patient had showed major personality changes after his frontal lobes had been virtually destroyed by a crowbar that had been accidentally driven upward through the front of his skull and emerged from the top of his head, without killing him.
Joseph Niepce, the French inventor, was also this age when he took the world's first photograph in the summer of 1826. The scene was the courtyard of his country house at Gras and the exposure time was eight hours.

Eleanor Roosevelt became Chairman of the UN Commission for Human Rights in 1946. She had been widowed by the President's death in the previous year and the new President, Harry S. Truman, had then made her a delegate to the UN. From that position she rose to become Human Rights Chairman at the age of 61 and continued to work vigorously in this field for the rest of her life. She played an important role in drafting and pushing through the Universal Declaration of Human Rights in 1948, and was widely acknowledged at the time as 'the world's most admired woman'.
Ferdinand von Zeppelin, the German inventor of the rigid airship, was nearly 62 in 1900 when he launched his first dirigible. He had retired from the army at the age of 53 to devote himself totally to the development of what he considered would be a brilliant new weapon of war. Zeppelins were, indeed, used in the first ever air raids on London during the Great War, but they were so easily shot down that they proved to be a military disaster, to his bitter disappointment.

Misfortune at 61 Willy Brandt, the West German Chancellor, was forced to resign at 61, in 1974, when one of his staff was arrested as a communist spy, ending an impressive period of rule during which Brandt achieved a great deal in his efforts to ease tension between East and West. A lively and charismatic politician, who had worked for the Norwegian underground during the Second World War, he had been awarded the Nobel Peace Prize in 1971, and was sorely missed from the German political scene after his departure.
Richard Nixon was also this age when he was forced to resign the presidency of the United States. It was in August 1974 that he left the White House in disgrace because of his involvement in the cover-up of the Watergate Scandal. This sleazy incident completely overshadowed the fact that he had taken the US out of the Vietnam War and had made a successful trip to China to ease East–West relations.

Life-span of 61 Beau Brummell, the English dandy who became the arbiter of high fashion at the end of the eighteenth century, gambled his fortune away, was imprisoned for debt and died in a charitable asylum in 1840.
Samuel Taylor Coleridge, the English poet who wrote 'The Rime of the Ancient Mariner' and 'Kubla Khan', was, for the last thirty years of his life, hopelessly addicted to opium. He claimed that he took drugs to stop pain rather than to cause pleasure, but when he died at 61 in 1834, the *post mortem* could reveal no physical cause for his long sufferings.
Paul Ehrlich, the German scientist who pioneered chemotherapy and revolutionized the treatment and control of infectious diseases, providing the earliest cure for syphilis, died at this age in 1915.

Thomas Gainsborough, the founder of the 'English School' of painting, and famous both for his portraits and his landscapes, died at 61 in 1788. His dying words, to his old rival Joshua Reynolds, were, 'We are all going to heaven and Van Dyck is of the party.' Reynolds commented that Gainsborough's 'regret at leaving life was principally the regret of leaving his art'.

George Hegel, the German philosopher, died of cholera at this age in 1831, after one day's illness during an epidemic that struck Germany.

Ernest Hemingway shot himself with his own gun a few weeks before his sixty-second birthday, in 1961. The act of killing was a central obsession of his. He wrote about it as a novelist, he reported it as a war correspondent, and he enjoyed it as a bullfight enthusiast. He also performed it as a big game hunter, slaughtering harmless wild life in Africa, and it was only appropriate that he should end his life with a violent suicide. There was a history of suicide in the family: his father had killed himself and his younger brother, Leicester, took his own life as recently as 1982. Like Ernest, Leicester shot himself with his own gun.

Ando Hiroshige, the Japanese print-maker who exerted an important influence on the French Impressionists, died in a cholera epidemic, at 61. In his will he quoted an old Edo verse: 'When I die/don't cremate me, don't bury me/just throw me in the fields/ and let me fill the belly of/some starving dog.'

Benito Mussolini, the Italian dictator, was this age in 1945 when he was caught fleeing to Switzerland, tried by partisans and shot, along with his mistress, Clara Petacci. With nine bullets in his corpse, he was dragged to the main square of Milan and hung up by his feet like a piece of meat, his mistress beside him. The angry crowd broke through police cordons and mutilated the two bodies. One woman shot Il Duce's corpse five more times, once for each of the sons she had lost in the war, and other women publicly urinated on his face.

Imre Nagy, the Hungarian Premier, was 61 when he was tried in secret by the Russians and executed in 1958. The son of a poor peasant, he rose to become the communist leader of Hungary and was the first such leader anywhere in Eastern Europe to introduce liberal policies, such as releasing peasants from previously obligatory collective farming. But he fell foul of the Russians and was replaced, only to be swept back into power by the October 1956 anti-Soviet uprising. He was made Premier again for ten glorious days of freedom before the Russians crushed the revolt and killed him. In his final days he wrote that there was no freedom in communism even for those who gave their whole lives to it.

Kurt Schwitters, the German artist who invented Merzism – art made from bits of rubbish – died at 61 in 1948. His collages of pieces of lino, used bus tickets, old shoe-laces, and other detritus, were a major influence in twentieth-century art, freeing artists from traditional materials and allowing them to experiment more freely with a wide variety of new techniques.

Walter Scott, the Scottish writer best known for *Rob Roy* and *Ivanhoe*, worked himself to death at 61, in 1832, attempting to avoid being declared a bankrupt. In 1826 the publishing firm in which he was a partner found itself in serious debt, to the tune of £130,000, and he was determined to prevent bankruptcy proceedings. His sense of honour demanded that he should repay all the money himself and he set to work to write himself out of debt. The strain and exhaustion of his efforts killed him, but the sale of his copyrights after his death did in fact complete the task he had set himself.

Julius Streicher, the German founder of the anti-Jewish German Socialist Party, one of the most brutal and fanatical racists in history, was hanged as a war criminal in 1946.

Diego de Silva Velasquez, the Spanish artist who was one of the world's greatest portraitists, was court painter to the Spanish Royal Family from 1623. He lived a highly successful luxurious life until 1660, when he died of a fever at 61 following a voyage with the court.

Stefan Zweig, the Austrian writer who was one of the first authors to apply psychoanalytical ideas to literature, became a target for Nazi oppression in the 1930s. As he was Austro-Jewish, his books were burned and his house was searched by Nazi police. He fled to England and then further still, to Brazil where, in 1942, in a fit of depression about the horrors of Nazism and the plight of the world, he committed suicide at 61 along with his faithful wife.

Zeppelin's invention reached its peak with the *Hindenburg*

Sixty-two is the year of the organizer. It is the age of the official, the committee chairman, the political leader, the administrator, the doer of 'good works'. It is a mentally active age when experience is made to compensate for the inevitable decline in creativity, and when the voice of orthodox authority replaces the cry of rebellious originality. An anonymous verse sums it up: 'At sixty-two/I cannot do,/But may decide/What others do.'

Life Expectancy Life expectancy at birth for males in Barbados, Costa Rica, Jamaica and Mexico, and for females in North Korea and Mongolia, is 62.

> Among the animals, 62 is the oldest recorded age for a horse. One called 'Old Billy' lived from 1760 to 1822.

Romance at 62 Aristotle Onassis, the Greek shipping magnate, was this age when he married Jackie Kennedy, twenty-three years his junior, in 1968, on his private Greek island of Skorpios. It was alleged that there had been a pre-nuptial contract running to 100 clauses, allowing for separate bedrooms at all times and relieving his new wife of any responsibility for bearing him children. It was supposedly a contract that gave her total security and privacy and him the world's most glamorous companion. All apparently went well until his son was killed in an air crash, at which point he went into a deep decline and began to tire of his marriage, turning to his old friend Maria Callas for comfort. He was said to be contemplating a divorce when he died of a muscle complaint in 1975.

Achievement at 62 Henrik Ibsen, the Norwegian dramatist, was this age when his greatest play, *Hedda Gabler*, appeared in 1890. He based the central character on an 18-year-old girl from Vienna called Emilie Bardach, with whom he was infatuated. He called her his 'May sun in a September life', but seems to have been too inhibited to develop the relationship into an active sexual one. It was probably his own complex, frustrated personality that led to his revolutionizing of the European theatre, replacing dramas based on external events with stories based on the psychological development of his characters. **Robert Koch**, the German bacteriologist, was 62 when he was awarded the Nobel Prize in 1905 for his amazing advances in the study of harmful bacteria. He isolated the anthrax bacillus and perfected a method of inoculation against it; he discovered the bacillus causing tuberculosis, and the one responsible for cholera; and he also carried out important studies on malaria, rinderpest, bubonic plague and sleeping sickness.
Harold Macmillan became Prime Minister of Great Britain at 62, in January 1957, bringing a cool-headed worldliness to Downing Street. At first

he had the magic touch, but then a series of embarrassments, followed by poor health, persuaded him to resign.
Carl C. Magee invented the parking meter at this age, in Oklahoma, in 1935. The machine's insidious spread across the globe had reached London by the year 1958.

Retirement at 62 Cary Grant, Hollywood's greatest exponent of sophisticated light comedy, remained at the top of his profession throughout his long career and then (unusually for an actor) ended it with a formal retirement while still a leading man. He started in films in 1932, opposite Marlene Dietrich in *Blonde Venus*, and ended in 1966 with *Walk Don't Run*. It was he who, in 1933, received Mae West's famous invitation to 'Come up and see me sometime.'

Life-span of 62 Tsar Alexander II was blown to pieces in the street by a terrorist bomb in 1881. As a young man he had met Queen Victoria, then 20, and they had become infatuated with one another, but the relationship was politically impossible. Later, when he was married and had eight children, he fell in love with a 16-year-old girl, Catherine Dolgorulci, by whom he had four children. Eventually, when his wife was dying of tuberculosis, he had his mistress and their four children installed in the Winter Palace above her bedroom, where she could hear them playing, much to her anguish. Forty days after she died he married Catherine and they remained passionate lovers until the day he died, at 62, in the eighth attempt on his life. When Catherine left Russia for Paris she took one of his shattered fingers with her.

Cary Grant   Aristotle

Aristotle, the world's most influential philosopher, died at 62 of a stomach illness in the year 322 BC. A pupil of Plato and a teacher of Alexander the Great, he founded his own school of philosophy in Athens, called the Lyceum, where he pioneered scientific thinking of the kind that still dominates the Western world today.
Clyde Beatty, the wild-animal trainer, died at this age after defying death almost every day since the age

of 16. He performed the most daring acts in the history of the circus, mixing forty lions and tigers of both sexes, and also using dangerous combinations of tigers, lions, leopards, pumas, hyenas and bears.

Giovanni Boccaccio, the author of the fourteenth-century masterpiece *The Decameron*, died at 62 in Tuscany in the year 1375. Ill-health and criticism had made him repeatedly depressed and on one occasion, in 1362, his friend Petrarch only just managed to prevent him from destroying all his works.

G.K. Chesterton, the English author, died at this age in 1936. He is best remembered for his *Father Brown* detective stories.

Hernan Cortés, the Spanish conqueror of the Aztec empire, was plagued by envy of his great success and when eventually he returned to Spain, his health already destroyed by his campaigns, he found himself the victim of vicious political moves which left him in dire circumstances. Instead of being fêted as a champion of Spain, he was forced to write to his ruler, 'I am old, poor and in debt.' He was eventually given permission to return to Mexico, but he died at 62, in 1547, miserable and disillusioned, before he had even managed to reach Seville.

Demosthenes, the greatest orator in ancient Athens, killed himself by taking poison after being tracked down by an enemy agent in the disastrous Greek war against the Macedonians, the Lamian War.

Edward the Confessor, King of England from 1042 until his death at 62 in 1066, was a pious but hopelessly ineffectual ruler. Although he had promised the throne to William, Duke of Normandy, he was persuaded on his deathbed to give it instead to the Anglo-Saxon Harold. Edward died in January 1066 and ten months later William killed Harold at the Battle of Hastings and took over the English throne.

Sid James, the British comedian, died of a heart attack while performing on stage in 1976.

Alfred Kinsey, the zoologist whose reports on human sexual behaviour swept aside a great deal of puritanical hypocrisy and created a more healthily open attitude to sexual matters generally, died at 62 in 1956, his work far from complete. He was planning several more volumes including studies of sex in prisons, in Europe and among animals.

Pierre Laval, the leader of France's Vichy Government during the Second World War, was executed after trying to poison himself, following a trial for treason in 1945.

Kwame Nkrumah, the first black African Prime Minister ever to attend the Commonwealth Prime Ministers' Conference (in 1957), became President of Ghana in 1960. With his skills as an orator he could have become the first president of the whole black continent, but he became totally corrupted by his power and developed into a megalomaniac tyrant with a belief in his own immortality. While on a trip to his communist friends in the East in 1966 he was deposed by a coup. This was followed by wild cele-brations in Ghana and the destruction of the huge statues of Nkrumah that he had erected in his own honour. He died at 62 six years later in 1972.

Nostradamus, the French physician and astrologer, died in 1566. Eleven years earlier he had written his famous prophecies – astrological predictions so successful that he was invited to court by Catherine de' Medici, and became physician to Charles IX.

Robert Oppenheimer, the American pioneer of the Atomic Bomb, died at this age in 1967. A tall, thin, thoughtful, chain-smoking physicist, he had been put in charge of the new Los Alamos laboratory in 1943. There the Atomic Bomb was designed and constructed. Later, when he opposed the development of the Hydrogen Bomb, he came under attack, was denied access to secret documents, and caused a split between the scientific and political factions involved. Accused of 'imprudence in his associations', by the Atomic Energy Commission, the blackening of his character caused much anger among other scientists, but his status was regained when, in 1963, he was awarded a high honour by the AEC. He retired in 1966 and died, at 62, shortly afterwards.

Chips Rafferty, the rough, tough, archetypal Aussie in many films from the 1930s to the 1970s, died of a heart attack in a Sydney street in 1971.

Damon Runyon, the American author of *Guys and Dolls*, who created a whole cast of low-life characters in a Brooklyn setting, including Harry the Horse, Little Dutch and Ropes McGonagle, died in 1946. His famous last words were: 'You can keep the things of bronze and stone and give me one man to remember me just once a year.' Such is the immortality of the artist.

Anwar Sadat, the enlightened President of Egypt who worked tirelessly for peace in the Middle East, was murdered during a military parade in Cairo on October 6, 1981. He took the unprecedented step of visiting Israel for peace talks in November 1977 and a Peace Treaty was signed in 1979. He shared a Nobel Peace Prize with Begin in 1978.

Evelyn Waugh, the master of the careful phrase and the economic comment, died in 1966 full of scorn for the decline of language and the moral and aesthetic coarseness of modern society. In particular, he had a growing disgust of the cult of the common man, which made him intensely unpopular in many academic and literary circles.

Ludwig Wittgenstein, the Austrian-born British philosopher, died at this age in 1951. He was Professor of Philosophy at Cambridge from 1939 to 1949, although during the Second World War he insisted on leaving the University and working as a humble porter in a London hospital during the Blitz.

Émile Zola, the French author of a huge series of naturalistic novels which concentrated on social horrors and moral decline, died of carbon monoxide poisoning in 1902. Officially it was an accidental death, but there were those who believed that he had been quietly murdered by political enemies.

Sixty-three is an easing-off year for those anticipating a retirement pension at 65. It is a time to avoid controversy or any action which may put that pension at risk. Even when a clash occurs at this age, the cautious employee will do his best not to cause any ripples as he coasts gently in to the end of his career. For the more forcefully-minded, this can create a severe inner conflict, and the need to suppress an explosion of feeling at work can sometimes give rise to seemingly inexplicable outbursts elsewhere.

In Britain it has recently been suggested that 63 should be the new retiring age for both men and women, removing the sexual inequality that exists at present, with women retiring at 60 and men at 65.

Life Expectancy

Life expectancy at birth for males in South Korea and for females in Guyana, Paraguay and Thailand, is 63.

Accession at 63

Pius XII became Pope at this age on the eve of the Second World War and remained pontiff throughout the war and until his death in 1958. To his critics he was an infamous Pope who knew of Nazi atrocities but found it expedient not to denounce them publicly, thereby rendering the Vatican a squalid bystander at Nazi crimes against humanity. To his defenders he was a valiant manipulator seeking desperately to retain the Vatican's independence from the pigsty of international politics.

Achievement at 63

Samuel Beckett, the Irish author of increasingly concentrated, distilled, brief plays, was this age when he won the Nobel Prize for Literature in 1969. Although he accepted the award, he refused to make the trip to Stockholm to attend the presentation ceremony and give his acceptance speech. This was in line with his lifelong rejection of any kind of personal publicity or public appearances.

Johannes Brahms, the German composer, fell in love at 21 with Clara, the wife of Robert Schumann. They did not marry when Robert died, but remained friends. On his sixty-third birthday, May 7, 1833, Brahms completed 'Four Serious Songs' while thinking of her as 'the most beautiful experience of my life, its greatest wealth and noblest content'. She was gravely ill and, when she died two weeks later, he travelled for forty hours by train to attend her funeral. Within a matter of months he too was dead.

Casanova, the Italian adventurer, spent most of his life travelling through Europe indulging in endless love affairs and other escapades. Eventually, in a more settled mood, working as a librarian to Count Waldstein, he sat down to write his infamous memoirs. They were published when he was 63, in 1788.

Benjamin Disraeli became Prime Minister of Great Britain at 63 in 1868. Everything had been against him – his Italian-Jewish origins, his debts, his flashy reputation, his dandyish way of dressing, even his exotic good looks, and when he finally achieved his goal he remarked, 'I have climbed the greasy pole.'

Francis Galton, the English geneticist, a cousin of Charles Darwin, was this age in 1885 when he devised a way of classifying fingerprints and established their use for identifying individuals. He named the three main patterns – whorls, loops and arches – and after further years of research had developed a system to store 700,000 sets of fingerprints. Later, the FBI improved on his system to allow filing of 200,000,000 prints.

Nikita Khrushchev

Nikita Khrushchev was 63 when he became Premier of the USSR in March 1958 and proceeded to confound the West with his sudden shifts of mood, from warm and boisterous to angry and belligerent, from diplomatic to eccentric, but the overriding goal of his reign was one of relaxing the tensions between East and West and of opening up contacts, instead of isolating Russia behind its Iron Curtain.

Richard Wagner was 63 when he saw his greatest masterpiece, the monumental opera cycle *The Ring of the Nibelung,* performed for the first time, in August 1876. He had set himself the task of converting classical opera into 'Music Drama' and he spent twenty-five years developing *The Ring* which required the construction of a specially designed opera house to present it. The Bayreuth Opera House was built in such a way that he could realize his dream of giving equal importance to the music, the words, the dancing, the mime and the elaborate scenic effects. Controlling and creating everything himself, he was able to achieve a unique unity that changed the face of opera when *The Ring* was eventually performed to an audience which included two emperors and a king.

Life-span of 63

William Bligh, famous as the sea captain whose crew mutinied against him on his ship *The Bounty,* died in Bond Street, London, in 1817, at 63, after having risen, step by step, to the rank of Admiral.

Cicero, one of Rome's greatest orators, died in the year 43 BC. Although he disliked Julius Caesar's dictatorship, he was not involved in his murder. He was violently opposed to Mark Antony's control after

Caesar's death and delivered fourteen brilliant orations against him. As a result he was captured and executed at the age of 63.

John Ellerman, the richest man in England in his day, was a reclusive shipping magnate who left the running of his vast business empire to others and devoted his life to the scientific classification of rats and mice. He died at this age in 1973.

Ekaterina Furtseva was the only woman ever to rise to the Kremlin inner circle. In 1960 she became the Soviet Minister of Culture. She died in 1974, aged 63.

Ulysses S. Grant, the eighteenth American President, died at this age in 1885, after considerable suffering from throat cancer.

William Holden, the Hollywood actor, died alone in his apartment in November 1981, after a fall following a heavy drinking bout.

Lord Hore-Belisha, the man who, as Minister of Transport, introduced Britain to a new Highway Code, driving tests for motorists, speed limits on roads in towns, and 'Belisha Beacons' for safety crossings for pedestrians, died at 63 in 1957.

Charles Laughton, the English-born American actor most famous for his portrayal of Captain Bligh in *Mutiny on the Bounty*, died at the same age as the real Bligh, succumbing to cancer in 1962. Laughton achieved the remarkable distinction of being the only fat, ugly man ever to become a top box-office star in the cinema. His success was due to his immense range and acting skill which enabled him to play everything from monsters to heroes and from butlers to despots. Nero, Henry VIII and the Hunchback of Notre Dame were among his most memorable roles.

Lin Piao, the Chinese communist statesman, was in 1968 officially declared to be Mao's heir and the next chairman of the party. Instead he became the victim of a mysterious air crash over Mongolia in 1971, at the age of 63, and was later said to have died while trying to escape China following an unsuccessful plot to remove Mao prematurely.

Archbishop Makarios, one of the last ethnarchs of the western world – a man who was at the same time head of both church and state, died of a heart attack in 1977. He wanted Cyprus to be joined to Greece and, with General Grivas, began the EOKA struggle against British rule after the Second World War. He was horrified when Grivas turned it into an armed conflict in 1955, as he had always been a man of peace. He was exiled to the Seychelles by the British but returned to become the President of Cyprus in 1960 on the understanding that union with Greece would not be pursued further. But the continuing presence of armed rebels alarmed Turkey so much that she invaded Cyprus in 1974 and took over the northern half of the island. Makarios fled the country just before this happened, to avoid assassination, but returned triumphant in December 1974 to rule the southern half of Cyprus. The strain of living in a divided island proved too much for him. His failure to have the unity of the country restored destroyed his health and he died at 63.

Airey Neave, the British MP, was killed by a car bomb which exploded under him as he was driving out of the House of Commons underground car-park in 1979 – a revenge murder carried out by the IRA.

Alfred Nobel, the Swedish chemist who gave the world dynamite, left a fund of $9,200,000 in 1896 when he died at the age of 63, to provide prizes for outstanding work in the realms of physics, chemistry, medicine, literature and peace. The first ones were awarded in 1901, on the anniversary of his death, December 10.

Robert Ryan, the American actor famed for his portrayals of hard, often cruel characters, but who in private life was a gentle, soft-spoken man, died of cancer in 1973.

August Strindberg, the greatest Swedish playwright, died of stomach cancer in 1912, after a prolific career in which he wrote over fifty plays, his most famous being *Miss Julie* and *The Dance of Death*. He suffered a tormented childhood, tortured by a gloomy hell-fire religious education, the death of his mother at 13, and the bankruptcy of his father which led to his being teased at school for his poor clothes. His plays were uncompromisingly realistic dissections of the battle between men and women, and he himself was consumed with jealousies that destroyed his three marriages.

Franchot Tone, the American actor who specialized in smooth-talking lounge-lizards or shallow sophisticates in many films in the 1930s and 1940s, died of lung cancer in 1968.

Florenz Ziegfeld, the American impresario famous for his *Ziegfeld Follies*, died at 63 in 1932. His stated aim was 'to glorify the American girl' and he did this in a long series of annual New York stage reviews starting in 1907 and ending in 1931. It is claimed that he alone was responsible for the introduction of slender figures as the epitome of female beauty in the western world, replacing the ample, fleshy shapes that predominated at the turn of the century.

Laughton as Capt Bligh

Benjamin Disraeli

64

Sixty-four is the year chosen by The Beatles to epitomize old age. In their 1967 song *When I'm Sixty-four*, the sad phrase 'Will you still need me, will you still feed me, when I'm sixty-four,' is repeated at the end of each verse, and the 64-year-old is characterized as losing his hair, 'handy, mending a fuse', 'doing the garden, digging the weeds,' scrimping and saving and having grandchildren on his knee. This picture of someone cosily pottering about in a condition of antique domesticity may apply to some 64-year-olds, but it is worth putting it into perspective by pointing out that Francis Chichester was this age when he set sail on August 27, 1966, in his 53-foot yacht *Gipsy Moth IV*, to travel completely alone around the globe. It is not too late at 64 to defy the traditional image of one's age.

Life Expectancy
Life expectancy for males in the Bahamas, Panama, Sri Lanka, Suriname, Trinidad and Tobago, and the USSR is 64.

Achievement at 64
Richard Burton, the nineteenth-century English explorer and author, is best remembered today as the translator of classics of Eastern erotica, including *The Perfumed Garden* and *The Kama Sutra*. His greatest triumph in this field was a huge sixteen-volume work entitled *The Thousand Nights and a Night* – his unexpurgated and sexually explicit rendering of the tales of the Arabian Nights. It first appeared in 1885 when he was 64, his obsession with the publication of erotic works having arrived late in his life, when he took the decision to introduce the West to the sexual wisdom of the East. His motivation seems to have been a hatred of what he called the 'immodest modesty' of Victorian society, and its gross hypocrisy. Reactions to his work ranged from praise for its robustness to attacks calling it the 'garbage of the brothels'.

James Callaghan, the British politician, was 64 when he became Labour Prime Minister in April 1976. An avuncular, stodgy ex-tax official, he was a socialist leader with a conservative personality.

Cecil Day-Lewis, one of the '30s poets' – along with W.H. Auden and Stephen Spender – who also wrote detective novels under the name of Nicholas Blake, became Poet Laureate in 1968 at the age of 64.

Edith Evans, the English actress, was 64 in 1952 when she gave her most memorable performance, as Lady Bracknell in *The Importance of Being Earnest*.

John Galsworthy was this age when he was awarded the Nobel Prize for Literature, in 1932, for his 'distinguished art of narration which takes its highest form in *The Forsyte Saga*'.

Bonar Law became British Prime Minister at 64 in October 1922, but his period of power was brief. Once he became aware that he was suffering from cancer of the throat, he resigned, after only 209 days in office. Five months later he was dead.

Ferdinand de Lesseps, a French diplomat, was able to celebrate his sixty-fourth birthday at a moment of great triumph, in November 1869. Fired with an ambition to create a great canal to serve all mankind, he had spent over a decade masterminding the construction of the Suez Canal. The technical, political and human problems involved were colossal and the cost rose to nearly $100,000,000 which, in the 1860s, was an astronomical sum and more than twice the original estimate. But despite everything, the canal was opened to traffic on November 17, 1869. Sixty-eight vessels made the inaugural trip, arriving in Suez safely on November 20. En route they had paused for special festivities on de Lesseps's birthday, the 19th. The event was also celebrated in many other countries and Verdi composed his opera *Aida* to commemorate the occasion.

John Napier, the Scottish mathematician, was this age in 1614 when he published his invention of what he called 'logarithms', after spending twenty years calculating the necessary tables. He was also responsible for the introduction of the 'decimal point'.

Laurence Olivier was 64 in 1971 when he took his seat in the House of Lords – the first and only actor ever to be accorded this honour in the whole of English history. Generally accepted as one of the greatest actors of modern times, he became world famous for his films of *Henry V*, *Hamlet* and *Richard III*, receiving Hollywood Oscars for the first and second of these, and a third such award for the 'full body of his work'.

Voltaire, the French genius of the eighteenth-century enlightenment, who hated priestcraft and fanaticism and called for tolerance, humanity, justice, freedom and the rights of man, was repeatedly in trouble with the authorities and was even imprisoned for his writings. In 1759, when he was 64, he published his most popular work, *Candide*. It is claimed that he wrote the manuscript in four days in July 1758, while staying at the palace of an old friend. When he arrived he is said to have locked himself in his room, opening the door solely to admit food and drink, until four days later he emerged with the work complete.

Conversion at 64
J.B.S. Haldane, the British geneticist, became converted to an eastern way of life at this age, taking Indian citizenship, adopting Indian dress and emigrating with his wife to India, where he became the head of the government's Genetics and Biometry Laboratory at Orissa. His conversion was so intense that he remained quite undaunted when, at one of his first important scientific meetings in his new home, he arrived in Indian robes, looking like an overweight Gandhi, to discover that every one of the huge assembly of Indian scientists was wearing a dark lounge-suit.

Life-span of 64
Barbarossa – 'Red-beard the Pirate' – who ruled the whole of the Mediterranean

in the early 1540s, was pictured by Christian leaders as a lecherous sadist who died in his 90s worn out by sexual orgies with his captive white slave-girls. In reality, he was an abstemious, multi-lingual patron of the arts who died exhausted at 64, in 1546, tired out from his incessant campaigning and attempts to maintain alliances and balances of power in the Mediterranean region.

left, Laurence Olivier as Hamlet; *right*, Edith Evans as Lady Bracknell

Béla Bartók, the Hungarian composer who combined the essentials of his native folk music with the more advanced forms of European music to become one of the most original and individualistic of modern composers, died of leukaemia in New York in 1945.

Edward Burne-Jones, the English painter, died suddenly at this age in 1898. Although associated with the Pre-Raphaelites, he was not truly one of them. His paintings can best be described as sickly sentimentality run riot, depicting dreamy, mystical, legendary scenes in a literary never-never land.

Claudius, the Roman Emperor, found himself thrust into a position of power when his predecessor, the odious Caligula, was murdered in AD 41. Claudius himself was murdered in AD 54, by his wife Agrippina who poisoned him with mushrooms, and was succeeded by the equally odious Nero. Views about Claudius vary. He seems to have been an ugly but scholarly man who wrote many books on Roman, Etruscan and Carthaginian history. Sadly, none of his works has survived.

Marion Davies, the American actress, was a mediocre talent raised to stardom by her rich protector. An ex-chorus girl, she became the protégé of the press baron William Randolph Hearst, and had the door to stardom pushed open by the weight of his cheque-book. When she became his mistress he formed a film company solely to produce her films and publicize her manufactured star status. This campaign cost him $7,000,000, but it succeeded in the end and she enjoyed playing host to Hollywood society at her 14-roomed studio bungalow, her 110-roomed Santa Monica beach house (with 55 bathrooms), her Beverly Hills mansion and Hearst's bizarre San Simeon Castle. She eventually died of cancer at 64 in 1961.

Edward III, King of England from 1327 to 1377, became prematurely senile in his early 60s, and the running of the country fell into the greedy hands of his mistress Alice Perrers and his son John of Gaunt. On June 21, 1377, when it became clear that he was dying, Alice removed all the rings from his fingers and left him. All his courtiers also walked out on him and only a single priest attended his deathbed. It was a small reward for a man who had ruled his country for half a century.

Alberto Giacometti, the Swiss sculptor famous for raising matchstick-men to the level of high art, died of a heart condition at 64 in 1966.

Lyndon B. Johnson, the thirty-sixth American President, died from a heart attack while on his Texas ranch, in 1973, five years after retiring from office. During his presidency he had achieved a great deal in the realm of civil rights and the war on poverty, but it was the other war – the one in Vietnam – that broke him. It was he who had escalated that war and who had been deeply hurt by the repeated chanting of 'Hey hey LBJ, how many kids did you kill today?' His vanity was such that he had to have his initials LBJ everywhere. For this reason, he called his wife Lady Bird Johnson, his daughters Lynda Bird Johnson and Lucy Baines Johnson, and his dog Little Beagle Johnson. For such a man, the open abuse from millions of young Americans was insufferable.

Karl Marx died of a lung abscess in March 1883, broken by the death of his wife in December 1881 and his eldest daughter in January 1883. He died an irritable man. When his housekeeper asked him if he had a final message for the world, he uttered his famous last words: 'Go on, get out! Last words are for fools who haven't said enough!'

Carry Nation, the American puritan campaigner, a large, formidable warrior against the evils of alcohol, became famous for her hatchet attacks on American saloons. She raged against booze, tobacco and sex and despised all men, calling them nicotine-soaked, beer-besmeared, whisky-greased, red-eyed devils. She was jailed more than thirty times for invading bars and destroying everything in sight with her favourite axe, while her female supporters stood outside singing uplifting hymns. She helped to pay her fines by selling souvenir hatchets. It may be significant that her first husband arrived drunk at their wedding and died of alcoholic poisoning within a year, and that her second husband was ugly and sexually repellent. She herself died at 64 in 1911.

Walter Raleigh, the Elizabethan adventurer who popularized both potatoes and tobacco, and was a great favourite of Elizabeth I, fell foul of her successor James I, who confined him to the Tower of London from 1603 to 1616. During this time he penned his great work *The History of the World*. He was then allowed one last expedition before finally being executed to please the Spanish, in the year 1618.

65

Sixty-five is the typical age for retirement. In most advanced countries it is the year when people become eligible for Social Security payments, or, in the old-fashioned terminology, become 'old age pensioners'. Exceptions to this rule are Japan, where official retirement comes earlier, and Denmark and Norway, where it comes later.

A few people still continue to work at this age, some because they want to and some because they have to. A study in France, for instance, revealed that, at 65, 10·6 per cent of all men and 5·0 per cent of all women are still employed.

The age of retirement is often greeted with wistful humour. James Thurber: 'I'm sixty-five and I guess that puts me in with the geriatrics, but if there were 15 months in every year, I'd only be forty eight.' (It also puts him at the bottom of the maths class, as he would be 52, not 48.) R.C. Sherriff: 'When a man retires and time is no longer a matter of urgent importance, his colleagues generally present him with a watch.'

Physically, 65 is the year when females start to show an appreciable decline in body weight, following the weight increase of middle age. (Men started this decline ten years earlier.)

A special female weakness that appears at this age is an undue susceptibility to 'deaths from falls'. From 0 to 64 males are more likely to die from falls, but the period 65 and over shows a massive reversal of this bias. The figures for 1976 in the UK were: male deaths 565, female deaths 1,689 – three times as many. The reason is that the human skeleton loses much of its strength at this point, and this loss is greater in women than in men.

The most encouraging news for the 65-year-old facing retirement is that it is not too late to start an entirely new career. Winston Churchill, for instance, was this age when he first became Prime Minister, in 1940, and started his epic struggle against Hitler. A heavy smoker, a heavy drinker and a heavy eater, he defied all the rules of the keep-fit brigade and proved that a good challenge is the best rejuvenator in the world.

Life Expectancy
Life expectancy at birth for males in Argentina, Colombia, Portugal, Singapore, St Lucia and Venezuela, and for females in Chile, Grenada, Lebanon, Mauritius and Samoa, is 65.

Accession at 65
John Paul I, the 263rd Pope of the Roman Catholic Church, was this age when he became pontiff in 1978. He was the first Pope in history to choose a double name. He nearly broke another record, being the Pope with the shortest reign in four centuries, for he was found dead in bed, having suffered a heart attack while reading a book, only thirty-four days after his accession.

Achievement at 65
Joseph Haydn, the Austrian composer, was this age in 1797 when he gave the Austrian nation the stirring song *God Save the Emperor Francis* which was used for more than a century as the National Anthem of Austria and the patriotic song *Deutschland, Deutschland uber alles* ('Germany, Germany above all else').

David Hume, the British philosopher, completed his curiously detached autobiography, *The Life of David Hume, Written by Himself*, on April 18, 1776, when he was 65. He died four months later, cheerfully defending his atheistic disbelief in immortality.

Life-span of 65
Peter Abelard, the French philosopher and teacher, was castrated by the vengeful canon of Notre Dame for seducing his 17-year-old niece, Heloise. She had been a pupil of Abelard at the time and had given him a child but, after his mutilation, they both took holy orders. They were reunited in death when he was buried alongside her in 1144.

Matthew Arnold, the British poet and literary critic, died of a heart attack at this age in 1888, when hurrying to meet his daughter.

Johann Sebastian Bach, the German musical genius, was immensely prolific, both as a composer and as a father, producing a huge output of music and twenty children. In his later years his eyesight failed and at the end he became blind. His health was undermined by two unsuccessful eye operations performed by John Taylor, an itinerant English quack who numbered Handel among his other failures, and he finally died of a stroke in July 1750.

Francis Bacon, the statesman, courtier, lawyer, politician and philosopher, was above all else a great writer, describing himself as 'naturally fitted rather for literature than anything else'. Some even claim that he wrote Shakespeare's plays, but this seems highly unlikely. He died of bronchitis at 65, in 1626, after catching a chill from leaving his coach in the snow to see if he could preserve a chicken.

Tallulah Bankhead, the American actress who raised the cutting remark to the level of an art form, became the 'dahling' of London society in the 1920s, before returning to America in 1931 to star in Hollywood films. Her greatest successes, however, were later, on Broadway, in *The Little Foxes*, *Private Lives* and *The Skin of Our Teeth*. She was more successful as a drawling, extravagant, amusingly bitchy personality than as a serious actress. The impression she created is best summed up by Mrs Patrick Campbell's remark: 'Tallulah is always skating on thin ice and everyone wants to be there when it breaks.' She died of Asian flu in 1968.

Sandro Botticelli, the Florentine master, died at this age in 1510. His *Birth of Venus* has a delicacy and freshness that has made it one of the most famous paintings in the history of European art. As a man he was described as passionate, careless, vehement, moody, unpredictable, indolent and reckless. In a letter to the tax-collector, his father wrote, 'He works at home when so inclined.'

Lewis Carroll, the English mathematician whose two tales invented to amuse a small girl – *Alice's Adventures in Wonderland* and *Through the Looking Glass and What Alice Found There* – outlived all his serious work, died at 65 in 1898. A shy, stammering man in the presence of other adults, he was at home only in the company of little girls, whom he entertained brilliantly with his amazingly fertile and child-like imagination. He had expressed strange ideas even as a small boy and it is recorded that 'he supplied earthworms with weapons in order that they might fight with more effect, fostered snails and toads, and inquired persistently the meaning of logarithms'.

Wilkie Collins, the English novelist whose mystery story *The Moonstone* was described by T.S. Eliot as 'the first, the longest and the best of modern English detective novels', died at this age in 1889, after his health had declined through opium addiction.

Walt Disney also died at this age, in 1966. The most original talent in the history of the cinema, his cartoon features, especially *Snow White and the Seven Dwarfs*, were masterpieces of animation skill and imagination.

Nelson Eddy, the American singer and actor whose sickly-sweet operettas with Jeanette Macdonald in the 1930s have now become rediscovered as 'high camp' entertainment, died at 65 in 1967, after being taken ill shortly after completing a night-club act in a Miami Beach cabaret.

Ghengis Khan, the founder of the Mongol nation, died following a fall from his horse in 1227. As a warrior-ruler he became the most famous conqueror in history, striking terror into his enemies with massacre and rape on a huge scale.

George Grosz, the controversial German artist, was always in trouble with the authorities. His problems began when he made a personal attack on the Kaiser in a magazine he had founded, and a warrant was issued for his arrest. He escaped, but then sent for exhibition a painting showing a crucified Christ wearing a gas-mask. For this, he was caught and condemned to be shot by a firing squad, but the Roman Catholic Church intervened and his sentence was reduced to being sent to the front-line trenches of the First World War.

Fritz Haber, the German chemist who discovered how to synthesize ammonia, which helped the Germans to make explosives and thus prolong the First World War, also played a leading part in the development of poison gas. The fact that he was awarded the Nobel Prize for Chemistry in 1918 should have been enough to persuade all other Nobel Laureates to return their medals in protest. He died at 65 in 1934.

Oscar Hammerstein, the brilliant American songwriter, also died at this age, in 1960. The basic rule governing his success was that no song should exist by itself in a musical: 'The song is the servant of the play', he said, and ensured that each song he wrote advanced the story and was part of it.

Oliver Hardy, who teamed up with Stan Laurel in 1926 to create the greatest comic double-act in the history of entertainment, died of a stroke at 65, in 1957.

Laurel and Hardy

Emil Jannings, the Swiss-born German actor, famed for his role as the professor in *The Blue Angel*, died of cancer in 1950, a lonely and embittered man, feeling disgraced at the defeat of Nazi Germany.

Anna Magnani, the powerful, passionate, Oscar-winning Italian actress, died following surgery for a tumour of the pancreas, at 65, in 1973. She was mourned in Italy by a vast crowd that turned out for her funeral 'as if she were a Pope'.

Eugene O'Neill, one of America's greatest dramatists and the first American playwright to be given a Nobel Prize for Literature, died at 65 in 1953.

Queen Salote of Tonga, the gigantic ruler of a group of tiny islands, endeared herself to the world when she visited London for the coronation of Elizabeth II in 1953 and insisted on braving the heavy rain to parade in an open carriage. She reigned for forty-seven years until her death at 65 in 1965.

Adlai Stevenson, the American politician who failed to become President because he was too much of an intellectual to engage in the barnstorming childishness demanded of a successful Presidential candidate, died at 65 in 1965.

U Thant, the Burmese diplomat who became the second Secretary General of the UN (1961–71), was humane, modest and open-minded, but sadly, as with all major UN attempts at peace-making, he was ineffectual. He was deeply depressed by his failure to halt the Vietnam War, and had to be persuaded to accept a second term of office. He died of cancer in 1974.

Donald Wolfit, the last of the flamboyant actor-managers, died at this age. A formidable Shakespearean performer with a glowering expression and a tendency to be over-dramatic, he kept his theatre open in London all through the Blitz, with lunchtime Shakespeare, and then went off to do Home Guard Service at night. He died of a heart attack in 1968.

Sixty-six is the year of the widow. It is the start of the phase of life when, in advanced countries, the widows outnumber the widowers by 3 to 1. The fact that women live longer than men has often been referred to as a 'natural' difference between the sexes, but this is not the case. It is cultural, and only occurs in post-industrial societies. In earlier, pre-industrial societies, the men on average lived longer than the women. In those earlier times, the hazards of child-birth and poor nutrition accounted for the shorter life-span of the females. In modern times, the stress of unhealthy city jobs accounts for the shorter life-span of the males. If tomorrow's women truly achieve sexual equality with men they will soon suffer from the shorter male life-span. In the meantime, women of 66 and over will continue to outnumber the men of the same age.

Life Expectancy
Life expectancy at birth for males in Hungary, Malaysia, Poland and Uruguay, and for females in Jamaica, Mexico and Sri Lanka, is 66.

If you lived in seventeenth-century London, you had only a 3 per cent chance of surviving to the age of 66, according to John Graunt's Life Table.

Achievement at 66
John Betjeman, the English poet and affable champion of nostalgia, became Poet Laureate in 1972 at this age. Although his light-hearted, mildly satirical verse had given pleasure to a wide audience, he was less than ideal as a proclamatory poet for grand occasions, such as coronations and royal weddings. But having become a national symbol of ultra-Englishness, he was easily forgiven.

John Betjeman

Edgar Rice Burroughs was 66 when he became the oldest war correspondent in the South Pacific, after witnessing the Japanese attack on Pearl Harbor in Hawaii. Already world famous for his Tarzan stories,

he set aside his book-writing for the duration of the war, to send back reports for the *Los Angeles Times*.
François Mauriac, regarded by many as the greatest French novelist since Proust, was 66 when he was awarded the 1952 Nobel Prize for Literature.
Lester Pearson, the Canadian politician who had won the Nobel Peace Prize for his attempts to solve the Suez crisis, became Prime Minister of Canada in 1963, when he was this age.
Niko Tinbergen, the Dutch-born zoologist and pioneer of naturalistic animal behaviour studies, was 66 when he was awarded the Nobel Prize for Medicine as one of the three founders of the science of comparative ethology (along with Konrad Lorenz and Karl von Frisch), in 1973. Tinbergen's unique contribution was to show that it was possible to carry out carefully controlled experiments with wild animals in their natural environment, rather than in the artificially sterile world of the laboratory.

Misfortune at 66
Aldous Huxley, the English author best known for his book *Brave New World*, moved to Southern California in 1937. In May 1961, when he was 66, he suffered the disaster of seeing his house burn down, destroying his entire library and all his papers. It made him, he said, 'a man without possessions and without a past'. To make matters worse, he had only just been told that he had cancer, from which he died two years later.

Life-span of 66
W.H. Auden, the most brilliant of England's modern poets, who began as a literary *enfant terrible* and ended as a grand master, died in 1963, shortly after returning to his old university home at Christchurch, Oxford.
Hieronymus Bosch, the greatest genius of imaginative imagery in the history of art, died in 1516. His surname is an abbreviation of the name of the Dutch provincial city from which he came, Hertogenbosch, but little is known about his personal life.

Detail from *The Garden of Earthly Delights* by Bosch

Pierre Brasseur, the larger-than-life actor who became one of the leading figures on both the Paris stage and in the French cinema, died at this age in

1972, while acting in a film.

Jacob Bronowski, the Polish-born scientist whose work on smokeless fuels led to the demise of the traditional London fog, was also erudite in the fields of literature, mathematics, biology and history. He combined his multi-faceted talents in 1973 to produce a major television series, *The Ascent of Man*. Although his presentation style was often irritating – a combination of verbal diarrhoea and postural constipation – he displayed an all-embracing clarity of mind that left a deep impression on his huge audience. He died in the United States in 1974.

Jack Buchanan, a British version of Fred Astaire, with a dry Martini voice and a sophisticated comedy bias, was immensely popular in inter-war stage and film musicals. After the Second World War, however, he suffered from spinal arthritis and soon faded from the scene, dying at 66 in 1957.

Joseph Conrad, the Polish-born British author of novels such as *An Outcast of the Islands* and *Lord Jim*, spent his early life at sea, travelling around the world. He then settled down to concoct stories based on his adventures, but he was already in ill-health as a result of a visit to the Congo where he contracted a recurrent fever. As he grew older he also suffered from gout and rheumatism, becoming increasingly irritable and short-tempered, before he died of a heart attack, at 66, in 1924.

Thomas Cranmer, the first Protestant Archbishop of Canterbury, and a good servant to Henry VIII in the business of removing unwanted wives and popes, met an inevitable fate when Mary came to the throne and Catholicism returned to favour. He was accused of heresy and burned at the stake in Oxford, in 1556. The spot where he died is marked today by a cross on the road surface in the middle of Oxford's Broad Street.

Thomas Cranmer   Edgar Rice Burroughs

Moshe Dayan, the Israeli soldier and statesman, who lost an eye fighting against the Vichy French in the Second World War in Syria, and afterwards always wore a distinctively piratical black eye-patch,

died at 66 in Tel Aviv in 1981. He masterminded and directed all of Israel's military actions in the 1960s and early 1970s, including the Sinai campaign and the Six Day War, either as Chief of Staff or as Minister of Defence.

C.S. Forester, the English novelist best known for *Captain Horatio Hornblower* and *The African Queen*, died in 1966 of arterio-sclerosis which had made him a semi-invalid in his later years.

William Hogarth, the English artist whose fame today rests largely on his moralistic series of engravings – *A Harlot's Progress*, *A Rake's Progress*, and *Marriage à la Mode* – died in 1764.

Joe Louis, the World Heavyweight Boxing Champion from 1937 to 1949, died in Las Vegas in 1981. He defended his title twenty-five times before retiring undefeated, but then unwisely returned in 1950 to fight Ezzard Charles, who took the title from him. Known affectionately as 'The Brown Bomber', he made a fortune of $5,000,000, but saw it all drain away, so that his last fights were for cash rather than glory.

Terence Rattigan was 66 when he died in 1977, having enjoyed one of the most successful careers of any playwright in the 1940s and 1950s. *The Winslow Boy*, *The Deep Blue Sea* and *Separate Tables* are among his most popular works.

Mark Rothko, one of the purest of the American abstract expressionist painters, killed himself by slashing his arms, in 1970. His self-mutilated body was found in his vast New York studio. It is hard to imagine why he should have wanted to kill himself, since he had been immensely successful with his art, and was the only American ever to have received the honour of a major exhibition at New York's Museum of Modern Art in his own lifetime. His great floating rectangles of colour either had a hypnotic fascination for the viewer, or drove him angrily from the gallery. His avowed aim was to remove from art 'the swamps of memory, history and geometry', and in that he certainly succeeded.

James Thurber, the great American humorist, who died in 1961, is reputed to have uttered the famous last words: 'God bless ... God damn ...'

Alexander Woollcott, Thurber's friend and colleague, and a fellow-member of the Algonquin Round Table, also died at this age, but earlier, in 1943. He was the most feared drama critic in New York in the 1920s and 1930s, when he could make or break a play with his comments.

Frank Woolworth, the American shopkeeper, died at 66 in 1919 after enjoying one of the greatest personal success stories in the history of high-street business. He became a shop assistant in 1873 and by 1913 he had seen the construction of the then tallest building in the world – the 57-floor skyscraper Woolworth Building – to house his central offices, the nerve-centre of a vast international chain of over 1,000 stores at which prices were kept at deliberately low levels.

Sixty-seven is the age of retirement in Denmark and Norway, two years later than in most countries. In view of the recent stretching of the active life-span, the authorities in these Scandinavian countries have probably made the better judgment. Many people at 67 now feel youthful and vigorous, as the following quotation from Mike Hepworth and Mike Feather-stone's book, *Surviving Middle Age*, reveals: 'I'm sixty-seven but I feel vital – anywhere from forty-five to fifty-five, and I resent being bound by the arbitrary measurement of time ... I have an appetite for life.' For such people, and they are many, Denmark and Norway are at least moving in the right direction.

Life Expectancy   The life expectancy at birth for males in Czechoslovakia, Italy, Kuwait, Luxemburg, Romania and Yugoslavia, and for females in the Bahamas, Barbados, Panama and South Korea, is 67.

Achievement at 67   **Josephine Baker**, the American-born French dancer and singer who was the toast of the Folies Bergère in the 1920s and 1930s, became preoccupied with racial discrimination after the Second World War. She gave up her American citizenship as a protest against racial unfairness there and undertook an 'experiment in brotherhood' at her private estate in France, where she created what she called her 'rainbow family' of adopted children of all nationalities. She retired from the stage in 1956, but the costs of her experiment were mounting so much that she had to make a comeback to pay the bills. This she did in 1973 at the age of 67, making a triumphant return to the Broadway stage, where she had first performed as a dancer more than fifty years before.

John Dryden, the English author who so dominated the literature of the late seventeenth century that it was known as the 'Age of Dryden', also undertook a challenging task at the age of 67. He was that age when he agreed to supply his publisher with 10,000 verses for the sum of £300. In the end, his *Fables Ancient and Modern*, his last great work, extended to 11,700 verses.

Grandma Moses, the American 'Naïve' artist, started to paint at this age, following the death of her farmer husband with whom she had enjoyed forty happy years. She began working as an artist, she said, because she did not want to become idle and 'sit back in a rocking chair'. Some years later, these paintings were to make her world famous (see ages 96 and 101).

Joseph Strauss, the American civil engineer, was 67 in 1937, when he saw the completion of his great Golden Gate Bridge at San Francisco.

Crime at 67   **Herod the Great** was this age when he ordered the 'Massacre of the Innocents' in Bethlehem. As King of Judea under the Romans, he had ruled in peace, with ambitious architectural achievements to his credit, but then, in his later years, suffering from arterio-sclerosis and intense jealousies due to the intrigue and deception taking place within his own family, he became increasingly brutal and murderously cruel. This behaviour culminated in the slaughter of the infants of Bethlehem shortly before his own death after an unsuccessful attempt at suicide.

Life-span of 67   **Ingrid Bergman**, the Swedish film star, died in the night following her sixty-seventh birthday. She had invited a group of close friends to a dinner party at her Chelsea apartment and, despite feeling weak, insisted on going ahead with her birthday celebrations. Her guests were horrified to learn that she had died shortly after they had left, but it was typical of her to ignore the pain of the cancer that had been threatening her for some time. She had had both breasts removed some years earlier, but kept on filming right up to the end of her life. She insisted: 'Time is shortening, but every day I challenge this cancer, and survival is a victory for me.'

Catherine the Great, Empress of Russia from 1762 to 1796, died at 67, two days after suffering a massive stroke. She enjoyed a vigorous sex life with a string of young lovers right up until the year of her death, when her current favourite was still in his 20s. Each new lover in her life had to pass three tests. First he had to be selected as suitable by one of her earlier lovers, Grigori Potemkin, then he had to undergo a medical examination for venereal disease from the Empress's personal physician and finally he had to prove his virility with a specially appointed lady-in-waiting. If he passed all three tests he was installed in a room below Catherine's, connected by a private staircase, and was rewarded with money and honours for his performances.

Paul Cézanne, the great French post-Impressionist painter, son of a rich banker, was a disappointment to his father. He was shy, shunned society and marriage, fathered an illegitimate child and wandered the countryside in old, rough clothes, painting rural scenes – the local eccentric whose works of art were laughed at. An awkward figure, he was hardly the ideal banker's son. Even he himself admitted that he was 'feeble in life' – still allowing himself to be bullied by his father at 47, when he married the girl with whom he had enjoyed a liaison for seventeen years. But his stubborn experiments with new ideas of perspective and form heralded the major upheaval that was to come in the revolution of twentieth-century art, and made him one of the pioneer figures in its history. He died at 67 in 1906, of a harsh chill, on top of serious diabetes.

E.E. Cummings, the most original and individualistic poet of the twentieth century, who experimented with poetic typography, presenting the verse on the page as a visual as well as a verbal composition, died at 67 in 1962.

Leonardo da Vinci, the giant of the Italian Renaissance, was not only one of the world's greatest artists, he was also one of the most inventive brains in the history of civilization. He nearly discovered the circulation of the blood, he devised the first armoured fighting vehicles, aircraft and helicopters, and the submarine. But he had one great failing – he hardly ever finished anything. Not one of his inventions was put into practice. And although he left thousands of sketches and drawings, we have only thirteen finished paintings and four unfinished paintings of his to enjoy. The famous last words of this illegitimate, homosexual, left-handed genius suggest that his problem was being too much of a perfectionist. He said: 'I have offended God and mankind because my work did not reach the quality it should have.'

W. C. Fields, the American comedian who was funny because he was so monstrous, died of dropsy, liver disease and heart failure, in 1946. He hated people, especially women and children and their pets. He also hated anyone in authority and distrusted bankers so much that he had 700 bank accounts, all over the world. He gained a huge following as an antidote to the wholesomeness of many of the early comedies. In private life he was an alcoholic as monstrous as his film image. He died in California on the day of the year he hated most – Christmas Day – and his famous last words were 'On the whole, I'd rather be in Philadelphia.'

Albert Fish, the American murderer, was electrocuted at Sing Sing prison in 1936, at the age of 67. An astonishing sado-masochistic cannibal who looked like a gentle old professor, he killed a 12-year-old girl in 1928 after persuading her parents to let him take her to a children's party. Taking her to his home, 'Wisteria Cottage', he cut her head off and made a stew of her body, adding onions and carrots to flavour. He then sat down and ate her. He was a house-painter with six children of his own who were encouraged to beat him. He also thrust sharp metal objects into his body and there was so much metal inside him that he short-circuited the electric chair and the executioners had to try a second time before killing him. The delay did not disturb him, since he described electrocution as 'the supreme thrill – the only one I haven't had.'

Robert Flaherty, the American documentary film-maker, was a passionate explorer from an early age and his 1920s field-studies of remote places became cinema classics. *Nanook of the North*, made in 1922, depicting the struggle for survival of an Eskimo fighting against the hostility of his northern world, was his most memorable achievement. Nanook's struggle was real enough, for he died of starvation shortly after the film was completed. Flaherty himself died at 67 in 1951.

George I, the unpopular King of Great Britain and Ireland from 1714 to his death in 1727, lived for sixty-seven years. The effective ruler of the country was Sir Robert Walpole as the King preferred to live in Hanover as much as possible.

George IV, the indolent, arts-loving King of Great Britain from 1820 to 1830, behaved in a curious manner as his death, at 67, drew near. He lay in bed all day, keeping his room at a fiercely high temperature and drinking vast quantities of cherry brandy. As his sight began to fade he became convinced that he had commanded a division at Waterloo and ridden a winning race at Goodwood. After he had died it was discovered that his wardrobe contained every item of clothing he had worn over the past fifty years and 500 pocket-books, each containing a small sum of money. There were also collections of women's gloves and women's locks of hair.

W. G. Grace   George IV

W.G. Grace, the most famous of Victorian cricketers, died at 67 in October 1915, having played his last game of cricket in July 1914 at the advanced age of 66. His score on that occasion was 69 not out.

Henry I, King of England from 1100 to 1135, died at 67 of a 'surfeit of lampreys'. His health had been poor for some time and his physician had urged him to be careful with his food. He ignored these instructions, feasted on lampreys (eel-shaped parasitic fishes) and then went out hunting in the Forest of Lyons. He became sick while on the hunt, developed a fever and died within a few days.

Spencer Tracy, the craggy, querulous hero of countless films, and the earliest master of 'naturalistic' film acting, died of a heart attack in 1967. He won two Oscars, the first in 1937 and the second in 1938.

Anthony Trollope, the English novelist who wrote the *Barchester* series about country life in Victorian England, died of a stroke in 1882. Persecuted as a child and bored as a young post office clerk, his early life was wretched. To earn more money he started writing, producing 2,500 words every day between 5.30 am and breakfast. He saw it merely as another job of work, requiring no inspiration or special talent. One feature of his life for which he is not remembered is that, having risen to importance in the postal service, he invented the pillar-box.

68

Sixty-eight is the average age at which men stop indulging in sexual activity. At this point in their lives roughly one man in four is impotent. Women, on average, stop sexual activity much earlier, at around 60. There are many exceptions, however, and both men and women are known to have enjoyed sex right into their 80s and even (with men) into their 90s. Their attitude is epitomized in the answer of an elderly actress when asked at what age a woman ceased to be interested in sex. 'How should I know?', was her reply.

For some, 68 is a sad year because of the loss of familiar friends. A poem by J.R. Lowell, called 'Sixty-eighth Birthday', reads: 'As life runs on, the road grows strange/ With faces new, and near the end/ The milestones into headstones change,/ 'Neath every one a friend.'

Life Expectancy Life expectancy at birth for males in Austria, Belgium, Bulgaria, China, Cuba, East Germany, Eire, Fiji, Finland and Malta, and for females in China, the Seychelles, and Trinidad and Tobago, is 68.

> Among the animals, 68 is the record longevity for an owl.

Accession at 68 Tomas Masaryk became the first President of Czechoslovakia at this age, in 1918. The son of a coachman and a maid, he had campaigned for years for an independent country for his people and worked tirelessly to carve a new state out of the Austro-Hungarian Empire. Once he had succeeded he proved himself to be a wise and liberal-minded leader and brought prosperity to his new nation until, at the time of his death, the Hitler threat arrived to disrupt the scene.

Achievement at 68 The Aga Khan, spiritual head of the Ismaili sect of the Muslims, celebrated sixty years as the Imam by having himself weighed against one million pounds' worth of diamonds, when he was 68.
Mrs Patrick Campbell, the actress friend of George Bernard Shaw, who dominated the London theatre for many years and who created the role of Eliza in *Pygmalion*, started a new career at the age of 68, appearing in her first film role, in the 1933 production *Riptide*.
Miguel de Cervantes, Spain's greatest author, was actively writing right up to his death in 1616. It was in the final year, at the age of 68, that he completed his work, *The Trials and Peregrinations of Persiles and Sigismunda*, penning the brief dedication for it while on his deathbed 'with one foot in the stirrup for the next world'.
Charles Grey, Prime Minister of Great Britain from 1830 to 1834, was this age when he pushed the 1832 Reform Bill through parliament, against furious opposition. It greatly extended and improved the voting powers of the middle classes and was a major step towards a truly democratic form of government.
Boris Pasternak, the Russian author, was 68 when he was awarded the Nobel Prize for Literature in 1958. He was accorded this honour for his novel *Dr Zhivago*, despite the fact that the book had been banned in Russia and had only been published after being smuggled out to Italy. Pressure was put on him to refuse to accept the award and he reluctantly agreed – the price he had to pay for continuing to live in his own country.

Life-span of 68 Edward I, King of England in the late thirteenth century, died in 1307 on his way north to do battle with Robert the Bruce in Scotland.
Edward VII, King of England for the first decade of the twentieth century, died in 1910 of a bronchial condition. It was reported of him that he 'walked little and ate and smoked much' and displayed 'a somewhat corpulent habit of body'. Not a life-style to encourage a ripe old age.
Wilhelm Furtwängler, the German conductor who worked with the Berlin Philharmonic throughout the whole Nazi period, including the Second World War, but was 'de-Nazified' after the war and formally exonerated of any accusations of complicity in Nazi crimes, died at 68 in 1954, of pneumonia.
Paul Hindemith, the German composer, left Germany in the late 1930s and became an American citizen. He refused an invitation to return to Germany after the war and remained in the United States teaching and composing. He was one of the most influential figures in the avant-garde of European music. His death came in 1963 at the age of 68.
Ito Hirobumi, the first Prime Minister of Japan under the newly created European-style cabinet system, was assassinated in 1909 by a gunman of the Korean Independence Movement. His dying words were 'He is a fool', meaning that the killer had removed the one man who was most likely to act fairly towards Korean policies.
Arthur Lowe, the English character actor famous for his role as Captain Mainwaring in the television series *Dad's Army*, the epitome of bumbling, well-meaning pomposity, died at this age in 1982.
Henry Luce, the American publisher who launched *Time* and *Life* and devised the *March of Time* dramatized newsreel, died at 68 in 1967. The secret of his success was embodied in his statement, 'Character is destiny.' In other words, it is important to stress the personalities in the news and to understand the nature of people who are influencing world affairs.
Max Miller, the last of the great music-hall variety performers, famous for his dirty jokes, which on occasion had him banned from the radio, died in 1963.
Nancy Mitford, the English novelist famous for her books *The Pursuit of Love* and *Love in a Cold Climate*, and for her enjoyment of the anger caused by her championing of the 'U' and 'non-U' class

distinctions, died at 68 in 1973. She made the social-climbing members of the middle class furious to discover that they were giving themselves away with almost every phrase they used, and her classification of 'U' and 'non-U' words was studied avidly. Here are some of the most notorious examples, with the non-upper-class word first in each case: bye-bye/goodbye; dentures/false teeth; home/house; ice-cream/ice; lounge/drawing-room; mirror/looking-glass; pardon?/what?; perfume/scent; perspire/sweat; serviette/napkin; settee/sofa; sweet/pudding; toilet/lavatory; wealthy/rich. Although some of these 'U' words have since been learned by the middle classes, many Mitford distinctions still apply even today.

Fridtjof Nansen, the Norwegian arctic explorer, died at this age in 1930. After his exploring days, he turned to good works and was in charge of the repatriation of 500,000 prisoners after the First World War. He also directed the Famine Relief of Russia in 1921 and was awarded the Nobel Peace Prize in 1922.

Ogden Nash, the American humorist who was the master of deliberately bad verse, died at 68 in 1971. He said he mangled the English language because 'you only tease the things you love'. He was a laughing philosopher who ridiculed the follies and foibles of his *New Yorker* world. His most famous poem, entitled 'Reflections on Ice-breaking' goes: 'Candy is dandy, but liquor is quicker.' In the 1960s he added 'Pot is not.'

Merle Oberon, the Tasmanian-born actress, died of a stroke, in 1979. Her cool, regal bearing made her a dramatic leading lady on both sides of the Atlantic in the 1930s and 1940s.

Joshua Reynolds, the English painter and first President of the Royal Academy, was the most fashionable portrait artist of his day. He painted every celebrated person of his epoch – 2,000 of them in all – and was a sociable, good-natured, well-organized man. He died at this age in 1792.

Dmitri Shostakovich, the Russian composer, died in Moscow in 1975. The greatest composer in Russia during recent decades, his work had a wide range of moods, from sombre solemnity to quirky gaiety. He managed to toe the party line while composing, but had to be rapped on the knuckles on occasion, when officialdom detected some bourgeois elements creeping in.

Stanley Spencer, the eccentric English artist with a haunting personal vision, a kind of primitive Pre-Raphaelite producing meticulously detailed scenes of imagined religious events in his native village of Cookham, died at 68 in 1959. The titles of his strange paintings sum up his oddity: 'Christ preaching at Cookham Regatta'; 'Resurrection, Cookham'. He could be seen, even after he had been awarded a knighthood, pushing his old pram, containing all his painting equipment, through the village streets. A unique artist, imitating no one and with no imitators.

Merle Oberon with Laurence Olivier

Joshua Reynolds, self-portrait

Stanley Spencer, self-portrait

Sixty-nine is the average life-span for the English-speaking male. This figure is a great improvement over those of earlier centuries and looks likely to rise again by the end of the present century. Surveys carried out in the late 1970s revealed that this average life-span figure for men is shared by the UK, USA, Canada, Australia and New Zealand – all the major English-speaking areas.

Life Expectancy In addition to the above, 69 is the life expectancy at birth for males in Albania, Cyprus, France, Taiwan and West Germany, and for females in Albania, Suriname and Venezuela.

Achievement at 69 **Neville Chamberlain** was British Prime Minister in 1938 when, at this age, he returned to England on September 30 having signed the Munich Agreement with Adolf Hitler. The event made him a national hero and he enjoyed the moment as a great personal triumph, but it was all to turn sour in a short period of time, when it emerged that he had been playing the role of a polite, old-fashioned gentleman to Hitler's ruthless gangster.

William (Buffalo Bill) Cody was still making public appearances at this age, in 1915, taking part in fancy shooting and the acting out of the capture of the Deadwood Stage, in his famous Wild West Show. He continued this vigorous routine to within a few months of his death at the age of 70.

Ronald Reagan became the fortieth President of the United States in 1980, at the age of 69. After making films such as *Going Places* (1938), *Dark Victory* (1939), *A Modern Hero* (1940), *The Winning Team* (1950) and *Law and Order* (1953) it should have been obvious that our handsome hero was trying to tell us something. When he finally entered the White House in 1980 we at last knew what it was. In addition to acting, he had also been a lifeguard (when he saved seventy-seven people from drowning), a baseball commentator, a radio sportscaster, a USAAF Captain in the Second World War, President of the Screen Actors Guild, a television host and a State Governor.

Noah Webster, the American lexicographer, was 69 when he published his famous dictionary in 1828. For the American population his work had the same impact as Johnson's dictionary in England. Single-handed, Webster updated the English language and all the differences between modern American spelling and modern English spelling are due to him. He claimed that he was extending and systematizing what was happening naturally anyway. Examples of his changes are: *behaviour* to *behavior*; *centre* to *center*; *colour* to *color*. Quaintly, and illogically, he did not change *Saviour* to *Savior* because he considered the word sacred and therefore untouchable.

Gilbert White, the English clergyman and naturalist, was 69 when he published his classic work, *The Natural History of Selbourne*, an early example of a good observer recording nature as he found it.

Brigham Young, the American Mormon leader, was 69 when he became a father for the fifty-sixth and last time. He had assumed leadership of the Mormon Church in 1844, when its founder was murdered, and agreed to obey the sect's rule about 'plural marriage'. It was felt that polygamy was necessary to obtain salvation and, although Young did not look forward to the idea, considering that it would mean 'great toil and labour for my body', he soon changed his mind and eventually took a total of twenty-seven wives. In the evening he would put a chalk mark on the door of the wife he intended to visit, then eat a large number of eggs to boost his virility, call on her, and later retire to his own room to rest. He was known as the 'Lion of the Lord' and the heart of his harem was called the 'Lion House'. (The Mormon Church officially renounced polygamy in 1890.)

Life-span of 69 **Clarence Birdseye,** the American Frozen Food King, died at this age in 1956, after starting a revolution in eating habits. As a young fur-trader in Labrador, he had watched Eskimos catching fish and letting them freeze solid. He noticed that when they thawed them out months later and cooked them, they tasted as fresh as if they had just been caught. This gave him the idea for quick-frozen food and he started a company to market it, in Massachusetts. Frozen fruits, meats, vegetables and fish were all soon selling well, enabling housewives to shop when convenient and then store food at home in their freezers. In 1955 the company introduced Birds Eye Fish Fingers, which rapidly became their most successful product. Today the British eat 710,000 tons of frozen food each year and the Americans 8,500,000 tons.

William Blake, the romantic, mystical English artist and poet, died at 69 in 1827. As a child, he claimed to see angelic visions and in 1783 he started writing poems. In 1784 he set up shop selling prints

Ronald Reagan in *Law and Order*   Buster Keaton

and engravings and later went on to illustrate all his books with his own hand-painted engravings. His subjects became increasingly mysterious, mythological and symbolic and, although he was largely ignored in his own lifetime, he retained his single-mindedness to the end.

Richard Burton, the English scholar-explorer, died at Trieste in 1890. During his extraordinary life he had not only explored many of the world's remote and dangerous places, but had also been a soldier, a consul, an author, a poet, a botanist, a geologist, an ethnologist, a linguist and a translator. He produced forty-three volumes on his travels and translated thirty volumes from a variety of languages. In his later years he became obsessed with pornography and immediately after his death his wife burned most of his diaries and journals, and his collection of erotica. She then wrote a biography of him, trying to turn him into a good Catholic, a faithful husband, and a refined and modest man for posterity. It is easy to understand her feelings in the prudish atmosphere of late Victorian England, but her destructive act was a literary tragedy.

Elizabeth I, the 'Virgin Queen' of England from 1558 until her death in 1603, was one of the greatest of all English sovereigns. Her reign saw England expand as a major power in politics, commerce and the arts. She inspired both loyalty and awe, and ruled with style, panache, knowledge and appropriate cunning. When she became ill early in 1603, she characteristically refused to take to her bed or to use any medicines. It has since been claimed that her death was caused or hastened by her use of lead-based cosmetics, employed for whitening the face.

Faisal, King of Saudi Arabia, was this age when he was assassinated at his Royal Palace in Riyadh, in 1975. The killer was his 27-year-old nephew, who believed in progress and saw King Faisal as an antiquated stumbling block to the westernization of Saudi Arabia. He shot the King three times in the face as the slim, saturnine monarch leant forward to have his nose kissed in greeting. The nephew, an Arab prince educated at Berkeley in California and familiar with the delights of marijuana and LSD, long hair and freedom, was held prisoner for several months, then taken into a public square in front of the Great Mosque of Riyadh, forced to his knees and decapitated with a single stroke of a huge gold-hilted sword. His head was then displayed on a sharpened stake.

Joyce Grenfell, the English comedienne who specialized in caricatures of dottily nice English ladies, bravely faded aristocrats and well-meaning schoolteachers, lost her fight against cancer in 1979, when she was 69.

Buster Keaton, the expressionless clown of silent films, known as the 'Great Stone Face', who represented the stubborn survival of dignity in chaos, died of lung cancer in 1966.

Marshall McLuhan, the Canadian director of the Centre for Culture and Technology at Toronto University, died at this age in 1980. A communications theorist and media analyst, he became famous for his catch-phrases, such as 'The medium is the message' – expressing his view that the mode of communication is as important as its content.

Nanak, the Indian guru and founder of the Sikhs, who preached a puritanical, monotheistic, tolerant form of Hinduism, and who rejected the caste system and all forms of power, died at 69 in 1538.

Ramon Novarro, the Mexican-born American film star who epitomized the 'Latin lover' and became the romantic idol of Hollywood silent films in the 1920s, second only to Rudolph Valentino, was murdered by two brothers in 1968. They beat him to death trying to find out where he kept his money, but were later caught and jailed for life.

Emmeline Pankhurst, the English leader of the Women's Suffrage Movement who founded the Women's Social and Political Union in 1903, soon discovered that arrests for misbehaviour brought enormous amounts of publicity and thereafter adopted public disruption as a deliberate weapon in her campaign for 'Votes for Women'. She started out by disturbing meetings and chaining herself to railings, then stepped up the pressure with destruction of property. In jail, she adopted the technique of refusing food. Repeatedly arrested and released, her health began to suffer from her series of hunger strikes and eventually, worn out by them, she died at 69 in 1928.

Stephen Potter, the English humorist who invented the suffix '-manship' (as in Gamesmanship, Lifemanship and One-upmanship), died in 1969. He described his approach as 'the art of winning without actually cheating', and gave advice on how to get ahead by exploiting the weaknesses of polite British manners. In the Foster Dulles era his invention became international with the coining of the word 'Brinkmanship'.

Seneca, the Roman statesman who was Nero's tutor, was ordered to commit suicide by his Emperor in AD 65, at the age of 69, which he did with great composure.

Achmad Sukarno, the Indonesian dictator, died in 1970 after being kept under house arrest for several years. A charismatic leader, he had fought for independence from Dutch colonial masters to the extent of welcoming the Japanese in 1943 as liberators. When the Allies showed signs of winning, however, he promptly declared independence, in 1945, and defied the Dutch until they officially handed over power to him, in 1949. His rule gave identity to his people, but suffered from extravagance and corruption.

Richard Wagner, the anti-Semitic, egotistical, disloyal, promiscuous, extravagant sponger, who also happened to be a musical genius, died of a heart attack in the arms of his young wife, in 1883. She clung to his body for twenty-four hours, then cut off her hair to bury it on his heart, in the coffin.

Seventy is the biblical age of man. According to *Psalms 90:10:* 'The days of our years are threescore and ten; and if by reason of strength they be fourscore years, yet is their strength labour and sorrow; for it is soon cut off and we fly away.'

Longfellow says that 'To be seventy is like climbing the Alps. You reach a snow-covered summit, and see behind you the deep valley stretching miles and miles away, and before you other summits higher and whiter, which you may have the strength to climb or you may not. Then you sit down and meditate and wonder which it will be.'

D.H. Lawrence views 70 as a brave age: 'I believe that one has to be seventy before one is full of courage. The young are always half-hearted.' The basis of this curious statement is presumably that, at 70, one has far less to lose, but others would argue that, at 70, caution tempers bravery more than it does with the young.

Emerson takes a more jaundiced view of the mood at 70: 'The man and woman of seventy assume to know all, they have outlived their hope, they renounce aspiration, accept the actual for the necessary, and talk down to the young.'

A happier note is sounded by Oliver Wendell Holmes, in a letter to Julia Ward Howe on her seventieth birthday: 'To be seventy years young is sometimes far more cheerful and hopeful than to be forty years old.' Carl Jung, the great psychiatrist, adds another reassuring comment: 'A human being would certainly not grow to be seventy ... if his longevity had no meaning for the species to which he belongs. The afternoon of human life must also have a significance of its own and cannot be merely a pitiful appendage to life's morning.' The answer, of course, is that in ancient, tribal times, the elderly members of the group were important as story-tellers, passing on the tribal knowledge, legends and traditions to the young; also as helpers in caring for the very young, either as grandparents or elderly relatives. Such roles were vital then, but today have been sadly eroded, leaving many older people unnaturally isolated.

A survey in 1970 revealed that there were 11,765,000 people of 70 or over in the world. Whether they feel defiantly young at this age, or wearily old, the fact is that their physical condition is considerably changed since they were 25, when they were at their peak of muscular vigour. The following alterations have taken place since that earlier age: body water content is down from 61 per cent to 53 per cent; cell solids are down from 19 per cent to 12 per cent; bone mineral is down from 6 per cent to 5 per cent; fat is up from 14 per cent to 30 per cent; body height is down by one inch; the brain is lighter in weight; responses to heat, cold and pain are reduced; body temperature is lower; reflexes are slower; joints are stiffer; metabolic rate is lower; bladder control weakens; pulse rate decreases; blood pressure rises; sense organs are weaker; speech becomes metallic and tremulous; memory fades. Sexually, 27 per cent of males are impotent at this age.

This unhappy inventory of physical decline reflects an inevitable ageing process that overtakes us all, but it still leaves the human being with enormous potential compared with any other species. The range of achievements of human beings aged 70 and over is staggering, even if the brain and the body are not working as efficiently as they once did.

Life Expectancy Life expectancy at birth for males in Spain, and for females in Colombia, Costa Rica, St Lucia and Singapore, is 70.

Late Start at 70 **Robert Broom,** the South African fossil-hunter extraordinary, whose first great task was proving that mammals had evolved from reptiles, set himself a new challenge at 70. In 1936, inspired by Raymond Dart's discovery of the skull of a young man-like ape, he began a search for man's ancient ancestors. Determined to find an adult version of the Dart skull, he was successful within six weeks, searching in the bone caves of the Transvaal. He continued to find more evidence all through his 70s and well into his 80s.

Alfred Wallis, the Cornish artist whose work has become famous in recent years, did not begin to paint until he was 70, in 1925. As a child of 9 he had gone to sea as a cabin-boy; later he became a fisherman; then a rag-and-bone man in St Ives; and finally an odd-job man until he retired with his pension. His wife having died he found himself alone and began to paint the marine scenes he knew so well. He became a virtual hermit, pig-headed, independent, proud, deaf and irascible. The artist Ben Nicholson discovered him and made his work known to the outside world.

Francis Chichester   Copernicus

Achievement at 70 **Copernicus,** the sixteenth-century Polish physician who is regarded as the father of modern astronomy, spent thirty years studying the then revolutionary idea that the sun and not the earth is the centre of the solar system. Although he finished his work in his late 50s, he was afraid to publish because of the hostile reaction that he felt his discovery would produce. In his late 60s he relented and his great study, *The Revolutions of Heavenly Bodies,* started to go through the presses.

The first completed copy was ready in 1543, when he was 70, but the great achievement came almost too late, for Copernicus was already dying. The volume was handed to him on his deathbed, where he lay already paralysed. It was placed in his feeble hands, to assure his fading brain that the work had been finished, and a few hours later he was dead.

Hilda Johnstone achieved the remarkable record of taking part in the Olympic Games at the age of seventy years and five days – the oldest ever British competitor. She was this age when she represented Great Britain in the equestrian dressage competition in the 1972 Games.

Golda Meir became Prime Minister of Israel at 70, on March 17, 1969. Russian-born and American-educated, she founded the Israeli Labour Party in 1967, and went on to become a greatly respected national leader, both at home and (especially) abroad.

Life-span of 70

Hans Christian Andersen, the Danish creator of *The Ugly Duckling*, *The Red Shoes* and *Thumbelina* and many other famous fairy-tales, died a virgin at 70, in 1875. His abnormally spindly body seems to have prevented him from gaining the love of the women he admired. When, as a last resort, he visited brothels he did no more than talk to the naked girls, being disgusted at the suggestion that he should actually make love to them. This may have been partly because he was a hypochondriac, riddled with irrational fears and phobias. He always travelled with a rope, for instance, in case he was trapped upstairs in a fire, and often left a note by his bedside which read 'I only seem dead', in case someone buried him by mistake. He died finally of cancer of the liver, having left instructions that an artery was to be severed before the coffin was closed, to ensure death.

Canaletto, the Venetian landscape painter obsessed by the beauty of the canals of Venice, died at this age in 1768. He was accused of being cold and mechanical by the Romantics, but his works have survived to be seen now as masterpieces of architectural landscape.

Benvenuto Cellini, the Florentine goldsmith and sculptor, was 70 when he died in 1571, after living a life that made him the prototype of the romantic image of the artist – passionate, loyal, egotistical, colourful, violent, but also technically brilliant. He was banished from Florence for fighting a duel and went to Rome. There he claimed he killed the Constable of Bourbon and the Prince of Orange during the sack of the city in 1527. Later he murdered a rival goldsmith, but was forgiven by the Pope until he was accused of stealing papal jewels. He escaped from prison and fled to France, eventually returning home to Florence, where he died.

Raymond Chandler, the American detective novelist who invented the character Philip Marlowe, hated contrived murder mysteries of the classic kind and brought both a new atmospheric toughness to the genre and a harsh sense of dramatic irony. His philosophy is summed up in the quote: 'Down the mean streets a man must go who is not himself mean.' Chandler died at 70 in 1959.

Francis Chichester, the English sailor who was knighted for his round-the-world solo sailing voyage when in his mid-60s, died at 70, in 1972. A small, quiet, modest man, he was stricken with lung cancer shortly before he won his first single-handed transatlantic sailing race in 1960. He had been given only a few months to live, but refused an operation and recovered. He had suffered a bleak and lonely childhood at the hands of his puritanical father. Once, for example, when he had been bitten by a viper, at age 11, his father had sent him off alone on his bicycle to pedal to an infirmary four and a half miles distant. Later, he was sent to Marlborough College which he hated for its appallingly deficient food and its brutality – its beatings, deprivations, cold, and petty restrictions. Eventually he simply told his form-master that he was leaving and went – emigrating to New Zealand at 18. It was the kind of determination that was to see him through his major voyages years later.

Honoré Daumier, the French artist, died in poverty, in 1879. Always a savage critic of society in his drawings and paintings, he was hated by the very people who enjoyed buying pictures, hence his penury. One of his cartoons enraged the King – Louis Philippe – so much that the artist was jailed for six months as a punishment.

Galen, the Greek physician who founded experimental physiology and influenced medical theory and practice for hundreds of years, died in AD 199.

Giotto, the Tuscan painter, sculptor and architect who was the founding father of the Renaissance, died at 70 in 1337. His fame rests on his daring break with formal Byzantine traditions. He gave his figures solidity and individuality – his faces had expressions that were not based on rigid traditions. He was the first European artist to allow nature and emotion to enter his works of art. In this he is the forerunner of most European art since his day.

Thomas Henry Huxley, the English scientist famous for his defence of Darwin's Theory of Evolution, died of a combination of influenza and bronchitis, in 1895.

James Robertson Justice, the Scottish actor who became famous playing bluff, booming bullies in a long series of film comedies, died in 1975. Unusually for a comedy actor, he had a doctorate from Bonn University and spoke ten languages including French, German, Italian, Dutch and Gaelic.

Thomas Rowlandson, the English master of caricature, died in 1827. It is thanks to the fact that he was a compulsive gambler that we today can enjoy his wonderfully humorous watercolours. He started as a minor, serious artist until his gambling lost him his inheritance and he was forced to make 'popular' works to survive, often producing his pictures on the spot to pay his debts.

71

Seventy-one is the year when people start to boast about their age, according to Irish-American humorist Finley Peter ('Mr Dooley') Dunne. Now that they are into their 70s they feel they have a right to brag about the fact that they have outlived the biblical three-score-years-and-ten. They cease to worry about the depressing aspects of growing old and begin to take pride in their staying power.

Some set out to prove that their staying power at 71 goes far beyond mere survival. In 1910, the American marathon walker Edward Weston was this age when he walked across the North American continent, from New York to San Francisco, a distance of 3,895 miles in 105 days. Four years later he walked 1,546 miles in fifty-one days. His secret, he said, was to be in bed by 2 am every night and up at 8 am, keep food simple, smoke and drink only in moderation, and walk as much as possible every day. It certainly worked for him. After his great walk at 71 he lived on for another nineteen years.

Life Expectancy Life expectancy at birth for males in Denmark, Holland, Israel and Switzerland, and for females in Argentina, Cuba, Fiji, Kuwait and Malaysia, is 71.

Late Start at 71 Leni Riefenstahl, the prize-winning documentary film-maker, took up scuba-diving at this age. Famous for her 1930s films *Triumph of the Will* and *Olympia*, she became Hitler's favourite film director and was an important figure in his propaganda machine. After the war she was imprisoned by the French for four years. Later, she attempted to revive her film career, but without success. Frustrated, she turned to still photography and produced several stunningly beautiful books about Africa. It was while she was there that she discovered, for the first time in her life, the magic of the underwater world, at Malindi in Kenya. In order to take a course in scuba-diving she had to lie about her age, putting her date of birth as 1922 instead of 1902. Even at '51' she was the oldest 'learner' in the group, and at her true 71 she would have been considered too great a risk to be accepted for training. With underwater cameras she was soon at work photographing marine life and the result was a book of amazing photographs called *Coral Gardens*.

Misfortune at 71 Sarah Bernhardt, the great French actress known as the 'Divine Sarah', who had an enormous range, playing everything from comedy to classics and melodrama, including male parts such as Hamlet, suffered an injury to her right knee during a tour of South America. She damaged it when jumping off the parapet in the last scene of *La Tosca* and the leg had to be amputated in 1915 when she was 71. This did not stop her from undertaking a sixteen-month tour of the United States in the following year, although it did reduce her acting range. To compensate for this she had new parts written specially for her, with her infirmity in mind.

Douglas MacArthur, the American general, who had been summoned back from retirement in 1941 when he was 61 and had achieved greatness as Allied Commander for the South-West Pacific in the Second World War, suffered a major insult at the age of 71. On April 11, 1951, he was relived of his command (of the United Nations' forces in Korea) by President Truman. He was sacked because of his increasing insubordination and unwillingness to wage a limited war. After a much publicized Senate investigation of his dismissal, his popularity waned.

Henri Matisse, the great French artist, was 71 when he underwent an intestinal operation in 1941. As a result, he became bedridden, but insisted on continuing to paint, and created many beautiful works in his final years. He was often forced to work from his bed with the aid of a crayon attached to a long bamboo pole but, despite his infirmity, his late works were often among his most daring and serenely optimistic.

Auguste Renoir, a slightly earlier genius of French art, also became severely handicapped at 71. In his case the problem was rheumatoid arthritis, which had started years earlier and gradually worsened until, in 1912, he was quite unable to move his gnarled and cruelly twisted fingers. His legs had also become paralysed and he was confined to a wheelchair. But, like Matisse, he refused to give up his work. He continued to paint even though he could no longer grasp the brush. Instead he had to slide it in between his crippled fingers, or strap it to his hand. To add to his discomfort he was also suffering from dyspepsia, bronchitis and a hernia, but he never stopped painting and never lost his cheerfulness, producing a total of 6,000 paintings during his lifetime, mostly of luscious, voluptuous nudes. Despite his physical condition, he was a painter of joy, warmth, colour, fecundity, light and beautiful human flesh.

Life-span of 71 Louis Armstrong, the greatest personality to emerge from American jazz, died of heart failure at this age in 1971, two days after his seventy-first birthday. His personal story was one of triumph over exploitation. He began as a traditional New Orleans jazz trumpeter, then horrified purists by allowing himself to be swept up into big-time Swing. Later, Hollywood encouraged him to commercialize fake-jazz in films that were often condescending and depicted him as a laughing clown with a trumpet. But his personality was so huge that he eventually transcended this imposed artificiality and rose to great heights as a much loved international entertainer, bigger than the music he played.

Enid Blyton, the immensely prolific writer of children's books who made a fortune from her work but who remained publicity-shy and reserved, died at 71 in a Hampstead nursing home in 1968. Altogether she wrote 400 books, often completing one in a single week, working in a trance-like state on the verandah of her house, with a portable typewriter on her knees and a shawl around her shoulders. Her writing was much loved by children but often much hated by adults. Noddy, Big Ears and Mr Plod in the village of Toyland came in for particular criticism. Noddy was once referred to as 'unwholesome, wet and odious'.

Daniel Defoe, the author of *Robinson Crusoe*, died at this age in 1731. It was recorded, rather oddly, that he 'died of a lethargy' in Ropemaker's Alley, Moorfields, London, on April 26.

William Laud, the much hated Archbishop of Canterbury who was religious adviser to Charles I, was beheaded on Tower Hill in 1645. He was a religious persecutor by nature and eventually came to be accused of high treason, but not before his repressive policies had done a great deal towards precipitating the Civil War in England.

Piet Mondrian, the Dutch abstract artist, died of pneumonia in 1944. His hard-edged, purist abstractions had a major influence, not only on twentieth-century art, but also on architecture, advertising, decoration and typography. He worked almost entirely in vertical and horizontal lines, using three primary colours.

Cole Porter, the American song-writer famous for such pieces as *Anything Goes, I Get a Kick Out of You, Night and Day*, and *Don't Fence Me In*, who was active from the 1920s to the 1950s, died in 1964.

Nicolas Poussin, the French grand master of stiltedly perfected composition, over-careful balance and super-controlled classicism, died at 71 in 1665. His theatrically gesturing figures combine in frozen artificiality that many find unappealing, but his technical brilliance cannot be denied.

Socrates, the Greek philosopher, was accused of impiety in 399 BC, because of his 'corruption of the young' and his 'neglect of the Gods whom the city worships' and 'the practice of religious novelties'. Sentenced to death, he 'drank the hemlock', spurning an escape that had been planned for him, on the grounds that the verdict, although contrary to fact, was that of a legitimate court and must, therefore, be obeyed. A man of grotesque appearance, short, stout and with prominent eyes, a snub nose and broad nostrils, he was nevertheless said to be 'all glorious within'. His dictum was 'know thyself', and he directed philosophical thought towards the analysis of human conduct and character.

Erich von Stroheim, the bullet-headed Austrian film actor and director, who specialized in aristocratic villains but was in reality the son of a Jewish hatter, died of a spinal ailment at this age in 1957. In early films he was cast as a cruel Prussian with a snarl and a monocle – the man you love to hate – but after the First World War he became a wildly extravagant film director of vast epics which always went so far over budget that they had to be taken out of his hands. This happened five times, largely because of his obsession with detail. He once insisted that his actors wore underwear appropriate to the period of the film, even though these garments were never seen on screen. In the end he had to return to acting, making a brilliant appearance with his old star from the early days, Gloria Swanson, in *Sunset Boulevard*.

Suleiman the Magnificent, the Sultan of Turkey, who brought the Ottoman Empire to its peak of success, expanding it into Europe and the Middle East, while his great fleet dominated the Mediterranean, died at 71 in 1566. At home he was said to be enlightened and progressive, reforming laws and erecting great mosques and other buildings, but in his family life he revealed his ruthless streak. He had fallen in love with a beautiful slave girl called Roxalana, and under her influence he had had his own son, his legitimate heir, strangled by a bowstring so that her son could succeed him.

Maurice Utrillo, the French artist who had been a lifelong alcoholic, nevertheless survived to this age, dying in 1955. Only the success of his paintings kept him going. A less creative man would probably have succumbed much earlier to his level of alcohol intake.

Tennessee Williams, the American playwright famous for his violent, emotional plays such as *A Streetcar Named Desire* and *Cat on a Hot Tin Roof*, was found dead in his Manhattan hotel suite in February 1983. His brother, who had had him committed to an institution in 1969 for a drug cure, was reported as saying that the playwright had probably returned to drug use again, shortly before his death. Although his earlier plays were immensely successful and won many prizes, his later works had failed badly and his last play, *Clothes for a Summer Hotel*, closed after a miserably short run on Broadway in 1981.

Louis Armstrong

Maurice Utrillo

Seventy-two is the age of prudence, according to Oliver Goldsmith, in *The Vicar of Wakefield* (1766): 'One virtue he had in perfection, which was prudence – often the only one that is left to us at seventy-two.'

Maurice Chevalier was this age when he was asked how it felt to be so old. His reply: 'Old age isn't so bad when you consider the alternative.'

Life Expectancy Life expectancy at birth for males in Greece, Norway and Sweden, and for females in Hungary, Portugal, Romania, Uruguay and Yugoslavia, is 72.

> Among the animals, 72 is the age of the oldest known bird – an Andean Condor.

Romance at 72 **The Marquis de Sade** was this age when he took his last mistress. Her name was Madeleine Leclerc and she was only 12 when he first met her, three years earlier, and became enamoured of her charms. Then, although he was incarcerated in an asylum, he managed to contrive that, at 15, she became his mistress, promising her mother that he would somehow launch the girl as an actress. The only detail of their relationship that has survived is that she was persuaded to shave off her pubic hair for him, presumably to make her appear even younger and to give him the thrill of one last sexual aberration. He certainly had no great opinion of himself at this stage in his life, for the last request in his will, when he died two years later, in 1814, was that: 'The ground over my grave shall be sprinkled with acorns so that all trace of my grave shall disappear so that, as I hope, this reminder of my existence may be wiped from the memory of mankind.'

Late Start at 72 **Dame May Whitty**, the British actress, after a long and distinguished stage career in England, made her first Hollywood film at the age of 72, in 1937. Her late début was in *Night Must Fall*, and she stayed on in Hollywood for eleven more years, enjoying her second career immensely. She came to stand for cheerful dignity and kindly common sense. Her greatest triumph was in *The Lady Vanishes*, for Hitchcock in 1938.

Achievement at 72 **Blondin**, the French acrobat and the most famous tightrope walker in the world, gave his last performance in 1896, in Belfast. He had been making his tightrope walks more and more difficult as the years went by. After crossing the falls at Niagara on a rope 1,100 feet long, 160 feet above the water, he repeated the act blindfolded pushing a wheelbarrow; then, later still, carrying a man on his back; then on stilts. After a period of retirement, he reappeared to perform his strangest ever act – carrying a stove to the middle of the line (in a display at Staten Island) and cooking and eating an omelette there. His final performance in Belfast at 72 was made at a time when he was both wealthy and famous and had no need to work. By this age, the dangerous act of walking a rope had become a personal challenge and an obsession.

Colette, the French novelist who wrote seventy-three books on the pleasure and pain of love, was 72 when she published her popular work *Gigi*, in 1945.

Robert Graves, the English poet, novelist and critic, was also this age when he published his controversial translation of the *Rubáiyát* of Omar Khayyám, with Omar Ali-Shah, in 1967.

Martin Heidegger, the German philosopher, was this age when he published *Being and Time*, his magnum opus, in 1962.

Jomo Kenyatta, known respectfully as 'Mzee' (Grand Old Man), was 72 when he became Prime Minister of Kenya on December 12, 1963. He then became the first president of his country and went on to live to a ripe old age and to govern his people with stability and common sense.

Jean de La Fontaine, the French poet, completed his masterpiece in 1694, at the age of 72, not long before his death. His great *Fables* – 230 of them, in twelve books – were largely based on those of Aesop, and he had been working on them for over a quarter of a century.

Konstantin Stanislavsky, the Russian founder of the Moscow Art Theatre best remembered for his 'method' acting system, was 72 in 1935 when he fulfilled a forty-year quest with the discovery of what he regarded as 'the unbreakable psycho-physical unity of human experience and its external expression'. Using his discovery, he developed his 'method of physical actions' – an analysis leading to the synthesis of experience and expression and the means to 'organized' emotions. He immediately began an experimental production of Molière's *Tartuffe*. His was the most influential theory of acting in the twentieth century and flourished later in the USA, especially in New York.

Life-span of 72 **Douglas Bader**, the legless British air ace of the Second World War, died at this age in 1982. He had been a war prisoner from 1941 until his release by the advancing Americans in 1945. In 1956 his story was made into a film called *Reach for the Sky*, with Kenneth More playing Bader.

Confucius, the Chinese teacher, philosopher and political theorist, died in 479 BC. One of the most influential men in world history, he was concerned with human ethics rather than religion. Confucianism is often referred to as a religious system, but God and the afterlife are missing from his teachings. His main concern was that people should honour natural relationships with mutual respect. His golden rule was: 'What you do not like when done to yourself do not do to others' – an idea that resurfaced much later on in Christianity.

Aleister Crowley, the British occultist who liked to be known as the 'Wickedest Man in the World', died at 72 in 1947. He devised his own brand of Satanism which attracted a group of bizarre eccentrics and involved various rituals – opium, cocaine and hashish were used, and there were 'Sacred Orgies'. His followers had to deny 'self' and were urged to gash their arms with razors whenever they inadvertently used the word 'I'. As a result of his magico-sexual ceremonies, several of his concubines ended up in hospital and a child died.

Bud Flanagan, the English comedian and the best-loved member of the Crazy Gang, died at 72 in 1968. Despite his Irish stage name, he came from a family of Polish Jews who arrived in London in 1896. During the depression years he wrote his famous song 'Underneath the Arches'.

D.W. Griffith, the American pioneer film director who has been described as 'the single most important figure in the history of American film' died a bitter, lonely, forgotten man at the age of 72 in 1948, after suffering a cerebral haemorrhage in a Hollywood hotel room. His last film had been made in 1931 and, although he was given a special Oscar in 1935 for his 'lasting contributions', he faded rapidly from the scene, unable to adapt to the latest trends in the medium which he had influenced so profoundly.

Sacha Guitry, the French playwright, who dominated the French theatre for a generation – the author of over 100 plays, also actor, producer and manager – died at 72 in 1957.

Barbara Hepworth, next to Henry Moore Britain's greatest modern sculptor, died tragically in a fire at her Cornish home in 1975.

Henry James, the Anglophile American author who spent the second half of his life in England, writing his tortuous and meticulously elaborate prose observations of human conduct, died at 72 in 1916. The year before his death he became a British citizen as an act of sympathy for the Allied cause in the First World War.

John Locke, the English philosopher whose great work *Essay Concerning Human Understanding*, was published in 1690, died at 72 in 1704.

Louis B. Mayer, the American film tycoon and head of MGM, ruthless, tyrannical, quick-tempered and anti-intellectual, died at 72 in 1957. A large crowd attended his funeral because, it was said, 'they wanted to make sure he was dead'. Throughout the 1930s and 1940s he provided the masses with escapist entertainment, enjoying immense financial success. He was the highest-paid individual in the whole of America during this period, earning well over $1,000,000 a year.

Malcolm Sargent, the most popular British orchestral conductor of his day, died at 72 in 1967. At 24 he had been the youngest ever Director of Music in England, and he was still actively working at the end of his life.

Arthur Schopenhauer, the German philosopher, died of a lung haemorrhage in 1860. His father committed suicide when Schopenhauer was only 17 and his mother rejected him and went to live with a young man whose company she preferred to that of her own son. When he himself became a lover he caught syphilis. Not surprisingly, he was the philosopher of pessimism, stressing the existence of human suffering. He blamed women for the emotional chaos caused by their sexuality and asked how men could give the name 'fair sex' to 'the undersized, narrow-shouldered, broad-hipped and short-legged race'.

Gertrude Stein, the American writer whose mannish appearance and eccentric style ('a rose is a rose is a rose') made her one of the personalities of the Paris avant-garde scene between the wars, died during surgery for cancer at this age in 1946. She was a lesbian who lived with her friend Alice B. Toklas for thirty-eight years. Her last words, spoken to Toklas as she was about to be taken in for her operation, were characteristic of her style: 'What is the answer?' she asked. Toklas said nothing. 'In that case,' she said, 'what is the question?'

Algernon Swinburne, the English poet who died at 72 in 1909, was a puny-bodied, large-headed, red-haired, falsetto-voiced masochist with an obsession for flagellation. He acquired the sexual taste for being whipped when he was at school at Eton and, as an adult, he attended flagellation brothels where it was his aim to be flogged by beautiful women. Many of his poems, and also his lifestyle, were attacked as obscene and debauched by the critics of his day, and *Punch* referred to him as 'Swine Born'. Rossetti once paid a whore to seduce him, but she failed. Despite his dissipated life and his aberrant obsessions, his writing often displayed great imagination and perception.

John Wayne, a football player turned actor, whose real name was Marion Morrison, died of cancer in 1979. Described as the Ultra-American, Super-patriot, Arch-hawk Fundamentalist, he was the President of the Motion Picture Alliance for the Preservation of American Ideals. He also happened to be a hard-working film star who made 250 films between 1927 and 1976, although he never gave the impression that play-acting was a suitable occupation for a 'real man'. He had suffered and survived major operations during his later years, having a lung removed in 1963, open-heart surgery in 1978, and then, finally, having his stomach removed in 1979. After his death he was accorded the unusual honour of having a Congressional Medal struck in his memory.

Walt Whitman, the first great, truly American poet, freeing himself from the 'tenacious and stifling anti-democratic authorities of Asiatic and European past', wrote for the 'common man'. Despite this, his greatest appeal has been to the American intellectual. A stroke paralysed him in 1873, he retired in 1884, and died at 72 in 1892.

Seventy-three is the average life-span of Americans when both sexes are considered together. A survey in 1980 gave the figure as 73·6 years, calculated from 69·8 for males and 77·5 for females.

Although many consider this to be an advanced age suitable only for passive retirement, some 73-year-olds have different ideas. Dr Benjamin Duggar, Professor of Mycology at Wisconsin University, after being retired from his post as too old, was offered a job as head of a large research department at one of America's biggest pharmaceutical companies. He took it eagerly and at the age of 73 made a major discovery that has since saved thousands of lives and shortened the illnesses of countless more. It was at this age that he isolated from a sample of Venezuelan soil the mould *Streptomyces aureofaciens*, from which it was possible to prepare the life-saving antibiotic drug 'aureomycin'. Seventy-three is not a year to be ignored, even in terms of human inventiveness.

Life Expectancy

The life expectancy at birth for males in Iceland and Japan is 73·4. This is the highest figure for males anywhere, and for those seeking a 'longevity habitat', it is worth pointing out that Iceland also holds the record for female life expectancy (79). What it is about Iceland that allows inhabitants of both sexes a long life-span remains to be discovered.

The life expectancy at birth for females in Bulgaria, Cyprus, Eire, Italy, Luxembourg and Malta is 73.

Achievement at 73

Konrad Adenauer, the German political leader, became Chancellor of Germany at this age in 1949 and remained in office until 1963, leading his country's economic recovery at an age when most people have long since retired. His stubborn common sense was largely responsible for West Germany's 'economic miracle'.

Barbara Woodhouse, English dog-trainer extraordinary, who achieved fame late in life through her teaching sessions on television, when she intimidated dog-owners even more than their pets, was 73 when she set off on a global dog-training crusade early in 1983. Visiting all the continents, with special stops in Japan and Australia, she uttered one of her classic comments on the eve of her departure, to the effect that 'The Japanese don't seem to know much about obedience.' Such a remark would come as a shock to any student of Japanese customs, unless they realized that Barbara Woodhouse always speaks in technical dog-jargon, where 'obedience' is short for 'canine obedience training'. In the same way, her advice that, if you want to reward a male, 'scratch him between the legs', was meant to apply only to male dogs.

Life-span of 73

Casanova, the Italian adventurer and libertine, died at this age, in 1798, after spending the last thirteen years of his life as a librarian in a castle in Bohemia, enjoying good food and good books, but no longer virile enough to continue his amatory affairs. His memoirs, 4,545 pages long, recount his sexual pleasures with a total of more than 132 women, the youngest only 11 and the oldest more than 50. His conquests included 24 female servants, 22 children under 16, 18 gentlewomen, 15 royal ladies and 2 nuns. Among other feats he lovingly recorded for posterity were the deflowering of 31 virgins and making love to one woman twelve times in a single day. His secret, he said, was that: 'There was not a woman in the world who could resist constant attentions.' But it would be wrong to portray him as a callous seducer, for he once remarked: 'Without love this great business is a vile thing.'

Coco the Clown, whose real name was Nikolai Poliakov, died at 73 in 1974. He started work in England in 1929 with Bertram Mills Circus and appeared annually for over forty years, with his big boots flopping and his red hair suddenly standing on end, giving endless pleasure to millions of children.

Noël Coward, the British playwright, actor, composer, novelist, director, singer, producer and professional wit, died in 1973, of a heart attack. His sophisticated, staccato style made him the toast of the London stage in the 1920s and 1930s, and he was responsible for the best of the smart comedies of the inter-war years. Although he appeared in films for over fifty years, he was highly selective in the roles he took, appearing in only eleven productions during all that time. He won an Oscar in 1942 for his war drama, *In Which We Serve*, and was knighted when he was 70.

Joan Crawford, the domineering, long-suffering heroine of many Hollywood melodramas, died at 74 in 1977, of a heart attack.

Bing Crosby, the most popular crooner of the 1930s, who sold 300,000,000 records of his songs, also died of a heart attack in 1977. His relaxed style – sleepy, dreamy, good-natured and modest – made him the second richest (to Bob Hope) entertainer in the world, with a personal fortune estimated at between $200,000,000 and $400,000,000. He suffered his heart attack after completing a round of golf and his last words were: 'That was a great game of golf, fellers.'

Charles Darwin also died of a heart attack at this age, in 1882. With his book *The Origin of Species*, he effectively buried the church and, rather surprisingly, the church returned the compliment and buried him – in Westminster Abbey.

Vittorio de Sica, the outstanding neo-realist Italian film director, who won four Oscars, including two for *Shoeshine* and *Bicycle Thieves*, his most famous films, died in 1974. In later years he became internationally popular as an actor, for usually lightweight comedy supporting roles, emanating worldly charm and ironic affability. He directed a total of twenty-five films and acted in over 150, before succumbing to surgery for the removal of a cyst from his lungs.

Bela Lugosi

Alexander Fleming, the Scottish physician who saved countless lives through his discovery of penicillin, lost his own with a sudden heart attack at his Chelsea home in 1955.

Antonio Gaudí, the extraordinary Spanish architect, was struck down by a trolley car on his way to Vespers and died later of his injuries, in 1926. The huge, unfinished Church of the Holy Family towers into the Barcelona sky, a monument to his idiosyncratic architectural originality.

El Greco, the Cretan-born painter christened Domenikos Theotocopoulos, who spent most of his adult life at Toledo in Spain, died at 73 in 1614. His extraordinarily intense religious masterpieces owe their unique style to the three influences in his life: first, the impact of Byzantine icons from his early years; second, the teaching he received from the aged Titian when he stayed in Venice as a young man; and third, the fanaticism and passion of Spanish Christianity, encountered in his mature years.

John Grierson, the founding father of the British documentary film movement, died at 73 in 1972. It was he who coined the word 'documentary', when writing about the early films of Robert Flaherty which made a great impact on him. At the start of Grierson's career, he made a statement which summed up his life's goal and its achievement: 'The motion picture can open for us a window on the world.'

Ibn Saud, the Muslim leader who united central Arabia and formed it into the modern state of Saudi Arabia, died in his sleep in 1953. It was he who initiated the exploitation of Arabian oil.

Omar Khayyám, the Persian poet and mathematician famous for his *Rubáiyát* poems, died at 73 in 1123. He was one of the great mathematicians of his period, reforming the calendar and making important advances in both algebra and astronomy.

Bela Lugosi, the Transylvanian-born American master of the horror film, died at this age from the effects of his increasingly severe drug addiction, in 1956. He played Dracula with such success in Tod Browning's 1931 film that he became typecast as the aristocrat of evil, the personification of the mad scientist or the dangerous megalomaniac, in dozens of Hollywood horror films. His Dracula image even invaded his private life and he was known to give interviews lying in a coffin. He was buried with his Dracula cape.

Billy Smart, the genial, cigar-puffing, twenty-stone circus showman, died in 1966 after collapsing while conducting the band at his circus zoo. The Billy Smart Circus was the most famous in Britain for many years.

Karl Walenda, another great circus figure, also died at this age, in 1978. He founded the Great Walendas, noted throughout Europe for their seven-man pyramid. There were many accidents in his daring aerialist troop, and he himself died when, at the age of 73, he attempted a walk between two hotels in Puerto Rico and fell to his death when winds exceeded 30 mph.

William Butler Yeats, the Irish poet and playwright, died of myocarditis, in 1939. He published his first book in 1889 and soon became one of the leaders of the Irish literary movement. He was deeply involved in the setting up of the Abbey Theatre in 1904. His plays and poems were strongly influenced by an Irish mysticism and in 1893 he wrote *The Celtic Twilight*, a title which has since passed into common speech. He was awarded the Nobel Prize for Literature in 1923.

W. B. Yeats

Seventy-four has been described as the year of the elderly who still scorn old age. When they reach the fast-arriving year of 75 they will finally have to accept that they are in the 'aged' category, but for the moment they resist this and resent any attempt to lump them with the more advanced group facing true senility.

Their scorn of old age is probably their best weapon against it, as this quote from Trader Horn emphasizes: 'Death is like any untamed animal. He respects a scornful eye.'

Life Expectancy
Life expectancy at birth for females in Czechoslovakia, East Germany, Poland, Taiwan and the USSR is 74.

Achievement at 74
Jean Cocteau, the French poet, painter, novelist, actor and film director, was busily decorating the Church of Saint-Blaise-des-Simples in Milly-la-Forêt at this age, but he did not live to see the task completed, dying while working there in 1963. His painter son continued with the work after his death. Jean Cocteau is best remembered today for his two brilliant films, *La Belle et la Bête* and *Orphée*.

Franz Liszt, the Hungarian composer and virtuoso pianist, gave his last performance at this age, in Luxemburg on July 19, 1886. Two days later he arrived at Bayreuth for the festival, but was forced to take to his bed with a high fever. His illness developed into pneumonia and he was dead by July 31. This ended a remarkable concert platform career spanning sixty-five years, during which he was hailed as 'the greatest virtuoso pianist of all time'.

Daniel Malan, the South African politician, achieved the status of Prime Minister in 1948, following the downfall of Smuts. Many people today regret his rise to power, for it was he who instituted the much criticized policy of *apartheid*. It is strange to think that, by calling, Malan was a Christian minister who must, in his time, have preached the message 'Love thy neighbour'.

Thomas Love Peacock, the English author of satirical novels in which conversation dominated plot, published *Gryll Grange* at this age, in 1860. It shows a serenity absent from his earlier works and reflects the mood of a man who has ceased to worry about political problems.

Jack Wingfield, the English athlete, took part in his last marathon run at this age, in 1982, the Robin Hood Marathon at Nottingham. He had been a long distance runner for fifty years, having won a bronze medal for the marathon event in the 1930 Empire Games. When asked for the secret of his unusual stamina at 74, he replied simply: 'Always been a steady sort of fellow and looked after myself.'

Life-span of 74
Armand Denis, the Belgian film-maker and naturalist, one of the pioneers of televized natural history, died in Nairobi, Kenya. With his wife, Michaela, he made a long series of *On Safari* programmes and in 1954 they won the award for the best television documentary.

André Derain, the French painter, who started out as a young avant-garde artist associated with both Fauvism and Cubism, but then later in life reverted to a more traditional, realist style, so that for some critics he had 'dried up', died at 74 in 1954.

Friedrich Engels, the German socialist philosopher and partner of Karl Marx in the famous *Communist Manifesto* of 1848, died of cancer at 74 in 1895. The irony of his life was that it was only as a capitalist textile manufacturer that he could make enough money to finance Marx and support himself in their crusade against capitalism.

Uffa Fox, the English marine architect, died in 1972. He was the pioneer of the modern racing dinghy and introduced the V-shaped hull in 1928 which revolutionized dinghy racing. He went on to design many other vessels, including the parachuted airborne lifeboat that saved countless lives when airmen were shot down in the Second World War.

Jean Honoré Fragonard, the master of French rococo painting, died at 74 in 1806 after completing a total of more than 550 paintings and several thousand drawings, all bathed in a delicate hedonism. The French Revolution ruined him and eliminated most of his rich patrons, so that he died in obscurity and his work was not appreciated again until many years later.

Frederick the Great, who was King of Prussia for nearly half a century, died from a chill caught when reviewing his troops in 1786. Although he had gained great honour and glory during his lifetime, as an old man of 74 he was a sad sight, known as 'Old Fritz', habitually wearing the threadbare blue coat of the Prussian uniform, soiled with snuff, his body haggard and bent, racked with rheumatism and other afflictions contracted over the years of stress and harsh living conditions. He supported himself with a silver-handled walking stick and carried a flute which he played until his teeth fell out.

Guiseppe Garibaldi, the Italian patriot whose dominant goal in life was the unification of divided Italy under Italian rule, died in self-imposed isolation on the island of Caprera. A great revolutionary in his younger days and one of the most skilful guerrilla generals of all time, he became a recluse in his later years, crippled by rheumatism and by his many wounds. On the barren island of Caprera, he grew his own food and had little regard for the ordinary comforts or conventions of life. He died at 74 in 1882.

W.S. Gilbert, famous for his collaboration with composer Arthur Sullivan, which produced the best-loved comic operas of the Victorian era, died at this age in 1911, when he suffered a heart attack during an act of heroism – he was attempting to

Handel

Stan Laurel

rescue a woman from drowning on his country estate. His dying words were: 'Put your hands on my shoulders and don't struggle.'

George Frederick Handel, the German-born composer who took up permanent residence in England in 1712, died there in 1759, after a series of misfortunes in his later years. In 1737 he suffered a mild stroke; in 1750 he was seriously injured in a coach accident; in 1751 he became partially blind and underwent eye surgery; in 1753 his sight was virtually gone. Despite this he maintained his interest in music until the very end of his life, attending a performance of his *Messiah* only two weeks before his death.

Stan Laurel, the Lancashire-born American film comedian who specialized in 'featherbrained naïvety' that made his audiences feel sympathetically superior, died of a heart attack at 74 in 1965. In 1926 he teamed up with Oliver Hardy to produce the best-loved comedy duo in the history of cinema. After Hardy's death at the age of 65, Laurel never acted again.

Fernand Léger, the French mechano–cubist artist who was deeply influenced by industrial technology and produced crude, boldly-coloured paintings, died at this age in 1955.

Basil Liddell Hart, the world's leading military historian and strategist, whose slogan was 'If one

Guiseppe Garibaldi

Nehru

wishes peace one should understand war', died at 74 in 1970. His two main concepts were the 'Expanding Torrent' (spearheads reinforced by immediately available reserves) and the 'Indirect Approach' (unexpected outflanking and striking at nerve-centres rather than the obvious main body of the enemy).

Jessie Matthews, Britain's greatest musical star of the 1930s, famous on both sides of the Atlantic, and one of the most graceful of the popular dancers of her epoch, died at this age in 1981.

Nehru, the first Prime Minister of India after the country became independent, died in 1964. A follower of Gandhi's Nationalist Movement, he was imprisoned repeatedly by the British in the 1920s and 1930s, but despite this ruled with restraint when he came to power in 1947, maintaining a balanced neutrality between East and West.

Herbert Read, the English art historian, poet and critic, who was the main interpreter of modern art for England from the 1930s to the 1960s, died at 74 in 1968. He was the gentle but stubborn defender of all forms of experimentation in twentieth-century art and was instrumental in establishing the Institute of Contemporary Arts in London, which he referred to as 'an adult play-centre'.

Tiepolo, the Venetian painter, also died at this age, in 1770. The most impressive of the rococo artists in Italy, he decorated palaces with gigantic frescoes of historical and mythological subjects, and could justly be described as the world's most brilliant decorator.

Mark Twain, the American author, whose real name was Samuel Clemens, died at 74, fulfilling his long-standing prediction that he had come into the world with Halley's Comet in 1835 and would go out with it, in 1910. His famous pen-name 'Mark Twain' meant 'two fathoms deep' and was a call used on riverboats when sounding water depths. He spent most of his early life on the great southern rivers, the Missouri and the Mississippi, and was able to draw on his personal experiences for his most famous writings, *Tom Sawyer* and *Huckleberry Finn*, which brought an earthy freshness to American literature. He doted on his wife, a highly respectable semi-invalid, and led a sexually simple life, for which he compensated by writing copious obscene poems, ballads and essays.

Clifton Webb, the waspish, punctilious, elegant, narcissistic snob of many films from the 1940s to the 1960s, died of a heart attack in 1966. His great achievement as an actor was to make the characters he played attractive despite their arrogance and pomposity.

Walter Winchell, the American broadcaster and journalist, died at 74 in 1972. For over forty years his syndicated newspaper articles formed the most read gossip column in America – full of inside information, candid comments, underworld revelations, show business talk and political chit-chat. He once said, 'Other columnists may print it – I make it public.'

Seventy-five is the start of the period of 'confirmed senescence or primary old age'. It ends the period that is called 'quiescent pre-senescence'.

At this age, 50 per cent of males suffer from erection impotence. The percentage of body fat, which has been increasing steadily with advancing age, at 75 is double that in the young adult. The incidence of accidental death shows a dramatic rise at this point.

Quotes from the famous reflect these facts of physical decline. Samuel Johnson: 'My diseases are an asthma and a dropsy, and what is less curable, seventy-five.' Horace Walpole: 'Nobody grows stronger at seventy-five.' Marlene Dietrich: 'Say I am seventy-five and let it go at that.'

One group who have preferred to ignore the weakening effects of being 75 are the Roman Catholic bishops. Alarmed that they might be so old for their jobs as to be making serious errors of judgment, the Pope issued an instruction recently insisting that all bishops must retire at the age of 75. They are, however, allowed to take part in Papal elections until they are 80.

The most famous remark, on reaching this age, comes from Winston Churchill. Interviewed on his seventy-fifth birthday, he said, 'I am ready to meet my Maker. Whether my Maker is prepared for the ordeal of meeting me is another matter.'

Life Expectancy Life expectancy at birth for females in Austria, Belgium, Greece, Israel, New Zealand, the United Kingdom and West Germany, is 75.

Achievement at 75 'Foxy Grandpa' Ed **Delano** was this age when he made up his mind to travel from his home in California to a reunion meeting in Worcester, Massachusetts. It was the fiftieth reunion of the Worcester Polytechnic Institute. His friends were worried that the trip might be tiring for him, but he arrived safely, having travelled the 3,100 miles in thirty-three and a half days – by bicycle.
Charles Stolfus, a retired mechanic living in Kansas, also defied his years, by taking up roller-skating at 75. He even invented a special kind of skate, with jogging shoes adapted to take polyurethane wheels. He skates everywhere in his home town and insists that it is the ideal form of locomotion.

Life-span of 75 **Archimedes**, the Greek scientist, mathematician and inventor, was killed by a Roman soldier at this age in 212 BC. He was living in Syracuse when it fell to the Roman general Marcellus, but his fame was such that the order was given for 'The sage, Archimedes' to be spared and his house left untouched. Sadly, in the confusion of the massacre that followed the capture of the city, he was stabbed to death. He was busy drawing a mathematical figure in the sand at the moment he was

Archimedes   Alexander Graham Bell

struck down, working to the very end on the kind of problem that had fascinated him all his life. Marcellus lamented his death and gave him an honourable burial, complying with the scientist's wish that his tomb be marked with a sphere and a cylinder, recalling one of his most important achievements.
Alexander Graham Bell, the Scotsman who invented the telephone and later, in 1882, when he was 35, became an American citizen, died at his summer home in Nova Scotia in 1922. A tireless inventor, for ever exploring new technological possibilities in many fields, his dying words were: 'So little done, so much to do.'
Leonid Brezhnev, the Russian leader from 1964 to 1982, who managed to maintain nearly twenty years of stable government, in a country where the removal or replacement of high officials at brief intervals has been the more common practice, died of a heart attack at this age in November 1982.
Raoul Dufy, the French Fauvist painter who died in 1953 at this age, was a prolific, high-speed artist whose deceptively simple compositions of pleasant scenes, such as race-meetings or regattas, were filled with light and colour, elegance and visual wit. In spite of severe arthritis, he painted until the day he died.
Duke Ellington, the American jazz band-leader who outgrew simple dance music to become an increasingly complex and ambitious composer, even giving sacred concerts in his later years, died at 75 in 1974.
Michael Faraday, the English scientist who developed from a simple bookseller to become one of the greatest experimentalists of all time, retired in 1858 to his home near Hampton Court and lived there quietly, in declining health, until his peaceful death in 1867.
Myra Hess, the concert pianist, who played lunch-hour concerts at London's National Gallery daily through the German Blitz, from Autumn 1939 to April 1946, died at 75 in 1965.
Samuel Johnson, the great dictionary-maker and wit, died at 75 following a paralytic stroke. One of the most quoted men in the English language, his life is known largely from Boswell's famous biography, where he is portrayed as 'the bear with the heart of gold'. He hated solitude, which induced a great mel-

ancholy in him, so that, in old age, despite an ailing body suffering from asthma and dropsy, he struggled to maintain his social contacts. His stroke came in June 1783 and even then he set off on visits to the Midlands and his old College at Oxford, before he succumbed in December of the following year.

Alfred Krupp, the eccentric German armaments tycoon, died at 75 in 1887. In many ways the founder of modern warfare, he had armed forty-six nations by the time of his death and virtually set the stage for the great holocaust of 1914. He was known as the 'Cannon King' and his guns became status symbols for nineteenth-century nations. His private life was bizarre. An insomniac who was terrified of fire, he built his castle entirely of stone and glass, with not a stick of wood in its structure. He dressed as a Field Marshal and worked in a permanent aroma of horse-dung. He believed that the smell of horses would stimulate his imagination, so he had his study built on top of his stables, with special ducts designed to waft the scent upwards.

Edward Lear, the English animal illustrator and nonsense poet, was a lifelong sufferer from epilepsy and spent the last fifty years of his life travelling and living abroad, mostly in the Mediterranean region, for the sake of his health. He returned to England a number of times, for commissions and exhibitions, and once to give drawing lessons to Queen Victoria, but his final years were spent at San Remo, where he died at 75 in 1888.

Duke Ellington

Chico Marx and Harpo Marx both died at this age, suffering from heart trouble. Chico died in 1961,

Chico and Harpo Marx in *Animal Crackers*

Harpo in 1964. As part of the Marx Brothers comedy team, they made a series of classic films in the 1930s, starting with *Animal Crackers* in 1930 and ending with *At the Circus* in 1939. Later films lacked the surrealist lunacy and intensity of their earlier work. Chico was the fast-talking swindler and piano-player. Harpo was the mute, fuzzy-wigged wild man and harp-player – an embodied id who leapt where others pondered.

S. J. Perelman, the American humorist who wrote the scripts for the most outlandish of the Marx Brothers films, such as *Monkey Business* and *Horse Feathers*, also died at this age, in 1979.

Jacopo Tintoretto, the Venetian mannerist artist, whose paintings gave the impression of being scenes from some religious opera, died at 75 in 1594.

Josef Von Sternberg, the archetypal temperamental, egotistical, jack-booted film director of the early days of talkies, whose main claim to fame was that he discovered Marlene Dietrich, died of a heart ailment in 1969. He was summoned to Germany in 1930 by Emil Jannings to direct the great actor's first talking film, *The Blue Angel*. Von Sternberg selected the little-known Dietrich to play the female lead and took her back to Hollywood in triumph to make a series of successful films with her in the 1930s. He has been described as arrogant, secretive, stubborn, volatile and autocratic, but also as a director of great intelligence.

Henry Wood, the English orchestral conductor, died at 75 in 1944. He was the founder of the famous 'Promenade Concerts' in London, the first of which he conducted in 1895. He continued to conduct them annually for half a century. His health was already failing when the time came for his fiftieth 'Prom', in the summer of 1944, but he managed to take the rostrum for the opening night, and then died soon afterwards. His memory was honoured by calling them the 'Henry Wood Promenade Concerts' after that date.

Seventy-six is the age when people find they have fewer 'wants'. For some this brings sadness, for others a feeling of contentment. The more cheerful view is summed up in the saying: 'The best thing about getting old is that all those things you couldn't have when you were young, you no longer want.'

Life Expectancy

Life expectancy at birth for females in Australia, Canada, Spain and Switzerland is 76.

If you had lived in seventeenth-century London, you would have had only a 1 per cent chance of surviving to this age, according to John Graunt's Life Table.

Accession at 76

John XXIII was this age when he became Pope in 1958. Although he held the office for less than five years, he inaugurated changes so far-reaching that his papacy has been considered as the start of a new era in Roman Catholicism. Short and stocky, simple and unaffected, the son of a peasant, he became one of the best-loved pontiffs, refusing to be the traditional 'prisoner of the Vatican'. His church reforms, his urge for direct contact with his people, and his wish to be able to say 'born poor, died poor' (he left his family his total personal fortune of $20), brought personal values to a leadership which, in the past, had seemed far too remote and often arrogantly antique.

Achievement at 76

Joseph Haydn, the Austrian composer, was honoured on his seventy-sixth birthday in 1808 at a special concert in Vienna, when *The Creation* was played in his presence by a group of eminent musicians. Already ailing, he was carried in on an armchair and seated among applauding members of the high nobility. At this, his last public appearance, Beethoven knelt down and kissed the hand of his former teacher.

Maggie Kuhn, the American founder of the 'Gray Panthers', was 76 when she celebrated the first decade of her organization and its success in gathering a following of 60,000 members, campaigning against 'ageism'. She accused the western world of 'gerontophobia' – a pathological fear of old people and of ageing – and attacked what she called the 'Detroit Syndrome' ('Only the newest model is desirable'). Her main platform was that 'old age should be the time of greatest personal liberation', and she demanded that younger adults should re-shape their ideas about the elderly, while the elderly should re-shape their ideas about themselves. 'Old people's attitudes have to change,' she said, if they are to avoid being isolated from the rest of society. 'The body may be tired but you can always reach out to new ideas and new ways of thinking.' Her own goal, 'until rigor mortis sets in, is to do one outrageous thing every week'. Although her crusade inevitably brings to mind the *Monty Python* 'Hell's Grannies', it deserves serious attention. As will become clear from the later pages of this book, exceptional individuals are capable of remarkable feats, both mental and physical, at great ages. If some can do it, why not more, especially if helped by a change in social attitudes?

Life-span of 76

Augustus, the first Roman Emperor, introduced an autocratic regime which brought stability, prosperity and peace to the Roman world. He was cautious about expanding the Empire and concentrated instead on trade, which flourished as never before. Roads, aqueducts and fine buildings blossomed everywhere. Only in his domestic life was he unsuccessful, so that when he died, at this age in AD 14, it was his stepson Tiberius who became his heir.

Lionel Barrymore, the American actor, who was one of the first stage stars to move into films, was already in leading screen roles as early as 1911. He is best remembered, however, for his many character parts as crusty old gentlemen in wheelchairs. The chair was not a film prop – he was crippled by a fall in 1937. He died in Hollywood, aged 76, in 1954.

Cecil Beaton, the English photographer and designer, best known for his scenery and costumes for ballet, opera and musicals, especially *Gigi* and *My Fair Lady*, died at this age in 1980. He was also renowned for his elegant portrait photographs of fashionable society.

Cecil Beaton with his portrait of Mick Jagger

Benjamin Disraeli, the British Prime Minister, died in 1881, after achieving considerable success both at home and abroad. He developed a particularly close relationship with Queen Victoria, and a few days after his burial in his family vault, she paid him the unusual honour of visiting it in person to lay a wreath.

Albert Einstein, the greatest mathematical physicist of all time, died at 76 in 1955. His contribution to our understanding of the universe was matchless, but his crusade in human affairs seems to have had no lasting impact. He himself anticipated such an assessment when he said 'Politics are for the moment. An equation is for eternity.'

Albert Einstein

Edward Elgar, the composer whose work stimulated a renaissance of English music, died in 1934. At his death he left unfinished a third symphony, a piano concerto and an opera. His most important works were his oratorio, *The Dream of Gerontius*, and his *Enigma Variations*.

T. S. Eliot, the American-born English poet who once worked in a bank and always looked like a bank manager, died in 1965. Fastidious, moralistic, intellectually severe, formal and detached, he gave the impression of having been born middle-aged. Described as 'stooping, sombre and shy' in personal life, he fostered the cult of impersonality and the anti-romantic. Yet, despite his earnest pessimism, he was capable of a poetic intensity that reveals powerful feelings and complex emotions barely submerged below the cultivated surface. His greatest work was *The Waste Land* of 1922. Only with cats did he seem to unbend, as in his delightful poems, *Old Possum's Book of Practical Cats*, in 1939. He was awarded the Nobel Prize for Literature in 1948.

George II, King of Great Britain and Ireland, died suddenly at 76 after a long reign of thirty-three years (1727 to 1760). He came to rely heavily on his ministers and his reign accelerated the development of a constitutional monarchy. In military matters, however, he remained active, being the last British king to command his army on the battlefield (at the Battle of Dettingen in 1743).

James Laver, the English art historian who was a genius at interpreting the social significance of changes in dress fashions, died at 76 in 1975. No fashion, he claimed, was ever arbitrary, and it was said of him that: 'He read omens in overcoats, portents in petticoats and sermons in stoles.' His own comment, summing up his psychological approach to the subject of human clothing, was: 'The hopes and fears of a whole society are reflected in the cut of a dress.'

Livy, one of the greatest of Roman historians, died at this age in AD 17. His history of Rome, from the foundation of the city, established itself as a classic in his own lifetime and exercised a profound influence on the style and philosophy of writing, down to the eighteenth century. He wrote a total of 142 books (five books to one papyrus roll), at a rate of about three a year.

Louis XIV, King of France, known as the 'Sun King', was loved by some but hated by many. Foreign journalists called him a 'bloodthirsty tiger' and the neglected French populace jeered when his body was carried to the Saint-Denis Basilica, in September 1715, when he died four days short of his seventy-seventh birthday. He has been accused of digging the grave of the French monarchy.

Osbert Sitwell, the English poet, satirist, novelist, essayist and art critic, the brother of Edith and Sacheverell, died in 1969 of Parkinson's Disease, in Italy. A man of taste rather than emotional feeling, he waged Sitwell-war on the 'philistine'.

Albert Speer, Germany's Minister of Armaments and War Production in the Second World War, died in 1981, in London – the city he helped to devastate between 1939 and 1945, when he was one of the masterminds behind Hitler's war machine.

Graham Sutherland, England's greatest modern painter, died at 76 in 1980. His vigorous, colourful and spiky plant and animal shapes and his alluring landscapes make him one of the most enjoyable artists to emerge from Great Britain in the twentieth century. In addition, he produced some fascinating religious works, including the world's largest tapestry – the *Christ in Glory* for Coventry Cathedral, hung there behind the altar in 1962. He was also a major portraitist, although this is a view not shared by one of his sitters, Winston Churchill. When his brilliant portrait of a tired but still pugnacious, elderly Churchill was formally presented to the great man at a special ceremony, the carefully considered and deliberately ambiguous Churchillian comment was, 'This is a remarkable example of modern art!' In reality, his vanity was seriously affronted by it and, on his orders, the magnificent canvas was later secretly burned by Lady Churchill.

Joseph Turner, the English landscape and marine painter who anticipated the French Impressionists with his daringly original studies of light and atmosphere, died at 76 in Chelsea, in 1851.

Brigham Young, the American Mormon leader, died in 1877 at Salt Lake City, the settlement he had founded thirty years earlier. He was largely responsible for opening up the West of North America. The region in which he settled had been a desolate place in the heart of the Rocky Mountains, but with his energetic leadership, he made the desert bloom. When he died, he left $2,500,000 to his twenty-seven wives and fifty-six children.

Seventy-seven is a dogmatic year. In *Journey to the Western Islands of Scotland*, Samuel Johnson records a visit he made with Boswell to an old minister of 77. The aged cleric was extremely assertive and rigid in his comments and, after they had left him, Boswell complained to Dr Johnson about the old man's dogmatism. Johnson replied: 'At seventy-seven it is time to be in earnest.'

Life Expectancy Life expectancy at birth for females in Denmark, Finland and the USA is 77.

Achievement at 77 **Clara Barton**, the American humanitarian and founder of the American Red Cross, known as the 'Angel of the Battlefield', was this age when she served in Cuba during the Spanish-American War, in 1898.

Anatole France, the French writer who won the Nobel Prize for Literature in 1921, when he was 77, was considered the ideal 'French Man of Letters' in earlier days, but has become less fashionable in recent times. He was a tolerant, undogmatic author, urbane and highly civilized, but with an ironic scepticism about established authority.

Retirement at 77 **David Ben-Gurion**, the Israeli politician who became the first Prime Minister when Israel was established in 1948, resigned unexpectedly in June 1963, at the age of 77, for unnamed 'personal reasons'. His retirement was in part due to the bitter internal controversy in his party, which resulted in his rivals rising against him for the first time. This was because of the political implications of the 'Lavon Affair', involving Israeli-inspired sabotage of American and British property in Egypt. He was probably also disappointed that the plans he had initiated for secret talks with Arab leaders, with a view to peace in the Middle East, had proved fruitless.

Martha Graham, one of the most influential figures in modern dance, retired as a dancer at this advanced age in 1970. She continued to teach and create dances, however, and over more than fifty years has produced nearly 150 works, from solos to large-scale presentations. Her aim was to make dance 'reveal the inner man' and to act as a visually aesthetic way of expressing the emotions.

Life-span of 77 **Cristobal Balenciaga**, the Spanish dress designer, born in 1895, became fashion's most influential innovator in the twentieth century. He ruled Paris from 1937 to 1968, his innovator's crown being handed on to his pupil Courreges. Balenciaga abolished the fitted line and this led to 'the sack', 'the chemise', and 'the shift'. He is said to have 'shaped the silhouette of the mid-twentieth-century woman'.

James Barrie, the Scottish dramatist and novelist, who achieved wealth and fame with his plays *Quality Street* and *The Admirable Crichton*, was most famous for his children's fantasy *Peter Pan*. When he died at 77 in 1937, he bequeathed all future royalties from this play to a children's hospital.

Robert Browning, the Victorian poet who avoided the more sickly romanticism of many of his contemporaries, and who based much of his writing on his own direct observation of the behaviour of ordinary men and women around him, died in Venice in 1889 after catching a cold.

Chou En-lai, the Chinese Premier from 1949 until his death in Peking from cancer, in 1976, was an affable, pragmatic, persuasive man who was a master at implementing policies. He became one of the great negotiators of the twentieth century, surviving purges and exercising restraint on extremists.

Le Corbusier, the Swiss architect, died suddenly while swimming at Cap Martin in 1965. He was a major influence on modern architecture, being the inspiration behind what has been called 'vertical living'. Had it not been for his creative innovations, we might all have been spared the horrors of impersonal tower blocks and high-rise developments. He was only indirectly to blame for these, however. His own work was always done with imagination, but when copied by inferior imitators it quickly became dreary and damaging to the social life of the humans condemned to live in it. Corbusier's most exciting building, and one of the masterpieces of twentieth-century architecture, is the Ronchamp church in France.

Joe Davis, the world's greatest billiards and snooker player, was 77 when he died in 1978. He became World Snooker Champion in 1927 and held the title until his retirement in 1946, a remarkable reign of nearly twenty years. He was also World Billiards Champion from 1928 to 1933.

Cecil B. DeMille, the American film director who died in Hollywood at 77, in 1959, was the master of the lavish epic and the over-blown spectacle – the man who put sex and violence back into the Bible.

George Eastman, the American inventor whose ingenuity converted photography from a specialist activity for professionals to an amateur pastime for millions, died at 77 in 1932. When he introduced the Kodak camera in 1888 with the slogan 'You press the button, we do the rest', it was the start of a multimillion-dollar success story. In 1896 he launched the box Brownie camera, which sold for one dollar, and as the years went by more and more innovations appeared. In later years he became a philanthropist and gave away more than $75,000,000 to technical and educational institutions. But when ill-health dogged him in his 70s, he decided to call a halt and took his own life, rather than wait for a slow end.

Henry Fonda, one of the major Hollywood figures, died in August 1982. In the previous year he had received an honorary Academy Award for his services to the film industry. On screen, he represented tough sincerity and dogged resourcefulness. In private life, his personality was more complex and there were

77

frequent family tensions. His wife, Frances, the mother of Peter and Jane Fonda, killed herself in 1950.

Galileo, the Italian mathematician and astronomer, who was the first man to use the telescope to study the skies, died at 77 in 1642. He came into conflict with the bigotry of the Roman Catholic Church over his support of the idea that the planets revolve around the sun and, at 69, he was sentenced to house arrest in Florence by the Pope. This remained in effect for the final eight years of his life, but it did nothing to reduce the scientific activity of his fertile brain.

John Edgar Hoover, the director of the FBI from 1924 until his death in 1972, rose to fame during his crack-down on the notorious gangsters of the 1930s. Under his ruthless leadership, the FBI became known for its integrity and freedom from political control. In later years, however, he became criticized for his authoritarian administration. He was so well established that no president dared to remove him.

Arthur Koestler, the Hungarian-born author most famous for his novel *Darkness at Noon*, which was a devastating exposure of conditions inside Stalinist Russia, committed suicide in 1983, at the age of 77, after suffering increasingly from both Parkinson's Disease and leukaemia. He was Vice-President of the Voluntary Euthanasia Society, formerly known by the name EXIT, and, in killing himself, was practising what he had been preaching for some time. Two days before he died, his 56-year-old third wife, Cynthia, took their pet dog to a local vet and had it killed. Then, although she was herself in perfect health, she joined her husband in a double suicide. On the night of March 2, they both took overdoses of barbiturates, sat down in their sitting-room armchairs, and went to sleep together, permanently.

Harold Lloyd, the American comedian who portrayed the worried, earnest, bespectacled 'ordinary' man who constantly and hilariously found himself faced with some cliff-hanging catastrophe, died of cancer in 1971. To many people he was funnier than either of his two great rivals in the 1920s – Chaplin and Keaton. Doing all his own stunts, he was once seriously injured when a studio bomb exploded in his hand and blew off his right thumb and forefinger.

Elijah Muhammad, the American black racist and leader of the Black Muslims, died at 77 in 1975. He claimed to be the messenger of Allah sent to rouse American blacks to set up their own nation – a new *apartheid*. Under his leadership, the organization flourished and grew, despite his imposition of a strict moral code, and built up assets of $60,000,000 and a membership of several hundred thousand – including the boxer Muhammad Ali.

Christabel Pankhurst, the English suffragette who, with her mother, founded the Women's Social and Political Union, died in Los Angeles in 1958. She was the driving force behind the militant wing of the women's movement before the First World War, and was its greatest orator, with a magnetic personal presence.

Paul Robeson, the American actor and singer, died of a stroke at this age in 1976. He possessed the world's most famous black voice and his rendition of *Ol' Man River* has remained popular for half a century. The son of an ex-slave, he had become a member of the New York bar, before Eugene O'Neill persuaded him to give up law and star in two of his plays. He was soon in Hollywood making *Sanders of the River* and *King Solomon's Mines*, and becoming a world star. Then, in the 1930s, he started visiting Russia and adopted left-wing politics as a way of helping the oppressed blacks of America. In the political witch-hunts of the 1950s he was condemned by McCarthy and his passport was confiscated. At the time of his death he was in poor health and living in almost total seclusion in Harlem.

Elsa Schiaparelli, the fashion designer, died in Rome in 1973. She was the first to use violent colours in high fashion (shocking pink, for instance), the first to use zips, and the first to use nylon and other man-made fabrics. She was also responsible for introducing padded shoulders in the early 1930s.

Edith Sitwell, the English poet and critic, died at 77 in 1964. Privately educated, from an aristocratic family, she was disliked by her parents because she was too introverted. An eccentric aesthete, full of airs and graces, she was a formidable, bejewelled and turbaned figure, parodied by Noël Coward as Miss Hernia Whittlebot. Despite much ridicule from literary circles, she survived the criticisms and ended her life with honours heaped upon her.

Marie Stopes, the British pioneer of birth control who wrote two best-sellers on the subject, *Married Love* and *Wise Parenthood*, used the money earned from them to open birth control clinics, against much opposition. She preached 'deliberate parenthood' and changed the sexual attitudes of her generation. It made her a 'notorious woman' at the time, but in the long term she won her important battle. She died at 77 in 1958.

Jules Verne, the French author who can justly be described as the father of science fiction, died at this age in 1905. His most famous 'scientific romances' were *Journey to the Centre of the Earth*, in 1864, *Twenty Thousand Leagues Under the Sea*, in 1870, and *Around the World in 80 Days*, in 1873.

The Duke of Windsor, who was briefly King of Great Britain in 1936, until he abdicated to marry Mrs Wallis Simpson, died at 77 in 1972. After his marriage he lived quietly in France for the rest of his life, except for a period during the Second World War, when he was carefully shunted out of the way, to be Governor of the Bahamas.

Darryl F. Zanuck, the American film mogul, on whose character all caricatures of film moguls have since been based, died at this age in 1979. As head of production at 20th Century-Fox Studios, he was the volatile taskmaster famous for his comment to subordinates: 'Don't say yes until I have finished talking.'

78

Seventy-eight is the year when the days seem to grow shorter. This strange, subjective phenomenon is first detected in middle age, but then intensifies in old age, especially in the late 70s, when many people get the feeling that life is slipping away faster and faster. This is elegantly summed up by Alice B. Toklas with the comment: 'Dawn comes slowly but dusk is rapid.'

Life Expectancy Life expectancy at birth for females in France, Holland, Japan, Norway and Sweden is 78.

Achievement at 78 **Thomas Beecham**, the British orchestral conductor, refused to allow his advancing age to interfere with his musical activities. Undaunted by the prospects of endless travel and the inevitable organizational stresses and strains, he set off at the age of 78 on yet another lengthy foreign tour with his orchestra. A man of great verve and elegance, he was the son of a Lancashire millionaire (who manufactured Beecham's Pills), and had already poured a great deal of his inherited fortune into the promotion of opera in England. He was an outspoken male chauvinist, who opposed the introduction of women into orchestras. His grounds for objecting were that if they were pretty they would distract his male players and if they were ugly they would distract him.

Ayatollah Khomeini, the Iranian religious and political leader, who acquired the nickname of the 'Biggest Shi-ite in the Middle East', was 78 when he started his revolution to overthrow the pro-western Shah. Despite his age, the momentum of his movement carried him along and swept him back to a power role that would have been arduous for a leader half his age.

H. G. Wells, the English author with a genius for stimulating the imaginations of others, had a secret ambition to be taken seriously as an academic scientist. Because of this he submitted a doctoral thesis to London University on 'The Quality of Illusion in the Continuity of Individual Life', at this advanced age of 78. He was awarded a D.Sc for it, but not, as he had hoped, an FRS. Although he had written over 100 books, including *The Time Machine, The Invisible Man, The War of the Worlds, The Shape of Things to Come, Kipps, The History of Mr Polly* and *An Outline of History*, and was justly world famous, he was still bitterly upset by the slur of 'pseudo-scientist' thrown at him by minor academics who envied his great gifts. For this reason he set out to obtain an academic title in his later years, to beat them at their own game. Despite his enormously fertile and inventive brain, he clearly suffered from self-doubt because, when asked what his epitaph would be, he replied, 'He was clever but not clever enough.' A short, plump, squeaky-voiced, querulous man, he found it easy to attract lovers because, as one of them recalled later, 'he smelt of honey'.

Retirement at 78 **Charles de Gaulle** resigned as French President at this age after failing to win a national referendum in April 1969 and retired to his home at Colombey-les-deux-Églises to write his memoirs. A career soldier with a profound contempt for career politicians, he was an aloof, remote leader who acted alone for 'Eternal France', and whose almost dictatorial attitude was only tolerated because of his immense personal presence.

Life-span of 78 **Bud Abbott,** the American comedian and partner of Lou Costello in a long series of immensely popular film comedies from the 1930s to the 1950s, died from cancer in 1974.

Harold Abrahams, the British athlete immortalized in the 1981 film *Chariots of Fire*, died at 78 in 1978. He won a gold medal for the 100 metres at the 1924 Olympic Games.

Charles Boyer, the Great Lover of the screen, continued his role in private. So great was his love for his wife of forty-four years that, two days after she died, he followed her to the grave by taking an overdose of barbiturates, in August 1978.

Alexander Calder, the American sculptor who invented *mobiles*, died in 1976. He had been influenced by the abstract paintings of Piet Mondrian, with their bright, primary colours in small patches of vivid punctuation, and managed to translate them into three-dimensional metal constructions. By the time of his death, Calder was recognized internationally as the most important of all modern American sculptors.

C. B. Cochran, the English theatrical impresario who staged Noël Coward's early successes and became the great showman of his generation, in the 1920s and 1930s, died in a tragic manner in 1951. At 78 he was so crippled by arthritis that he found himself unable to turn off the hot tap in his bath one day, and was so badly scalded that he died a few days later.

Jean-Baptiste-Camille Corot, the French landscape artist, and the most forged painter in art history, died in 1875. Over 2,000 fake Corots were found in one collection alone. He was a quiet, unemotional man, kindness and soft-heartedness personified – so much so that when purchasers of forged Corots took them to him in distress at having lost their investment in his art, he would sign the forgeries to cheer them up. He also did the same for poor students, to help them out of financial difficulties. He never married – he was in love with his landscapes.

Dwight Eisenhower, the thirty-fourth President of the United States, died of heart failure at 78 in Washington in 1969. Despite criticism of his administration he remained an immensely popular father-figure, partly because of his cheerful personality and partly because he was still loved as the 'Ike' who, as Supreme Commander in the Second World War, had masterminded the invasion of Hitler's Europe and the destruction of the Nazi menace.

Jacob Epstein, the controversial sculptor, died at this age in London in 1959. Born in New York of Russian-Polish parents, he moved to England in 1905, where his sculptural work developed in two distinct directions. On the one hand, he became a respected bronze bust portraitist, with sitters such as Shaw, Einstein and Churchill; on the other, he became a notorious 'distorter' of the human form in a number of monumental pieces, which aroused great hostility in the 1930s.

Euripides, the Athenian playwright of the fifth century BC, wrote more than ninety plays, but only eighteen survive today, including *Medea*, *Electra*, *Hecuba* and *The Bacchae*. His view of life was full of gloom, doom and violence – especially within the family – and he was fond of portraying women as particularly vicious and evil. He died at 78, in the year 406 BC.

John Ford, the greatest director of Westerns in the history of the cinema, who won six Oscars, died in 1973. Basically a simple, romantic man, he was a poet of the cinema, if a somewhat violent one.

Mahatma Gandhi, the 'Father of India', who used a variety of non-violent methods, such as non-cooperation, civil disobedience and, in his own case, 'fasts until death', to destroy the British rule and gain independence for India, was assassinated at 78 in 1948. The saintly Gandhi had been preaching unity between Hindu and Muslim with such success that it alarmed Hindu extremists to the extent that one of them shot him dead at a prayer meeting.

Mahatma Gandhi

Henrik Ibsen, the Norwegian playwright, had become world famous by the turn of the century, when he suffered two strokes, one in 1900 and a second in 1901, which left him a helpless invalid until his death at 78 in 1906.

Kublai Khan, who conquered China to become its first Mongol Emperor, died at 78 in 1294. Unlike his brutal grandfather, Genghis Khan, he was a liberal-minded ruler who encouraged international trade and supported the arts and architectural developments.

Ben Lyon, the American actor and comedian, died of a heart attack in 1979. A popular leading man in the films of the 1920s and 1930s, he is best remembered for his wartime radio show *Hi Gang*, with his wife Bebe Daniels. After the war he became head of Fox talent-scouting and was responsible for discovering Marilyn Monroe.

Trofim Lysenko, the notorious Russian biologist who was virtual dictator of communist biological research during the Stalin regime, died at Kiev, at this age, in 1976. His wild promises of major improvements in Soviet agriculture assured him his dominant role, but, as they were based on false genetic arguments, his days were inevitably numbered, and he fell from grace in the mid-1950s. He made the cardinal error of allowing political doctrine to colour scientific theory – suggesting that acquired characteristics could be inherited. Unfortunately for him, the plants and animals about which he was talking were not members of the Communist Party.

Vladimir Nabokov, the Russian-born novelist who irritated many readers with his literary mannerisms and his mandarin posturing, was nevertheless a technically brilliant writer in both Russian and English. He is best known for his erotic novel *Lolita*, of 1959, about a middle-aged intellectual's infatuation for a 12-year-old nymphet. He died at 78 in 1977.

Sylvia Pankhurst, the English suffragette, who was repeatedly imprisoned for her vehement feminist views, died in Addis Ababa, in 1960. During the First World War she had split with the rest of her suffragette family because they had helped the war effort, while she adopted intensely pacifist views.

Juan Perón, the Argentinian President, died at 78 in 1974, after a final triumphant return to power, following many years in the wilderness. In 1943 he had led a revolt against President Castillo and then, as Vice-President, he had fought for social welfare reforms in order to gain support from the unions. He was imprisoned in 1946, but released when his wife Eva campaigned for him. He was then made President and ruled as dictator for nine years. During this time, he favoured the working classes in such a way that financial chaos resulted and he was exiled in 1955 for eighteen years, returning for his last taste of glory in September 1973 – until his death in July 1974.

Tiberius, the second Roman Emperor, was an efficient administrator and a good soldier, but also aloof, cruel, pedantic, austere and suspicious. In his 60s he retired to Capri where, it was rumoured, he presided over unspeakable orgies. He died at 78 in AD 37.

Paolo Uccello, one of the most brilliant and original artists of the fifteenth century, died in the winter of 1475, at the age of 78, lonely, melancholy and poor. A few years earlier he had complained to his tax inspector: 'Being old and out of work, I am unable to support myself, and my wife is ill, too.' It was an unfitting end for the Florentine master whose paintings show more daring and experimentation than any of his contemporaries. His masterpieces, such as *Hunt by Night* and *The Rout of San Romano*, which rank among the greatest paintings of all time, are full of strangely rearing horses, running animals and tilted lances, creating some of the most exciting visual compositions ever conceived.

Seventy-nine is at present the highest figure for 'life expectancy at birth', male or female, in any country. The record is held by Iceland, for females. Iceland also has the highest figure for males (73). It remains an intriguing mystery why Icelanders should prove to be the longest-lived population in the world.

Achievement at 79

George Cayley, the English inventor, was this age before he saw his great invention, the glider, make a successful flight carrying a man, in 1853. He had begun studying aeronautics as early as 1796 and his writings in the early nineteenth century effectively make him the father of the science of aerodynamics. But his first great success did not come until a few years before his death, when he persuaded his coachman to soar across a valley near his family home, Brompton Hall in Yorkshire. The flight covered 500 yards and made Cayley's servant the first human being ever to leave the ground in free flight. He did not, however, appreciate the honour, saying to his excited employer after the flight: 'Please, Sir George, I wish to give notice. I was hired to drive, not to fly.'

Edith Evans, the English actress, was given the New York Critics Award in 1967, for her performance in *The Whisperers*, when she was 79.

Life-span of 79

Ethel Barrymore, the female member of the 'Fabulous Barrymores', died of a heart attack in Hollywood, in 1959. The sister of John and Lionel, she was a renowned stage actress who also won fame for her 'grand old lady' roles in Hollywood films. She won an Oscar in 1944 for her part in *None but the Lonely Heart*. In private she was known for her morbid sense of humour, her huge library and her fanatical love of baseball.

Pierre Bonnard, the French painter who was a master of freely applied paint creating a sensation of light and colour, died at 79 in 1947. He was a member of the post-Impressionist movement – in particular, an 'Intimist', who transformed ordinary objects and scenes into things of beauty.

Nikolai Bulganin, who was Premier of the USSR for a few years before being ousted by Khrushchev, spent his final years in a long period of quiet obscurity, before dying at 79 in 1975.

Alan Cobham, the English pilot who pioneered long-distance air travel and won the Britannia Trophy in 1926 for his flight to Australia and back, died at 79 in 1973. He also numbered among his achievements the invention of mid-air refuelling.

Charles de Gaulle, the French President, died of a heart attack in 1970. His personal philosophy was: 'Authority requires prestige and prestige requires remoteness.' He is generally regarded as having rescued his nation from disaster and restored it to its former rank in world affairs. When he died at his French village home at Colombey-les-deux-Églises, he was less than two weeks away from his eightieth birthday.

Charles de Gaulle inspecting French cadets

Anthony Eden, the British Prime Minister whose time in office was shortened by ill-health and the Suez crisis, died in 1977, twenty quiet years after his resignation.

Grock, the Swiss clown who perfected an act with musical instruments and was honoured with the name 'The King of Clowns', died at this age in 1959.

William Harvey, the English physician who discovered the circulation of the blood, died at 79 in 1657, from cerebral thrombosis. A few years before his death he was rich enough to present a new college building, with a library and his own collection of books, to the Royal College of Surgeons, and it was perhaps fortunate that he did not live long enough to see it all destroyed by the Great Fire of London.

William Harvey demonstrating the circulation of blood

Ho Chi-Minh, the President of the Democratic Republic of Vietnam, died in Hanoi, in 1969. Had he lived four more years he would have seen the end of the Vietnam War, the fall of Saigon and its renaming as Ho Chi-Minh City, in his honour.

Lancelot Hogben, the English scientist who described himself as having 'a sheer genius for making enemies', was a brilliant popularizer as well as being a skilled geneticist and endocrinologist. He died at 79, in 1975. His books *Mathematics for the Million* (written during a spell in hospital), *Science for the Citizen* (written on trains at the weekends), and *From Cave Painting to Comic Strip*, were immensely popular, unlike their author, who was a quarrelsome, awkward individual lacking in social graces and filled with a bitter resentment against the condescension he had suffered from society in his younger days.

Louis Mountbatten, the English statesman, grandson of Queen Victoria and uncle of Prince Philip, was murdered by the IRA in August 1979, when his holiday boat was blown up in the Republic of Ireland. Among his many roles, he was Supreme Commander in South-East Asia in the Second World War, the last Viceroy of India and the first Governor General of that country after the transfer of power; later, he was Commander-in-Chief of NATO forces in the Mediterranean and First Sea Lord of the admiralty.

Louis Mountbatten

Edward G. Robinson, the most famous movie gangster in Hollywood history, whose portrayal of 'Little Caesar' in 1930 was the start of a long career playing attractive villains, died of cancer at 79 in 1973. Short-bodied, broad-headed, ugly-faced and harsh-voiced, he succeeded purely by the force of his personality and sense of authority. In private life he became one of the great art collectors of California, but was forced to sell his huge collection to finance a divorce settlement. He was accused, but cleared, of communist contacts in the political witch-hunts of the late 1940s. His film career spanned forty years and his last film, *Soylent Green*, contained a final, moving death scene which was to be echoed in real life within a few months.

R. C. Sherriff, the English author who wrote one of the most famous plays about the First World War, *Journey's End*, after being a volunteer soldier in the war himself, died at 79 in 1975. He used the fortune he made from his play to go up to Oxford to read History, after which he went to Hollywood to write scripts.

Matthew Smith, the most French of English painters, with rich colours, broad, swirling brushstrokes, voluptuous nudes and colourful still-lifes, died at this age in 1959.

Edgar Varèse, the modern French composer, whose avant-garde compositions produced reactions of hostility and incomprehension in many who heard them, died at 79 in a New York hospital, in 1965. An original, experimenting, exploring composer, he was one of the first to use electronic sources for sounds.

Horace Walpole, the English writer who set the fashion for 'Gothic horror-romances' with his 1764 novel *Castle of Otranto*, died at this age in 1797. He is best remembered as an industrious letter-writer, whose huge output provides us with a valuable social history of his time, complete with gossip and scandal.

Bombardier Billy Wells, the British heavyweight boxing champion from 1911 to 1919, lived a long and happy life, dying eventually in 1967. Although his career ended in 1925, such was the impact of his personality that he was still fêted as a guest of honour at sporting occasions forty years later.

Henry Williamson, the English author and naturalist who became famous especially for his books *Tarka the Otter* and *Salar the Salmon*, died in 1977. He was a meticulous writer – *Tarka* took him four years to complete and was rewritten no fewer than seventeen times.

Leonard Woolley, English archaeologist, who excavated Ur of the Chaldees from 1922 to 1934, to uncover some of the most beautiful ancient objects ever seen, died in London, at 79, in 1960. It was one of the most exciting excavations in the history of archaeology. Ur was part of ancient Sumer and is now in modern Iraq, south of Baghdad.

William Wyler, the American film director, who won three Oscars, for *Mrs Miniver*, *The Best Years of Our Lives*, and *Ben Hur*, died in Beverly Hills, California, at 79, in 1981. He was a perfectionist and taskmaster, known as '90-take Wyler'. He had furious rows with Bette Davis and other major stars, but in the end it was his tyranny that won them their Oscars.

Edward G. Robinson playing the gangster in a scene from *Little Caesar*

Eighty is the age of frailty, according to a nineteenth-century French print called *Degrés des ages*. It shows a human couple ascending a staircase of age and then descending again as old age overtakes them. On the step labelled *80 Ans*, it shows an elderly couple with heads slightly bowed, and the name *Age Caduc* – the Age of Frailty.

Certainly the body is more fragile at this stage of life, but a great deal is still possible, as Longfellow points out in *Morituri Salutamus*: 'Cato learned Greek at eighty; Sophocles/Wrote his grand *Oedipus*, and Simonides/Bore off the prize for verse from his compeers,/When each had numbered more than fourscore years.'

Charles Cooley, the American sociologist, records the optimistic attitude of his eighty-year-old grandmother who, while on a visit to the East, kept sending things home she had bought for her house. Her philosophy was: 'I don't suppose I shall live for ever, but while I do live I don't see why I shouldn't live as if I expected to.'

Putting forward two other comments on being eighty are Jean Renoir, the French film director: 'The advantage of being eighty years old is that one has had many people to love'; and Joyce Carey: 'A man of eighty has outlived probably three new schools of painting, two of architecture and poetry, a hundred in dress.'

Achievement at 80

Henri Breuil, the French priest who saw more of the insides of prehistoric caves than of churches, and became the world authority on Stone Age art, was immensely prolific. On his eightieth birthday, his bibliography listed more than 600 items.

Levi Burlingame, the American jockey, broke the record for the oldest competitive rider when he rode his last race, at Stafford, Kansas, in 1932, at the age of 80.

George Burns, the American comedian, was the oldest person ever to win an Oscar – at 80 in 1976 for *The Sunshine Boys*.

Marc Chagall, longest-lived of all modern artists, who was still active in his 90s, was a mere 80 in 1967, when he created the sets and costumes for a Metropolitan Opera production of Mozart's *Magic Flute*. His dreamlike paintings of rooftop violinists, multicoloured farm animals and floating brides, are among the most personal and original works of twentieth-century art.

Lucas Cranach the Elder, the brilliant German artist whose charming, elongated female nudes have a strange beauty all their own, despite their remarkably plain faces, was 80 years old when he was given a new contract as court painter to Emperor Charles V, in 1552.

Donatello, the Florentine sculptor who introduced the 'heroic' style – forceful figures slightly larger than life-size – was still actively working at the age of 80, in 1466, designing twin bronze pulpits for San Lorenzo, in Florence.

André Gide, the French author, was 80 in 1950 when he published the last volume of his *Journal*, which took the record of his life up to his eightieth birthday (November 1949). He had kept the journal from 1889 and in more than 1,000,000 words he recorded his experiences, impressions and moral crises over a period of more than sixty years. After its publication he decided to write no more and died less than a year later.

Victor Hugo, the French romantic poet and novelist, achieved great success in his own lifetime. In 1878 he was struck down by cerebral congestion, but lived on for some years at his home in the Avenue d'Eylan. On his eightieth birthday he was honoured by having it renamed the Avenue Victor Hugo. When he died he was given a national funeral, lay in state in the Arc de Triomphe, and was buried in the Pantheon, having been carried there, at his express wish, in a pauper's hearse.

Retirement at 80

Winston Churchill resigned his position as Prime Minister on April 5, 1955, when he was this age, but remained in the House of Commons to become 'Father of the House'. Four years later he fought and won another election. Earlier, in 1953, he had collapsed with a stroke that brought on partial paralysis, but he managed to make a recovery which, even for a younger man, would have been remarkable.

Life-span of 80

Jack Benny, the droll American comedian with brilliant timing and a skilled economy of action, died of cancer at 80, in California in 1974. Self-mocking to a fault, especially concerning his age, his miserliness, his vanity and his lack of musical talent, he became the master of the 'meaningful pause'. In private life he was modest and generous, although his generosity sometimes took an odd form: he once gave every child in his home town $39 invested in trust funds and not to be cashed until they were 39.

Busby Berkeley, the American film director of twenty-one musicals, who revolutionized this type of film by introducing immensely elaborate sets, with such items as waterfalls using 20,000 gallons of water a minute, 100 grand pianos played simultaneously by 100 girls, and the 'Berkeley top-shots' for which he drilled holes in the studio ceilings. He died at 80 in 1976, a quarter of a century after directing his last film.

The Buddha died of food-poisoning at 80 in the year 483 BC. After his enlightenment he had undertaken forty-five years of continual wandering and preaching. This came to an end shortly after he ate a meal at the house of one of his followers, the goldsmith Cunda. The man had prepared various delicacies, including a pork dish called *sukara-mad-*

dava. For some reason, the Buddha instructed that only he should eat the pork dish, and that his disciples were to be given other foods. At the end of the meal he asked Cunda to bury the leftovers, so that no one else could eat them. He then became sick and suffered agonizing pains, but set off again on his journeying. After a while it was clear that he was dying and he sent a message back to Cunda to say that he should feel no remorse.

Wyatt Earp, the American marshal who wore a black stetson, black frock-coat, black string-tie, and black boots, and kept law and order in Tombstone, Arizona, died quietly in his sleep, at his Los Angeles home, in 1929. His main interest in keeping order in Tombstone was to ensure the smooth running of the town casino, in which he had an interest. He earned $250 a month for being the marshal and $1,000 a month for keeping an eye on the saloon.

Erle Stanley Gardner, the American author who created Perry Mason, was one of the most successful authors of fiction in modern times, having sold 170,000,000 copies of his books – usually mystery stories based on court cases. An energetic, ranch-dwelling, archery fanatic, he wrote 1,000,000 words per year at his peak. He travelled widely, accompanied by a battery of secretaries, assistants and drivers. As a champion of good causes, he became an American hero. He died at 80 in 1970.

Francesco Guardi, the Venetian painter, in his lifetime could not rival Canaletto's high prices for views of Venice, but today he is preferred by many because of his poetic use of light and the brilliant deftness of his brush-strokes. He died at this age in 1793.

Alfred Hitchcock, the pear-shaped, sad-faced master of film suspense, who suffered from passionate fixations on his favourite ice-cool blonde stars, died in 1980. He was one of the first directors to become a public personality in his own right – a bigger attraction than even his stars. Like Churchill, he is proof that 'success defeats obesity' in its role as an early killer.

Elsa Maxwell, the American hostess who described herself as 'the arbiter of international society and the most famous hostess in the world', died at 80 in 1963. What she lacked in talent, beauty and social background, she made up for with energy and determination, the latter stemming from a childhood anger at being refused permission to attend a party because her parents were too poor. She vowed to give the best parties when she grew up – and she did.

Golda Meir, the Prime Minister of Israel from 1969 to 1974, and the founder of the Israeli Labour Party, died at this age in 1978.

Franz Anton Mesmer, the Austrian physician, died in 1815, after an extraordinary career. He claimed that his patients could use 'Animal Magnetism' to cure themselves. Called 'Mesmerism', his technique was, in reality, hypnotism. Condemned as a charlatan and driven out of Austria, he set up a clinic in Paris, with soft music and ornate decorations, where young ladies came for his fashionable treatment. It consisted of sitting them in a circle with their legs apart and making them hold on to magnetic rods while young male attendants massaged their thighs, breasts and buttocks until they experienced convulsive 'fits', after which they collapsed in delighted heaps, feeling greatly improved by their experience. Marie Antoinette and Madame du Barry were among his satisfied customers and, although he was again investigated for fraud, he amassed a fortune and, it is said, died with a smile on his face.

Thomas Love Peacock, the English author of satirical novels, was burnt to death at 80, in 1866. He had acquired a country residence, constructed out of two old cottages, and when it caught fire he refused to give up in his efforts to save his library from destruction. His dying words were 'By the immortal God, I will *not move*!'

John Ruskin, the English art critic, died in 1900 after repeatedly suffering from mental illness in his later years. He was a manic depressive. As an art critic he exerted a powerful influence on Victorian attitudes to beauty. He was also a social reformer, the first to advocate the old age pension, universal free education and improved housing for all.

Alfred von Tirpitz, the German pioneer of submarine warfare and the father of the U-boat, died at this age in 1930. He had been dismissed in 1916 because of the outrage caused by his sinking of unarmed vessels, including the *Lusitania*. He was ahead of his time – in the Second World War his callous policies were adopted without hesitation.

William Wordsworth, the English poet who opposed formality and artifice and turned to nature for his inspiration, died in Westmorland in 1850. He heralded the romantic revival in English poetry.

A Busby Berkeley 'top-shot'

Eighty-one is the age of reminiscence, the age when, according to an old proverb, 'Everything reminds you of something else.'

Sean O'Casey expresses a similar thought: 'When one has reached 81 one likes to sit back and let the world turn by itself, without trying to push it.'

In a study made at Duke University in the United States, it was found that 81 was the oldest age at which a woman was still enjoying sex. Some men, however, continued to do so for the rest of this decade.

Achievement at 81
Benjamin Franklin, the American statesman and scientist, was this age in 1787, when he helped to frame the US Constitution. His call for compromise and unanimity prevailed over deep disagreements and made possible the adoption of the Constitution. It was his final act in an official capacity.

Johann Goethe, the German literary giant, was 81 when he finally finished his masterpiece, *Faust*, in the summer of 1831. He had been toiling at it for over half a century and when it was done, he felt that his life's work was completed. Within a few months he was dead.

Life-span of 81
Phineas T. Barnum, the greatest showman on earth, became famous for his self-confessed acts of hokum, humbug and wild publicity. Penniless in 1835, he borrowed $1,000 from friends to buy an old black woman. He then exhibited her as George Washington's nurse, aged 161, and made $1,500 a week from his display. When she died, the *post mortem* revealed that she was under eighty years of age. Undaunted, he then opened a museum of freaks including a famous mermaid that consisted of the body of a monkey plus a fish's tail. In 1842 he acquired a dwarf, renamed him General Tom Thumb and toured with him across America and Europe where he met both Queen Victoria and the King of France. Barnum built a home for himself which was a replica of the Brighton Pavilion and had the fields nearby ploughed by elephants. In 1871 he opened his travelling circus – 'The Greatest Show on Earth' – and ten years later joined his great rival to create Barnum and Bailey's Circus. He was a multimillionaire when he died at 81 in 1891, the world's greatest expert, as he put it, of the 'Philosophy of Humbug'.

Constantin Brancusi, the Romanian sculptor, was one of the most famous, if one of the most austere sculptural artists of the modern movement. His work foreshadowed that of the minimal artists and he was a major influence on such figures as Moore and Hepworth. In 1927 his 'Bird in Flight' attracted customs duty in the USA because it was too 'minimal' to be considered a 'work of art'. He died in Paris in 1957.

Georges Braque, the French painter and, with Picasso, the co-inventor of Cubism, was the 'quiet one' of the pair, lacking Picasso's theatrical personality. Studious, modest and retiring, he was totally devoted to his art and created paintings which, if less rebelliously dramatic than Picasso's, are thought by many to be more beautiful and of more lasting worth. He died at this age in 1963.

Joe E. Brown, the American film comedian, died at 81 in 1973. The possessor of the widest mouth and the biggest grin in show business, his face was his fortune. He became the perennial innocent victim in a long series of films covering nearly forty years of Hollywood's history.

Georges Carpentier, 'Le Gentleman-boxeur', or 'Gorgeous Georges', as he was known, became the world boxing champion who was respected by people from all walks of life. Arnold Bennett said of him: 'He might have been a barrister, poet ... Fellow of All Souls, but not a boxer.' He was handsome, debonair and, despite terrible beatings, lived to a ripe and intelligent old age, dying at 81 in 1975.

Colette, the French novelist, was greatly honoured in her final years, despite the scandals of her youth, but by this time was already severely crippled by arthritis. She ended her days a legendary figure, confined to her Paris apartment, surrounded by her beloved cats. She died at this age in 1954.

George III

Marcel Duchamp, the French anti-artist, who outraged the art world by exhibiting such items as a Mona Lisa reproduction with a moustache painted on it, and a white porcelain urinal labelled as a sculpture by a Richard Mutt, abandoned the art world altogether in 1923 to devote the rest of his life to playing chess. He died at 81 in 1968.

Joseph Kennedy with his 'Clan'

John Fastolf, the English soldier immortalized through Shakespeare's character Sir John Falstaff, was in real life very different from the dissolute, clowning, cowardly fiction. He served with distinction in the second phase of the Hundred Years War, at Agincourt, Verneuil and Rouvay where he used barrels of herrings to shield his troops in the Battle of the Herrings of 1429. He died in Norfolk at 81 in 1459.

Gracie Fields, the English actress/singer, was so popular at the peak of her career that the British Parliament once adjourned early in order not to miss one of her radio shows. Her unquenchable spirit was used to cheer people up during the depression, but she suffered the fate of falling in love with an Italian just before the Second World War broke out. She moved to America to be with him and lost her immense popularity by marrying him during the war when Italy was part of 'the enemy'. She moved to Capri after the war and opened a restaurant there, often returning to England in her later years, when all was forgiven, to give nostalgic performances that kindled memories of the time when she was the highest-paid film star in the world. She died in Italy at 81 in 1979.

George III, King of Great Britain and Ireland from 1760 to 1820, suffered from an agonizing metabolic defect in later life. To his ministers he was mad, and for the last ten years of his life he was permanently deranged, tortured by pain and his well-meaning physicians, in a living death. He finally succumbed at Windsor Castle in January 1820, at the age of 81.

Boris Karloff, the British actor who became the master of horror in Hollywood films, retired in 1959 to an English village in Sussex, to spend the last ten years of his life 'enjoying cricket'. Towards the end he was confined to a wheelchair with arthritis, dying eventually of a respiratory ailment in 1969.

Joseph Kennedy, the American financier and head of the Kennedy 'Clan', who had been US Ambassador to England from 1938 to 1940, suffered a stroke in 1961 from which he did not fully recover. He died at 81 in 1969.

Antonio Salazar, Prime Minister of Portugal for thirty-six years, was a discreet dictator. He rescued his country from financial chaos and kept it neutral in the Second World War, but used unduly strict methods to suppress opposition. He managed to act in a dictatorial manner without arousing major criticism by avoiding any overt displays of power. In 1968 he had a severe stroke and was not even aware that he was no longer Premier at his death ten years later.

Walter Sickert, the British painter who was the leader of the 'Camden Town Group', used French Impressionist techniques to which he added dark, sombre tones. Described as the most important of the British Impressionists, he died at 81 in 1942.

George Stubbs, the world's greatest horse portraitist, who was almost entirely self-taught, also died at this age, in 1806. He learned about his subjects by making extensive dissections of horses and in 1766 published a classic work, *The Anatomy of the Horse*.

J. R. R. Tolkien, the English creator of *The Hobbit* and *The Lord of the Rings*, who invented a whole language, culture and mythology for his fictional world, died at Oxford in 1973. Like Dr Johnson he had a horror of going to bed and was always at his best after midnight. He was an Oxford professor from 1925 to 1959.

Queen Victoria

Victoria, Queen of Great Britain for sixty-three years, seven months and two days – the longest reign in English history – died at 81 on January 22, 1901, at Osborne on the Isle of Wight. Despite her epoch's prudery and piety, she enjoyed a great, passionate love – for her husband Albert. When her favourite, Disraeli, was dying, he refused a visit from the Queen, saying: 'She will only ask me to take a message to Albert.'

Eighty-two is the age at which Victor Hugo addressed the French Senate with the following words: 'It is difficult for a man of my years to address such an august body – almost as difficult as it is for a man of my years to make love three, no four, times in an afternoon.'

> Nobody knows precisely how long a large fish may live, under favourable circumstances, but a Lake Sturgeon which was still growing was recorded at the impressive age of 82.

Achievement at 82

Winston Churchill was this age when he published the first part of his four-volume work, *A History of the English-Speaking Peoples*, in 1956.

William Gladstone, the British politician and leader of the Liberal Party, was Prime Minister on four occasions during his sixty-year career in parliament. He was what Queen Victoria called an 'old, wild, and incomprehensible man of $82\frac{1}{2}$' at the time of forming his fourth cabinet in August 1892. She disliked his pomposity, but he achieved many important reforms and is responsible to a large degree for the increased 'liberal-mindedness' of all later political parties. Despite his advanced age he remained in office for a further two years, finally retiring when he was 84.

William Gladstone

Bill Kane, the American cowboy, was still riding rodeo at the age of 82. A rancher from the age of 10, he had ridden competitively in more than 3,000 rodeos, and was still winning prizes at the age of 80. His secret was never to miss his daily workout. He claimed it was fatal to stop, even for a short period: 'If you do, you're a goner.'

Misfortune at 82

Haile Selassie, Emperor of Ethiopia for forty-three years, known as the Power of the Trinity, the Lion of Judah, and the King of Kings, was deposed in 1974, at this age. He was made a prisoner in his own castle in September and was dead in less than a year. He had been an absolute monarch and was reputed to be a direct descendant of Solomon and the Queen of Sheba. People fell down and kissed his feet. He travelled with an entourage of seventy-two, and a pet chihuahua called 'Chicheebee' who wore a diamond-studded collar. Despite his ancient style of rule, he was a reformer who suppressed slavery and promoted education. His educational policies were so successful that his more intelligent subjects learnt enough to take over the government of the country and throw him out. Since his day there have been an estimated 30,000 executions.

Life-span of 82

Aldrovandus, the Italian Renaissance naturalist, whose natural history museum was the largest in Europe and whose beautifully illustrated volumes, covering the whole range of life forms, were of major importance in the history of natural science, died at this age in 1605.

H. M. Bateman, the comic genius whose cartoons summed up hideous social gaffes, with such titles as 'The Man who Lit His Cigar Before the Royal Toast', and 'The Boy who Breathed on the Glass in the British Museum', died in Malta, in 1970. His simple aim in life was to make people laugh. In later years he worked hard, but failed, to establish a 'National Gallery of Humorous Art' in London.

Hilaire Belloc, the French-born British poet and historian, best remembered for his *Cautionary Tales* and other light verse for children, died in 1953.

Hoagy Carmichael, the prolific American songwriter who composed and performed some of the most popular songs of all time, and who made a new career for himself in musical films when he reached middle age, died at 82 in 1981.

René Clair, the French film director, who specialized in comedies full of bitter irony, also died at this age in 1981. He failed his exams at school and became an ambulance driver during the First World War. When he was injured he retired to a monastery, but emerged later to join the French film industry.

Gladys Cooper, the leading English pin-up girl of the First World War, died of pneumonia in 1971. She became a star of the London stage in the 1920s and then moved to Hollywood for a thirty-year career in films, in the 1940s, 1950s and 1960s, where she played gracious, dignified English ladies.

Francisco Franco, the Spanish dictator, managed to survive unharmed into old age. Even after his health declined in the 1960s and he became a more benign ruler, he still held the reins of power in Spain and did not give up his position as Premier until he was 80. Even then, he remained Head of State and Commander-in-Chief of the armed forces, until his death at 82 in 1975.

Francisco Franco

Francisco Goya, one of the great Spanish artists, who was court painter to Charles IV, died at 82 in 1828. In middle age he suffered an illness which left him permanently deaf and his later works showed an obsession with darker, more sinister themes. His series of etchings called *The Disasters of War* are among the most haunting works of art of all time.

David Lloyd George, the British Prime Minister from 1916 to 1922, suffered a long twilight to his career which was a melancholy anti-climax for the great Liberal leader. He died in 1945, after refusing a place in Churchill's war cabinet on the grounds of health and age.

Leonide Massine, the Russian choreographer, who was also the principal dancer of the Diaghilev Ballet in Paris from 1914 (when he replaced Nijinsky) until 1920, died in 1979. He created a number of new ballets, including *La Boutique fantasque* and *Le Sacre du printemps.* He was one of the boldest pioneers and innovators in the history of ballet.

André Maurois, the French writer who held a prominent place in French literary circles for half a century, died at this age in 1967.

John Nash, the Welsh-born architect responsible for London's most beautiful Regency terraces, and who designed the exotic Brighton Pavilion for the Prince Regent, died at East Cowes Castle, which he had built for himself. He was the favourite of George IV to such an extent that he had many enemies. Victorians hated his plaster surfaces, as this 1826 couplet reveals: '... But is not our Nash, too, a very great master? He finds us in brick and leaves us in plaster.' It was he who redesigned the West End of London, including Regent Street, Regent's Park, Trafalgar Square, St James's Park, Buckingham Palace and Marble Arch. He died at this age in 1835.

Helena Rubenstein, the Polish-born cosmetics tycoon, died in 1965, in New York, after amassing a vast fortune, all starting from a family formula for face-cream. She is commemorated in a striking portrait by Graham Sutherland.

Hannen Swaffer, the British journalist who was the founder of the 'gossip column', died at this age in 1962. 'Swaff' was one of the great characters of British journalism, a flamboyant personality resplendent in black hat, high collar and cravat, and long flowing locks.

Leo Tolstoy, the author of the two greatest novels to come out of Russia, *War and Peace* and *Anna Karenina,* died of a chill in 1910, at 82. In his middle years he was increasingly troubled by an inner conflict which eventually resolved itself in the later 1870s when he became a fanatical Christian of a non-orthodox type. He took to wearing peasant dress, worked as a farm labourer and tried to dispose of his estate, his property and all his belongings. This led to conflict with his wife, with whom he had produced fifteen children. He finally left her, at the age of 82, abandoning his large estate and setting off for a new, purified life. He was too old for the adventure and only managed to reach the railway station at Astapoyo. There he collapsed with a severe chill and died seven days later. As he lay dying, he refused to see his wife.

Sophie Tucker, the Russian-born American entertainer who was billed as 'The Last of the Red-hot Mommas', was still performing vigorously in 1965 at the age of 81, belting out her famous songs, such as *Some of These Days, Life Begins at Forty* and *My Yiddisher Momma.* Late in 1965 her health failed and she died early in the next year at 82.

Cornelius Vanderbilt, the American financier, also died at this age, in 1877. He had been a wilful child who had refused to attend school after the age of 11, and had gone into business when other children of his age were still playing with toys. By the time of his death he was one of the richest men in the world and the classic advertisement for America as the 'land of opportunity'.

Maurice de Vlaminck, the French Fauvist painter, who was also an avid racing cyclist and a café violinist, died in 1958. He became famous for his dark, forbidding landscapes of stormy weather, with wild skies and dark shadows. His savage brush-strokes and violent application of paint were the result of Van Gogh's influence. He was an impetuous 'young giant' who boasted that he had never been in a museum in his life. Once, in 1928, he discovered that there was an exhibition of forged paintings carrying his signature, and he burst in to the Paris gallery and attacked them with a knife, slashing them to bits.

Dame May Whitty, the distinguished British actress who made a second career for herself late in life, in Hollywood, died at 82 in 1948, shortly after the death of her husband, the actor Ben Webster, from whom she was inseparable. Throughout their fifty-five-year marriage, he brushed her long hair every night, until the day he died. Earlier in her life she had been a non-stop fighter for people's rights and the British actors' union, Equity, was founded in her London home, at Covent Garden.

Eighty-three is the year when 'things are not what they used to be'. Somerset Maugham was this age when he arrived for a stay at one of his favourite hotels, the Dorchester in London, in 1957, and there was a crisis when he found that they were no longer serving crumpets: 'When I was young one could have crumpets and muffins for tea. One cannot any more in this hard life we lead', he complained bitterly.

Marian Hart would have had little time for such trivia – she was busy making one of her several solo crossings of the Atlantic in a single-engined plane at this age, having learned to fly at 54.

Achievement at 83 Agatha Christie, the
immensely successful English detective novelist and playwright, whose books sold more than 100,000,000 copies, became a record-breaker at the age of 83. Her play, *The Mousetrap*, ended the longest continuous run in a theatre of any play in history, when she was this age. It ran, at the Ambassadors Theatre in London, from November 25 1952, to March 25, 1974. Even then it did not close, but merely moved 'down the road' to the St Martin's Theatre, where it was still running when Dame Agatha died in 1976.

Alexander Kerensky, the Russian revolutionary leader, was 83 when he wrote *Russia and History's Turning Point*. An opponent of the Tsarist government, he briefly became Head of State, having proclaimed the Russian Republic on September 15, 1917. He also assumed command of all the Russian armed forces, but by November 8, the Bolsheviks had staged a *coup d'état* and he was forced to leave Russia for permanent exile, first in London, then Paris and finally the United States.

Life-span of 83 Max Beerbohm, the British
author, caricaturist, dandy and wit, died at this age in Italy, in 1956, a few weeks after marrying his secretary-companion, Elizabeth Jungmann. Described by Shaw as 'the incomparable Max', he is best remembered for his brilliant novel of Oxford life, *Zuleika Dobson*, published in 1911.

Arthur Bliss, the Master of the Queen's Musick from 1953, whose rather astringent compositions included the music for the ballets *Checkmate* and *Miracle in the Gorbals*, died at 83 in 1975. His last composition, called *Shield of Faith*, a choral work, was performed a few weeks after his death, at the 500th anniversary celebration at St George's Chapel, Windsor.

Andrew Carnegie, the Scottish-born philanthropist who grew up in poverty but went on to become a multi-millionaire steel magnate in the United States, spent the latter part of his life giving his money away, through various Institutions and Foundations. In a famous essay of 1889, called *The Gospel of Wealth*, he was the first man to propose that the rich should dispose of their surplus wealth for the general welfare of society. He practised what he preached, to the tune of $350,000,000 before he died at 83 in 1919.

Maurice Chevalier, the French entertainer, died of a heart attack in 1972. His long career began in earnest when he became Mistinguett's dancing partner at the Folies Bergère, in 1910, and ended (as a voice only) in Disney's *Aristocats*, in 1970. In between he became the epitome of the French cabaret star, with the rare quality of being able to create an immediate atmosphere of intimacy, even with a huge audience. When his death was reported, the *Times* obituary commented sadly, 'Paris has lost another part of its history and its legend.'

Thomas Cook, who invented 'tourism', died at 83 in 1892. He began as a Baptist missionary, but then introduced the idea of the 'conducted tour' and founded Thomas Cook and Son, which became a world-wide travel agency. In 1856 he led the first Grand Tour of Europe, but it is doubtful that he could ever have imagined that his ideas would spread to become the vast tourist industry of modern times.

Edgar Degas, the French Impressionist painter who became the master of depicting the human figure in motion, died at 83 in 1917.

Walter de la Mare, the oil company employee who retired at 35 to give his full time to writing, produced many fine poems, novels, short stories and reviews, before dying at 83 in 1956.

Norman Douglas, the Scottish novelist and essayist, who grew up in a castle at Tilquhillie to become an eccentric, restless, wittily critical, hedonistic writer, died at 83 in Capri in 1952. He particularly disliked Christianity, socialism and any other form of puritanism. Among his provocative sayings are: 'Leisure is the curse of the poor in spirit'; 'All sentimentalists are criminals'; 'Socialism has its roots in envy and nothing else'; 'I always feel as if I needed a bath after talking about religion.'

Henry Ford, the American motor-car magnate, was a man with a complex personality. He was a diet-faddist, a runner, a folk-dance enthusiast, an Americana collector, a practical joker, and an unsuccessful politician. In 1915 he was a pacifist who tried to stop the First World War with a 'Ford Peace Ship'. Later, he published virulent anti-Semitic articles and in 1938 was awarded a medal by Hitler, whom he greatly admired. When he was a millionaire he still wore darned socks – mended for him by his wife. He doubled his workers' minimum wage, but employed informers to spy on them and used terror tactics against the unions. As a benefactor, his Ford Foundation became one of the major philanthropic institutions in the world. This complicated, contradictory man died in Dearborn, Michigan, at 83, in 1947.

Sigmund Freud, the Viennese father of psychoanalysis, felt that he would die young, but survived until he was 83. His later years were, however, a misery of suffering. In his late 60s he had a growth removed from his palate. It was cancer but he was not told. From then on he was in ever-increasing pain and during the last sixteen years of his life he underwent thirty-three operations, during which

83

most of the roof of his mouth was removed. In order to eat and talk he had to wear a false palate. He moved to London in May 1938, in flight from the Nazis who were burning his writings as 'Jewish pornography' – a former patient of his paying a £20,000 bribe for his safe passage. He died there the following year, with a little help from his doctor. On September 21, 1939, he reminded his physician that: 'You promised me you would help me when I could no longer carry on.' He was given 'adequate sedation' and died three days later.

Paul Getty, the President of the Getty Oil Company, was one of the world's richest men. Totally lacking in charisma and with no social graces, as he himself admitted, he was a shy, quiet, sad-faced man with a great business talent. Like Henry Ford he was a mixture of meanness and generosity. He installed a pay-phone for his house-guests at his English country mansion, but the Getty Museum in California was so richly endowed that it has an income of more than $1,000,000 per week. At his death at 83, in 1976, he was worth between $2,000,000,000 and $4,000,000,000.

Hippocrates, the father of medicine, died at this age in the year 377 BC. The Hippocratic Oath is an ethical code attributed to him and adopted by the medical profession ever since, all over the world.

Thomas Jefferson, the third President of the United States, from 1801 to 1809, and the principal author of the Declaration of Independence, died on July 4, 1826 – the fiftieth anniversary of the adoption of the Declaration.

George Jessel, the American entertainer known as 'America's Toastmaster General', whose show business career spanned seven decades, from the time he teamed up with Eddie Cantor in vaudeville when he was only 11, died at 83 in 1981.

Augustus John, the bohemian British painter who loved gypsies and spoke their language, and was the living proof that large quantities of alcohol do not always shorten the life-span, managed to live to the ripe old age of 83. In later life it was said that he needed two mistresses to lift him into bed after a night's imbibing. A gifted portrait painter, he was the archetypal artist of romantic imagination – robust, bearded, swashbuckling, free-loving and living from day to day.

Robert Menzies, the Australian politician, held the longest continuous tenure as Prime Minister in the history of Australia – from 1949 to 1966. He retired at 72 and died at 83, in 1978.

Ludwig Mies van der Rohe, the German-born architect who became the arch-priest of 'minimal' architecture in which all detail is eliminated on the exterior to create huge towering slabs of characterless monotony, died at 83 in 1969. He built the first 'all-glass' skyscraper in 1921 and in his hands the very daring of his austerity sometimes worked, but as an influence on lesser architects it was an aesthetic disaster, which created a minimal wasteland. His axiom was 'less is more', but for his untalented imitators it became transformed into 'less is a bore'.

J. Arthur Rank, the British film tycoon, died in 1972. Heir to a milling fortune, he virtually took over the British film industry in the 1930s. By the mid-1940s he owned more than half the British studios and more than 1,000 cinemas. Called 'King Arthur', he led the Rank Organization and dominated the British cinema for more than a quarter of a century. Although he was a Victorian Methodist in personality, he had the good sense to give creative talents a free hand, which was the secret of his success.

Alfred, Lord Tennyson, the English poet who succeeded Wordsworth as the Poet Laureate and became the most popular of Victorian poets, died at 83 in 1892. His popularity was such that the demands for his work made him over-prolific and he published many second-rate works as a result. His reputation did not survive the rapid change of tastes in the twentieth century.

Voltaire, the French author who championed tolerance – famous for his comment: 'I may disapprove of what you say but I will defend to the death your right to say it' – made a triumphant return to Paris, the city of his birth, in 1778. At 83, the excitement was too much for him and he died shortly afterwards. His famous last words, to a priest who was hoping for a deathbed conversion, were: 'For the love of God, let me die in peace.' Then, when the flame of the lamp suddenly flared up, he asked, 'The flames already?' and died.

James Watt, the Scottish engineer whose Condenser Steam Engine heralded the industrial revolution, died at this age in 1819. The unit of electrical power, the watt, was named after him.

Henry Ford

Eighty-four is the ideal age at which to make a parachute jump over Paris – one has so little to lose. Bernarr Macfadden, the American publisher, carried out this feat on his eighty-fourth birthday – October 12, 1953 – wearing spongy shoes to deaden his impact on the ground and red flannel underwear to protect himself from the cold. Seven ambulances and some 200 reporters stood by to watch him make his jump, but winds carried him half a mile off target. He landed well outside the giant police cordon that had been set up to protect him from the crowd that had gathered, but he was not at all put out, dancing a jig as soon as he touched the ground and announcing to everyone that he felt 'damn good'. In his childhood he had been a sickly orphan, worried that he might catch a disease and die like his mother. He had therefore started building up his body. As a young man he became the publisher of his own magazine, *Physical Culture*, in which he was able to spread his theories about curing diseases with fasting, health foods and exercise. He bought up other magazines and made a fortune of $30,000,000, giving some of it to a special foundation to perpetuate his ideas. As an old man he started a series of stunts, such as the Paris jump, to publicize his own health as proof of his theories.

Achievement at 84

Henri Matisse, the inspired French master of colour and the leader of the Fauvist painters, was still vigorously creating at the age of 84, in the year of his death. Living in a sun-bathed apartment in the Hotel Regina, overlooking Nice, cared for in his bedridden state by a faithful Russian woman who had once been his model, he defied his age and his ill-health to produce, at the very end of his life, some of his most daring and beautiful pictures. Even at this age, he was capable of transforming pieces of coloured paper into great works of art, tearing them into simple shapes to make collages of breathtaking colour and design.

Somerset Maugham was 84 when he published a series of essays written 'for his own pleasure' called *Points of View*, in 1958. One of the most successful authors in the world, he claimed to have sold a total of 64,000,000 copies of his books.

Claude Monet, the French Impressionist, was this age when he completed his last great work, a series of murals for the Orangerie of the water-lily pool in his garden. He had started on them nine years earlier, in 1916, and they constituted his crowning glory. He had built the garden himself many years earlier, diverting a stream from a nearby strip of marshland he had bought, next to his house. Once the lily-ponds had matured, he painted them over and over again, despite his failing sight, catching every condition of light and shade at different times and in different seasons. The final pictures are huge shimmering patches of colour, devoid of form and detail, forerunners of the style of art known as Abstract Expressionism.

The Snail by Henri Matisse

Life-span of 84

Elizabeth Arden, the American beauty expert, lied about her age, but, if we are to believe her entry in *Who's Who in America*, she was born in 1891, making her 75 when she died in 1966. More objective reports place her true life-span as 84 – a credit to her products. She was a tough, independent woman whose real name was Florence Nightingale Graham. A grocer's daughter, she ended up as one of the richest women in America, owning one of the world's biggest chains of beauty salons. A racehorse owner who won the Kentucky Derby, she had her horses massaged with one of her beauty creams. Despite her success as the Cosmetics Queen of America, she was a pessimistic conservative who saw decline everywhere.

Nancy Astor, the first woman to sit in the House of Commons, who devoted much of her parliamentary effort to improving the lot of women, and whose speeches were 'never dull, always brief and sometimes brilliant', died at Grimsthorpe Castle in Lincolnshire at the age of 84, in 1964.

Clement Attlee, the Labour Prime Minister of Great Britain from 1945 to 1951, died in his sleep in 1967. His three major achievements were to preside over the establishment of the welfare state at home, the granting of Independence to India, and the conversion of the British Empire into the Commonwealth of Nations. He was also instrumental in the nationalizing of major industries such as coal, steel and railways, and the Bank of England. A modest man, he abandoned the state apartments at 10 Downing Street and converted the servants' attics into a family flat.

A.J. Cronin, the Scottish physician and novelist who wrote *The Citadel* and *The Keys of the Kingdom*, died at 84 in Switzerland in 1981.

The Dalai Lama – the first in the long line of

84

leaders of the dominant order of Tibetan Buddhists – died in 1475. His name was Dge-'Dun-Grub-Pa and it was believed that he was reincarnated in the form of his successors. His fourteenth embodiment was forced to flee when China annexed Tibet in 1959.

Thomas Edison, the American inventor described as the 'genius of technology', who held over 1,000 patents, died at this age in 1931.

Thomas Edison's laboratory

Max Ernst, the German artist, served with the German army in the First World War, became a dadaist in the 1920s, a surrealist in the 1930s and then moved to America for the 1940s. He returned to France in 1949 and stayed there until his death at 84 in 1976. Described as having the 'most magnificently haunted brain in the world of art', his best paintings are among the purest examples of surrealism, both of the 'automatic' and the 'dream fantasy' kind.

José Iturbi, the Spanish concert pianist who was so precocious that he was already playing the piano at the age of three and teaching at seven, went on playing all his long life, until he died at 84 in 1980.

Joseph Lister, the English surgeon who pioneered the use of antiseptics and saved countless lives in the process, had the happiness of knowing that his methods had gained almost universal acceptance, before he himself sank into the terrible isolation of combined blindness and deafness in the final years of his life. Before this happened, he had found himself heaped high with honours, in which he had little interest. He asked for no personal reward, being a shy and unassuming man who believed himself directed by God. The end finally came for him, at 84, in 1912.

Oswald Mosley, the British Fascist leader, also died at this age, in 1980. As a young Member of Parliament in 1918 he had a great future ahead of him, but already acute observers were beginning to smell a rat. Socialist Beatrice Webb wrote in her diary: 'Here is the perfect politician who is also the perfect gentleman ... So much perfection argues rottenness some-where.' The rottenness finally came to the fore when he founded the British Union of Fascists in 1932 after abandoning both the Conservative Party and the Labour Party.

Isaac Newton, the English physicist and one of the greatest figures in the history of science, died in Kensington, at this age, in 1727.

Sean O'Casey, the Irish dramatist, who was born in a slum and whose writing presented life in a harsh, realistic light, but with a poetic use of words, died at 84 in 1964. He is best remembered for his play *Juno and the Paycock*, of 1924.

Jean Piaget, the Swiss child psychologist, attended Jung's lectures, read Freud and then set about applying his new knowledge to an understanding of child development. He wrote fifty books and became the world authority on the stages of concept-formation during childhood. Surprisingly, this was not his first academic interest. As a 15-year-old he had become an authority on molluscs and had published several papers on them. His papers were so impressive that he was offered a post as Curator of Mollusca at the Natural History Museum in Geneva – until it was discovered that he was only a schoolboy. He died at 84 in 1980.

Noah Webster, the great American lexicographer, who laboured for twenty years on his *American Dictionary of the English Language*, which contained entries for 70,000 words, died at 84 in 1843.

Clement Attlee

Eighty-five is an age for contemplation. Half-way through their 80s, people pause and consider the fact that somehow they have beaten the system. Most of their friends and loved ones are dead now, but they have miraculously managed to survive. Eighty-five is an age for quietly relishing this victory over death. Each morning brings a renewed pleasure from observing the simple, familiar sights and, as Sophocles said in the fifth century BC, 'Nobody loves life like an old man.' Indeed, for those 85-year-olds who have retained their health, a long life suddenly seems like a huge reward. There are drawbacks, as one retired professor of 85 put it: 'The sad thing about a long life is that you have to grow old to achieve it.'

Achievement at 85

Coco Chanel, the great fashion queen, was still ruling her empire at this age, when a Broadway musical based on her career was staged, with Katharine Hepburn in the leading role. It was a celebration of the way in which Chanel had liberated the female body after the First World War. It was she who was responsible for the death of the corset, the introduction of the chemise dress, the sweater, the shoe-string shoulder strap, the floating shoulder scarf, and costume jewellery, among many innovations that matched the new, more lively mood of the modern female.

Pope Pius IX broke all papal records at this age in 1878, when his pontificate became the longest ever (before or since) at thirty-two years. Although he was an amiable and witty pontiff, his long reign was not a peaceful one, involving as it did an uprising in Rome, the murder of several members of the administration and the Pope's eventual flight to Neapolitan territory. Returning later, he ended his rule with a period of self-imposed imprisonment in the Vatican, after refusing to recognize political changes in Italy.

Life-span of 85

Lord Beaverbrook, the Canadian-born press baron, died in 1964, after nearly half a century as one of the major figures in Fleet Street journalism. Although deeply involved in politics, especially during wartime, the greater part of his energy was given to the almost dictatorial running of his newspapers, the *Daily Express*, the *Sunday Express* and the *Evening Standard*.

Thomas Carlyle, the Scottish historian and essayist, died at 85 in 1881. His greatest work was his *History of the French Revolution*, and he is remembered especially for his interpretation of history in terms of great hero figures and heroic acts.

Ivy Compton-Burnett, the English novelist who wrote about family relations in middle-class society around the turn of the century, died at 85 in 1969.

George Cruikshank, the satirical and humorous artist, famous for his biting caricatures and his illustrations of Dickens's *Oliver Twist* and Grimm's fairy-tales, died at this age in 1878.

John Evelyn, the English diarist, died at his family estate, at Wotton in Surrey, where he had been born eighty-five years earlier. His diary, an important record of life in seventeenth-century England, was not found at Wotton until over 100 years after his death.

Haakon VII, King of Norway, a Danish prince who became a much loved monarch of Norway, was a true 'people's king'. A tall, thin, friendly man, he suffered a fall at the age of 83 and broke a thigh. It was the first time he had been to hospital in his long life. He died two years later in 1957.

Ernst Haeckel, the German evolutionist, 'more Darwinist than Darwin himself', lived until the age of 85, when he died at Jena in 1919. His major work, *The Evolution of Man*, which first appeared in 1874, raised a storm of controversy and was described by an outraged German theologian as 'a fleck of shame on the escutcheon of Germany'. But he lived long enough to see his work widely accepted and he turned, in later life, to a philosophical attempt to bridge the growing gulf between religion and science. His philosophy was called Monism and he defined it as 'the belief in the fundamental unity of all things'. It can best be described as a form of evolutionary pantheism.

Hermann Hesse, the German author who wrote *Steppenwolf* and *The Glass Bead Game*, in which the main theme is the tension between people's material needs and their spiritual longings, died at 85 in 1962. He was awarded the Nobel Prize for Literature in 1946.

Carl Gustav Jung, the Swiss psychiatrist, died at this age in 1961. He introduced the concept of a *complex* – such as the inferiority complex, the persecution complex, the Oedipus complex and the guilt complex – which had a major impact on psychological thinking. He defined a complex as 'a group of ideas unconsciously linked together as a result of repressed emotions or fears'. His Analytical Psychology also introduced the controversial idea of a 'collective unconscious'. In his 30s he worked closely with Freud until they had a disagreement and went their separate ways. Freud thought that Jung was too mystical and Jung thought that Freud was too sex-obsessed – and they were both correct. Ironically it was Jung who had the more involved sex life, maintaining a sexual triangle of wife and mistress for almost forty years, and justifying it by saying that a many-faceted gem needed more than a simple cube. Freud's sexual complexities were all in his head. Once when the two great men visited New York together, Freud complained to Jung that he couldn't get to sleep because he kept thinking about prostitutes. When Jung suggested that he do something practical about it, Freud was horrified.

Henry Kaiser, the American industrialist whose company became the largest and fastest of the world's shipbuilders, died at 85 in 1967. During the Second World War he pioneered prefabrication techniques that enabled him to launch one $1,000,000 ship per day, each taking only five days to build. He supplied one-third of all ships ordered by the US Government in the Second World War.

Lamarck, the French naturalist who coined the word 'biology' in 1802, died a scientific outcast, embittered, solitary, blind and in poverty, in the year 1829. His central idea – that it is possible to inherit acquired characteristics – has since been discredited.

Queen Mary, who had spent a quarter of a century as George V's queen and many more years as the Queen Mother, died at 85 in 1953. She had become Britain's grandmother figure, always stiffly erect, the personification of dignity and self-discipline. To some, she was an austere and formidable grandmother, lacking in spontaneity and charm, but she compensated for this with her ramrod decorum and restraint.

Queen Mary

Lord Nuffield, the British industrialist, died in 1963, after establishing himself as one of the greatest philanthropists of modern times. Starting out with a small bicycle shop in Oxford and building a vast motor-car industry, he found himself a rich man with no interest whatever in the luxuries of life. His reaction was to give his money away to deserving causes, especially in medical and education fields and, by the time he was 80, he had already handed over more than £27,000,000. As an employer, he paid high wages, provided holidays with pay long before others had thought of doing so, made high-quality sports grounds available and instituted welfare schemes. He disliked socialism and trade unionism, seeing them as reflecting failure on the part of employers.

George Raft, the American actor who appeared in more than 100 films (but never saw one of them), died of leukaemia in 1980. His best roles were as sinister, sleek-haired, almost expressionless gangsters, and the roles he played on the screen had a faint echo in real life. He was raised in New York's 'Hell's Kitchen' and grew up in the company of racketeers. Later in life he became intricately involved in gambling house ventures. Castro closed down his Havana gambling casino without compensation and, in the 1960s, he was refused entry into the United Kingdom to manage a gambling club in Berkeley Square, because of his underworld connections.

George Robey, a comic genius of the traditional music hall, half-way between gesturing clown and talking comedian, became a national figure in England and was the star of the first Royal Variety Command Performance, held before King George V. He was most famous for his sentimental duet during the First World War: 'If you were the only girl in the world and I was the only boy.' He was knighted in 1954 (a rare event for a comedian), about six months before his death at the age of 85.

Harriet Beecher Stowe, the American author, whose first novel, *Uncle Tom's Cabin*, in 1852, brought her international fame and became the best-selling book of the nineteenth century, died at 85 in 1896. She sentimentalized and exaggerated her characters in true Victorian fashion, but her work was important in its attack on the institution of slavery.

Richard Strauss, the German composer and conductor, at his best in his operas *Salome, Elektra* and *Der Rosenkavalier*, enjoyed great success and made a fortune from his music. His old age was ruined by the Second World War. His daughter-in-law was Jewish and he had to crawl to Nazi officials to save his family. He lost his fortune and his public standing, but the greatest blow came when Allied bombers destroyed his beloved opera houses at Berlin, Munich, Dresden and Vienna. He never fully recovered from this and, when the war ended, hid himself away, penniless, in Switzerland, refusing all calls to attend de-Nazification proceedings. Despite this, he was found guiltless and was invited to London to conduct a concert of his music in 1947, when he was already 83. He died two years later.

Serge Voronoff, the Russian-born French surgeon who attempted to rejuvenate human beings by grafting monkey glands into their bodies, died at 85 in 1951. He also grafted thyroid from monkeys into mentally defective children in attempts to make them normal again. On the French Riviera he had a zoo of monkeys used for grafting. His results were officially described as 'of little value' to the aged.

Mortimer Wheeler, the English archaeologist, whose most important field-work was on Roman Britain and the Indus Valley civilization of ancient India, died in 1976. He was the first serious archaeologist to attempt to popularize his science on radio and television, and in papers, magazines and books.

Vaughan Williams, the first composer since Elizabethan times to capture the spirit of England in his music, died at 85 in 1958.

Eighty-six is an age when work keeps you going. Those who enjoy life at this age, rather than merely exist, are usually those who are still working hard at their favourite projects. William Hocking, the American philosopher, was 86 when he said: 'I find that a man is as old as his work. If his work keeps him moving forward, he will look forward with the work.' Nothing could be more true of British archaeologist Margaret Murray, who was this age when she published her important work, *The Splendour that was Egypt*, and whose passion for her subject kept her going for another fourteen years and through several more books.

Achievement at 86

Jean Auguste Dominique Ingres, the leading French neo-classical painter of the nineteenth century, famous for such pictures as *The Turkish Bath* and *Odalisque with Slave*, was still actively working at this age – until within one week of his death in January 1867. All his life he had been fascinated by technique and his drawings and paintings show an almost clinically cold precision that makes them at once fascinating and daunting. Despite two happy marriages he died childless and left the contents of his studio, including many brilliant drawings, to the people of his birthplace, Montauban, where they are housed in the Musée Ingres.

Retirement at 86

Elizabeth Blackwell, the American gynaecologist, did not retire until a serious accident disabled her at this age, in 1907, ending an active career that lasted for fifty-eight years. The first woman to qualify as a doctor in the United States, she moved to London in middle age and became Professor of Gynaecology at the School of Medicine for Women, teaching and writing right up to the time of her accident.

Life-span of 86

Giacomo Balla, a founder member of the Futurist art movement in Italy in 1910, and the painter of one of the few important pictures to come out of that short-lived 'ism' of modern art, *Dog on a Leash*, in 1912, died at this age in 1958.

Giovanni Bellini, the father of the Venetian High Renaissance style, teacher of Titian and Tintoretto, died in 1516. His genius made Venice a centre for Renaissance art comparable with Florence and Rome. He died after a serene and prosperous career in both his artistic and personal life.

Édouard Daladier, the French Prime Minister at the time of the Munich Agreement when, along with Neville Chamberlain, he adopted a policy of appeasement towards Hitler, died at this advanced age in 1970. Unlike Chamberlain, when he returned to his native France after that fateful meeting, he was fearful and anxious about the response. To his amazement he was met by cheering countrymen. But the cheering did not last long and by 1940 he was arrested by the Vichy government and spent the rest of the Second World War in prison.

Naum Gabo, the Russian-born artist, one of the twentieth-century pioneers of abstraction who was a member of the original Constructivist group in Russia, died at 86 in the United States, in 1977. He disliked the idea that the individual must be suppressed for political ends in art, so he left the Russian group and moved to Berlin in 1922, then to England in 1935 and finally to America in 1946, where he remained for the rest of his long life.

Walter Gropius, the German-born architect who founded the Bauhaus in 1919, where many new ideas about art and design were fostered in the 1920s and 1930s, until Hitler's rise to power led to its closure, died at 86 in Boston. Gropius had exiled himself just in time and moved to the United States where he became Professor of Architecture at Harvard University, a post he held for many years. He had made a special request that his funeral should not be mournful, but marked in a festive manner. His Harvard friends accordingly gave a champagne party in his honour two days after his death.

Calouste Gulbenkian, the American oil magnate and art collector, known as 'Mr Five Per Cent' because he owned 5 per cent of the Anglo-Iranian Petroleum Company, died in Lisbon in 1955. He left £300,000,000 for an international trust – The Gulbenkian Foundation – to foster educational and artistic activities.

Frans Hals, the great Flemish portrait painter, most famous for his picture of *The Laughing Cavalier*, 1624, lived a long life in which he kept working until the very end. It is in the paintings he produced in his old age that his genius for portraying character is fully revealed. His last portraits are his masterpieces, despite the fact that they were painted at a time when he was worried by both financial and family problems.

Martin Heidegger, the German philosopher, also survived to this age, dying eventually in 1976. A leading exponent of existentialism, he supported Hit-

Édouard Daladier

ler in the early 1930s and was investigated by the occupying powers at the end of the Second World War. He was judged to have been only a 'passive' supporter and was allowed to return to his native Black Forest, where he spent his later years in retreat. He died in the room in which he was born.

Paul von Hindenburg, the German President from 1925 to 1934, who ruled with more dignity than skill and more prestige than cunning, died at 86 in 1934, allowing Hitler to take control. Hindenburg's victory over the Russians in the First World War had made him a national hero, but by the time Hitler was threatening, he was already a weak old man. He did everything he could to delay Hitler's rise to power, but by 1933 he could put off the evil day no longer and reluctantly made him Chancellor.

Paul von Hindenburg   Jomo Kenyatta

Jomo Kenyatta, the outstanding Kenyan President, died at this age in 1978. Some quarters feared a bloodbath would follow swiftly, as it had done in certain other countries in black Africa, following a presidential change, but Kenyatta's rule had left a country sufficiently stable to avoid this – a great credit to his administration.

Ivan Pavlov, the pioneer Russian physiologist, was still working regularly in his laboratory and his clinic until his death in 1936. A splendidly bold non-conformist, who once wrote to Stalin to castigate him for his bad treatment of intellectuals, he was the son of a priest and, despite popular opinion, was never at any time a communist. He went so far as to denounce communism, attacking its leaders by saying: 'For the kind of social experiment you are making, I would not sacrifice a frog's hind-leg.' He is most famous for his development of the *conditional reflex* (usually mis-

translated as the 'conditioned reflex'). In his now classical experiments he trained dogs to salivate at the sound of a bell previously associated with the appearance of food. This subject occupied him for over thirty years, but in later life he began to apply his laws to the explanation of human psychoses. He assumed that the massive inhibition typical of withdrawn psychotics was a defence mechanism shutting out the external world and its damaging stimuli. This thought became the basis for the Russian method of treating psychiatric patients in quiet, non-stimulating surroundings. Pavlov would be horrified if he had lived to see the terror treatments being used in Russian mental institutions today. (He has sometimes been accused of being the 'father of brain-washing', but such criticism is grossly unfair.)

Emily Post, the American etiquette expert, died of pneumonia in New York, in 1960. Her first book of etiquette, published in 1922, ran to ninety-nine printings over forty-seven years and sold 4,000,000 copies. Her basic message was 'to make the other person feel comfortable'. She hated affectation and loved good manners. According to her son, her goal was to 'tell the new rich how to behave like nice people'. Later on, she gave her attention to the new problem of 'how to be gracious without servants'.

Georges Rouault, the French Expressionist painter whose pictures looked like stained-glass windows – with bright colours edged in broad black lines like 'leading' – died at 86 in 1958. A decade before, when he was 76, he had made a bonfire of 315 of his canvases. They were pictures considered by him to be substandard, but which would have been worth millions today, had he left them in his studio.

Arnold Toynbee, the English historian whose major work was the massive, ten-volume *Study of History*, died at this age in 1975. His general review of his subject is an oddity of twentieth-century historical research, which has become increasingly narrow and specialized. By contrast, Toynbee's *magnum opus* has an enormously wide scope and reflects an incredible breadth of knowledge. It also, his critics claim, reflects an increasing obsession with certain religious themes that devalues it as objective historical writing.

Jack Warner, the English actor who began as a variety comedian, then became a radio comedian, then became a character actor in British films of the 1940s and 1950s, and finally ended his long career with the title role of the long-running television series *Dixon of Dock Green*, died of pneumonia following a stroke, in 1981. Through all phases of his work he epitomized the good-natured common man.

Jack L. Warner – another Jack Warner – also died at this age, in 1978. He was the American film producer who revolutionized the cinema by bringing sound to the screen. With his brother Harry, he founded Warner Brothers Pictures in the early 1920s and in 1927 they presented the first talkie – Al Jolson in *The Jazz Singer*. It created a sensation and heralded a new era in mass entertainment.

Eighty-seven is the oldest age at which anyone has been awarded a Nobel Prize. Francis Peyton Rous was this age when he shared the 1966 Nobel Prize for Medicine, for his research into the cause of cancer.

American senator Theodore F. Green gave his recipe for long life, at 87, with the following words: 'Most people say that as you get old, you have to give up things. I think you get old *because* you give up things.'

Achievement at 87
Bernard Berenson, the American art critic and the recognized world authority on the Italian Renaissance painters, abandoned America as a young man for an eighteenth-century mansion near Florence, where he lived for over half a century. In his old age he wrote an autobiographical trilogy which he completed when he was 87, in 1952. The three volumes were called *Sketch for a Self-Portrait, Aesthetics and History* and *Rumour and Reflection.*They brought him to the notice of a much wider public. At the same time, in 1952, his major work, *The Italian Painters of the Renaissance*, was reprinted in a new edition. In the introduction he concludes with words that sum up his life: 'No artifact is a work of art if it does not help to humanize us. Without art ... our world would have remained a jungle.'

George Burns, the American comedian, was 87 in 1983 when he gave a special performance for the Queen and Prince Philip on their first visit to California. At a relaxed dinner held in the Fox film studios, he proved that his memory, sense of timing and stage presence were as impressive as ever, even if one or two of the jokes were older than he was.

Misfortune at 87
Ferdinand de Lesseps, the French diplomat famous for building the Suez Canal, also undertook to construct the Panama Canal. Suez had been a success, but this time he failed and the company formed for the Panama project was liquidated in 1889. In 1893, when he had reached the advanced age of 87, following an official enquiry into the débâcle, de Lesseps was sentenced to five years' imprisonment. But he was so old that an appeals court reversed the decision and he was allowed to go free. He died the following year.

Retirement at 87
Arturo Toscanini, the Italian conductor, gave his final performance in April 1954, a few weeks after his eighty-seventh birthday. His farewell concert to the world was a gruelling programme of Wagner at Carnegie Hall. At the end, in a dramatic gesture, he dropped his baton, walked out of the concert hall, and did not return for the applause of his devoted fans.

Life-span of 87
David Ben-Gurion, the father of Israel, who had retired from all political activity at 84, to write his memoirs in a kibbutz in the Negev Desert, finally died at 87, in 1973.

Clive Brook, the British-born actor whose stock-in-trade was stiff-upper-lip suavity and stern authority and who was a famous leading man both on stage and on film in the 1920s and 1930s, died in 1974.

Avery Brundage, the American President of the International Olympic Committee, which directs the Olympic Games, died at 87 in 1975. He was the colossus of the modern Games, dominating the scene for twenty years – from 1952 to 1972. He was a dedicated amateur athlete, determined at all costs to keep professionalism out of the Games. He also fought a ruthless battle against other kinds of abuses, such as the use of drugs and the inclusion of sexual oddities, such as intensely masculine women. In private life he was a builder who claimed to have constructed half of Chicago. He married a 36-year-old princess in 1972, when he was in his mid-80s.

Chiang Kai-shek, the ruler of China from 1928 to 1949, died at this age in 1975, on his island retreat of Formosa, renamed Taiwan. A professional soldier, who received his military training not only in China, but also in Japan and Russia, he spent most of his rule fighting one enemy or another. First, it was the war-lords of Northern China, then the Chinese Communists, then the Japanese in the Second World War, and finally the Chinese Communists again, after the war. Losing this last conflict, he withdrew to Formosa and converted it into his final Nationalist stronghold, where he stayed until his death.

Chiang Kai-shek

Cicely Courtneidge, the Australian-born actress who was on stage for an astonishing total of seventy years, from 1901 to 1971, died in 1980 a few weeks after her eighty-seventh birthday. A vivacious comedienne, she was the toast of the London stage for many years with her husband Jack Hulbert, and was created a Dame in 1972.

Jimmy 'Schnozzle' Durante, the American comedian who did not need to wear a false nose to make people laugh at his face, died of pneumonia at 87, also in 1980. The Cyrano of Hollywood and Broadway musicals, with his gravel-voiced, honky-tonk piano-playing performances, he was one of the great originals of American show business.

Thomas Hardy, the West Country novelist, died at 87 in 1928. His ashes were placed in Westminster Abbey, but his heart was buried near his birthplace in his beloved Dorset. Sadly he wrote no novels during the last thirty years of his life because of his susceptibility to the attacks of the puritanical moralists of his day. *Tess of the D'Urbervilles* and *Jude the Obscure* came under heavy fire from the church. The Vicar of Wakefield, having announced that he had thrown his copy of *Jude* on the fire, then proceeded to have it banned from bookshops. Deeply hurt, Hardy spent the rest of his life writing poetry.

Robertson Hare, the English comic actor who was a star of a string of popular farces during the 1920s and 1930s and was still acting on stage and in television comedy series in his 80s, died at 87 of bronchial pneumonia, in 1979. Small, bald and bespectacled, he had only to open his plummy mouth to make audiences laugh, with such antique phrases as 'Double-dealing most damnable', 'Oh calamity' and 'Indubitably'.

Julian Huxley, the English zoologist who described himself as 'The only FRS who ever won an Oscar', died at this age in 1975. He was the first serious academic to attempt to popularize zoology, which he did with magazine articles, radio programmes and films. He was a distinguished evolutionist – which ran in the family – and a pioneer conservationist, in addition to being one of the first zoologists to study animal behaviour in the field.

Helen Keller, the American author and lecturer who was blind and deaf from infancy, but refused to give in to her disabilities, and forged an active life for herself against all the odds, eventually died at this age in 1968.

Groucho Marx, the sloping, loping comedian with the Hitler moustache, the flicked cigar, the flashing eyebrows and the comeback wisecrack, outlived his brothers Chico and Harpo by many years, dying from pneumonia in 1977. He tried going solo in his later years, but the magic of the great days of the Marx Brothers films could never be recaptured.

Ben Nicholson, the foremost British abstract artist, died in 1982, after half a century devoted to the subtle balance of circles and rectangles, and other simple shapes. A master of refinement, the whispers of his quiet paintings could easily drown out the shouts of noisier artists.

Thomas Hardy

Gillie Potter, the English comedian who lectured his audiences wearing a straw hat, a blazer and grey flannel trousers, holding a cane and a notebook, died at 87 in 1975. His favourite subject was the impoverished nobility, such as Lord Marshmallow at Hogsnorton whose drive gates were tied up with string.

Ezra Pound, the avant-garde American poet who turned to fascism during the Second World War, died in Venice in 1972. He had been living in Italy and stayed there during the war, making broadcasts for the fascists. After the war he was taken to the United States and charged with treason. Declared insane, he was put in a Washington asylum from 1945 to 1958 and was then returned to Italy. Although he was a major influence on modern poetry, his misdeeds have led to divided opinions on his lasting merit.

Marshal Tito, the moderate communist President of Yugoslavia, died at 87 in 1980. The son of a locksmith, he helped to set up the Communist Party in Yugoslavia and became a partisan leader in 1941 when the Germans overran his country. After the war he exiled King Peter and declared a communist state. He became its president in 1953 and maintained a clever balance between East and West – following an independent form of communism, and trading in all directions. He wisely allowed the peasants to keep their land and always opposed the stricter, Soviet form of collective communism.

Giuseppe Verdi, the Italian composer who represents the full flowering of the Italian operatic style, died at Milan in 1901, at this advanced age. *Rigoletto* in 1851, *Il Trovatore* and *La Traviata* in 1853, and *Aida* in 1871, helped to make him the giant of Italian opera.

John Wesley, the founder of Methodism, the breakaway Christian sect that scorned high church ritual and formality, and appealed in particular to the working classes, died at 87 in 1791.

Groucho Marx in *Animal Crackers*

88

Eighty-eight is not too late. It is a year when certain great eccentrics have simply refused to allow old age to interfere with their travels and their field-work. Dame Freya Stark was this age when she set off on yet another intrepid journey, this time with Sherpas and a television crew, pony trekking in the Himalayas. She was fulfilling a passion begun when she was a girl, and inspired by her godfather, to 'gaze upon mountains'. She ridiculed any suggestion that her age might create problems for her. Another determined field-worker, King Gustav VI of Sweden, took part in the excavations at the Etruscan ruins in Acquarossa, in Italy, when he was 88.

Achievement at 88

Konrad Adenauer, who had resigned as Chancellor of West Germany only a few weeks before his eighty-eighth birthday, was this age in 1964 when he started work on his memoirs. Despite his age, he also remained intensely active politically and created a storm by his interventions over the question of German reconciliation with France.

Pablo Casals, the great Spanish cellist, was still giving concerts at this age. His perfectionism and total devotion to his music seemed to endow him with an extra source of energy, enabling him to undertake exhausting world tours in his late 80s, the effort of which left him 'undisturbed'.

Michelangelo, the colossus of the Renaissance – painter, sculptor, architect and poet – who decorated the ceiling of the Sistine Chapel and was the chief designer of St Peter's in Rome, was working feverishly at the age of 88. As he neared his end, the man whose artistic achievements were already immense said in his last confession: 'I regret that I have not done enough for the salvation of my soul and that I die just as I am beginning to learn the alphabet of my profession.' Driven on to the very end, he was still working on his marble statue, known as 'The Rondanini Pietà', a few days before his death in 1564.

Life-span of 88

Robert Bunsen, the German professor of chemistry who invented the spectroscope, with which he discovered two new elements, Caesium and Rubidium, and developed the new science of spectrum analysis, died at this age in 1899. He is better remembered, however, as the man who gave his name to the Bunsen Burner, used today in laboratories all over the world.

Charles Chaplin, the comic genius of the cinema, died in 1977. Within four years of starting his film career he was capable of earning a million dollars in a single year (1917), but his professional successes were not matched by his progress in other spheres. Several marriages collapsed, there was a sex scandal and his political views led to a life in exile from 1952. Accused of being pro-communist, he replied 'I am a citizen of the world,' and made his home in Switzerland. In 1972 all was forgiven and he returned to an emotional reception in America and the presentation of a special Academy Award for 'the incalculable effect he has had on making motion pictures the art form of this century'. But instead of returning to California, he lived out his life in his Swiss villa. He never gave up his British citizenship and in 1975 was knighted by the Queen. After his death, his grave was robbed and his body mysteriously stolen but it was later returned.

Auguste Escoffier, the French master chef whose sixty-two years in the kitchen won him a worldwide reputation, died at 88 in 1935. His fame grew, in particular, from his time as director of the kitchens at the Savoy and Carlton Hotels in London. At the Savoy he invented his famous Pêche Melba in honour of Dame Nellie Melba, the opera singer. At the Carlton Hotel, one of his assistants was Ho Chi-Minh, the future President of North Vietnam.

Edith Evans, one of the finest actresses on the English stage for half a century, died in 1976, after a short illness following a stroke and a heart attack. In later life she became a much revered figure and provided a number of eccentrically brilliant cameos in films, right up to the year of her death.

Robert Frost, one of the greatest poets of his generation, died at 88 in 1963. He was an American farmer at heart – a nature poet – a man close to the land, and he defined good poetry with the following words: 'Like a piece of ice on a hot stove, a poem must ride on its own melting.'

William Gladstone, the Liberal leader and British Prime Minister, died in 1898. Together with his opponent, Disraeli, he had dominated British politics throughout the later Victorian age.

William Randolph Hearst, the American newspaper tycoon, died in Beverly Hills in 1951. He was the pioneer of tabloid, sensationalist journalism, whose famous witch-hunts exposed many scandals. As a collector he had few equals, as far as sheer quantity was concerned. His art collection was housed in his vast, $30,000,000 San Simeon Castle, which stood on 200,000 acres and had a six-mile driveway. The two acres of cellars contained most of his art objects. His 'buying period' lasted for forty years and, at its peak, he purchased one-quarter of all the art objects sold in the world. When he died he left his Castle to the University of California, but they refused it and the state took it over.

Arthur Keith, the British anthropologist who was one of the earliest students of human evolution and who wrote The Antiquity of Man in 1914, died at 88 in 1955.

Otto Klemperer, the German conductor who was an early champion of the modern composers, died in Zurich in 1973. Although he had been converted to Roman Catholicism, he was a Jew by birth and was forced to leave Germany in the 1930s. The rest of his long career was spent largely in the United States and England. A fall from the rostrum in 1933 led to a brain tumour, the treatment for which left him lame and partly paralysed, but nothing could stop him from conducting – even if he was unable to hold a baton.

Anita Loos, the mistress of the wisecrack and the clever line, died at 88 in New York in 1981. Her most famous work, *Gentlemen Prefer Blondes*, had a long and successful history. It began in 1925 as a novel which was turned into a stage production. Then in 1928 it became a film and was refilmed with Marilyn Monroe in 1953.

John Masefield, the British Poet Laureate best known for his poems of the sea, especially *Salt Water Ballads* of 1902, died at this age in 1967.

Henry Miller, the controversial author of *Tropic of Cancer* and *Tropic of Capricorn*, died at 88 in 1980. He wrote little in his later years but his reputation survived on the basis of his earlier books. They had been banned for many years and copies had to be smuggled out of Paris where they were published. Despite this they were widely read and Miller became a major 'underground author'. Then, with the greater sexual freedom of the 1960s, they appeared in the United States officially for the first time, a quarter of a century after their first publication. Miller was promptly prosecuted for obscenity, but in 1964 a Supreme Court decision went in his favour, and for the first time in his already long life (he was then 72) he became rich. In retrospect his writings seem slightly pretentious, but they undeniably broke new ground and have since given many serious authors the courage to deal frankly with sexual matters.

Igor Stravinsky, the Russian composer, died at this age in 1971. One of the most influential figures in twentieth-century music, his earlier works also caused considerable controversy. When his *Rite of Spring* was first performed in 1913, its savage moods and pagan fertility dances caused a riot which ended with police having to eject its first-night Parisian audience.

Vesta Tilley, the English music hall entertainer, who was one of the great characters of the Victorian and Edwardian music hall, died at 88 in 1952. She was already top of the bill when she was only 15, in 1879, and remained there for the next forty years. The best male-impersonator of them all, she always appeared in male attire.

Harry S. Truman, the thirty-third President of the United States and the only world leader, so far, to have ordered a nuclear attack, took office in April 1945, when Roosevelt died. At the time he was not even aware of the existence of atomic weapons. Four months later he had ordered the dropping of two atomic bombs on Japan to bring a quick end to the war. In 1950 he initiated the Korean War and a year later dismissed Douglas MacArthur when the general began to meddle in politics. He was also responsible for setting up NATO to contain the communists. He was, indeed, a man who never shirked difficult decisions and the famous sign on his desk (The Buck Stops Here) was one he took seriously. At home he

promoted the 'Fair Deal' policy to help the under-privileged. He attributed his long life, which did not end until 1972, thirty years after he ended his presidency, to his habit of taking a 'daily constitutional', which he continued to do well into his 80s.

Mae West in *Myra Breckinridge*

Mae West, the American actress who gave her name to a wartime inflatable life-jacket, also lived to the age of 88, defying time to age her. She was the daughter of a heavyweight boxer and was already an entertainer at the age of 5. Sex figured in her career even as a child, when she was billed in burlesque as the 'Baby Vamp'. After she had matured, she wrote, produced and directed her first play, in 1926, called simply *Sex*, for which she was jailed for ten days for obscenity. The show was closed by the police. The following year she directed her next play, *Drag*, on the subject of homosexuals. When she went to Hollywood she once again clashed with the censors, but solved her problem by adding humour to her erotic roles, thus defusing them. In her early 60s, she started a night-club act of stunning vulgarity, surrounding herself with huge muscle-men. She reappeared in films in 1970 at the age of 78 in the notorious *Myra Breckinridge*, and capped this, in 1978, at the age of 85, with an offering called *Sextette*, ending her career as lasciviously as she had begun it. She died of a stroke two years later, in 1980.

Eighty-nine is the year of the oldest hang-glider. Eighty-nine-year-old Londoner, Thompson Horan, the veteran of a dozen successful flights, was outraged when teachers at his local hang-gliding school refused to give him further lessons because of his advanced age, and he immediately lodged a protest with the British Hang-Gliding Association.

Achievement at 89

Mary Baker Eddy, the American founder of the Christian Science Movement, was still in control of 'The Church of Christ, Scientist', as she called it, at this advanced age in 1910. Her longevity and her ability to remain in charge of a large organization at 89 were excellent advertisements for her creed, which, stated simply, was that strength of mind could control the weaknesses of the body. In her earlier years she had suffered from ill-health – seizures and nervous collapse – but when she applied the principle of spiritual healing, she cured herself and decided to teach others to 'will themselves' better. Psychosomatic medicine has come a long way since her day and there is good evidence to support her basic belief, although our modern understanding of psychogenic influences on physical states does not require the religious setting she chose to give it.

Alexander von Humboldt, the German naturalist, explorer and pioneer ecologist, was still actively working on the final volume of his great five-volume *Cosmos* at the age of 89 – the year of his death. Yet another sickly child who overcame his youthful weaknesses, he spent five years exploring Central and South America, covering 6,000 miles on foot, in canoes and on horseback, between 1799 and 1804. For an encore, he went mountain climbing in the Andes to a height of 20,651 feet. When he returned to Europe, he produced thirty volumes of scientific reports on his expeditions, but then, in the last twenty-five years of his life, turned his attention to his *Cosmos*, an ambitious attempt to give a comprehensive account of the universe as it was known in the early nineteenth century. In 1859, when he was 89, he was busily preparing the fifth volume, with hardly diminished enthusiasm and vitality, but sadly did not live to see its final publication.

Arthur Rubinstein, the Polish-born American concert pianist, was 89 when he gave one of his greatest recitals at New York's Carnegie Hall.

Albert Schweitzer, the French missionary doctor, who devoted most of his life to caring for the sick at a hospital he established in Lambaréné in Gabon, West Africa, was still in charge of it at this age, in 1964.

Retirement at 89

Adrian Boult, the English conductor, retired at this age in 1979. It was as conductor of the BBC Symphony Orchestra that he made his name, his policy being to bring 'Music to the Masses'. Later he became conductor of the London Philharmonic and after that of the Birmingham City Orchestra.

Life-span of 89

Frank Brangwyn, the Welsh artist, died at this age in 1956. He was born in Belgium, at Bruges, where a Brangwyn Museum was opened in 1936. He was apprenticed to William Morris at the age of 15, ran away to sea at 17, and exhibited for the first time at the Royal Academy at 18. As a cabin-boy he was able to visit many countries and gain experience for his later paintings. In his mature work he became famous as a mural-painter on a large scale. He produced murals for the Skinners Hall in the City of London, the Canadian House of Parliament, Ottawa, the Court House at Cleveland, Ohio, and 'Radio City', New York. In personality he was described as 'the boy who had never grown up'.

Karl Doenitz, the commander of the German U-boat campaign in the Second World War, died at this age in Hamburg in 1980. He had served in U-boats in the First World War also, on which occasion he had been captured and placed in a Manchester mental hospital after exhibiting signs of insanity. In the Second World War it was he who devised the 'wolf pack' strategy, in which a group of U-boats together preyed upon convoys of Allied ships. His successes resulted in the rank of admiral in 1942. The following year he became Grand Admiral of the entire German navy. On the day Hitler died he proclaimed himself the new Reichsführer, but during the twenty-two days of his 'rule' as Nazi leader, his only major act was to surrender to the Allies on May 6, 1945. He was arrested on May 23.

Frederick Edridge-Green, one of the leading authorities on colour vision and colour blindness, died at 89 in 1953. He was the inventor of the colour perception spectrometer, used throughout the world for testing colour blindness, and wrote several important books on the subject, including *Colour Blindness and Colour Perception*, *The Physiology of Vision* and *Colour Vision*.

Friedrich Flick, the German multi-millionaire industrialist, died in 1972. A modest recluse, he was also an autocratic ruler of his vast iron, steel and coal empire, which was of great help to Hitler in the Second World War. After the war, he lost 80 per cent of this empire in the partitioning of Germany. He himself was imprisoned as a war criminal at the Nuremberg trials, but after he was released he set

Arthur Rubinstein

about rebuilding his businesses.

William Russell Flint, the Scottish watercolour artist, died in London, in 1969. Technically brilliant, he produced decorative watercolours that provided scenes of beautiful young naked girls, with sufficient art to avoid accusations of creating high-class pin-ups.

Mary Garden, the operatic soprano who was born at Aberdeen in Scotland and who died there eighty-nine years later, but who spent most of her life in Chicago, enjoyed a unique moment in 1921. It was then that she was appointed General Director of the Chicago Grand Opera Company, with complete artistic and business control – the first time that a Prima Donna had ever been given such a position in the history of opera. Her earlier claim to fame had been that in 1902 she had created the part of Mélisande in Debussy's opera, on the occasion of its première at the Opéra-Comique in Paris.

Otto Hahn, the German chemist who was the discoverer of nuclear fission, died after a fall in 1968.

Hokusai, the Japanese artist most famous in the West for his picture *The Breaking Wave of Kahagawa*, from his Mount Fuji series, died at 89 in 1849. He was not ready to go, but his appeals to heaven for 'yet another decade, nay, even another five years' for 'the old man mad with painting', went unheeded. He had become, for the West, the best known of all Oriental artists.

Hetty King, one of the two great male-impersonators of the Edwardian music hall (the other being Vesta Tilley, who lived to 88), died at 89 in 1972. The main characters she portrayed were 'the guardsman, the swell and the drunk'.

John L. Lewis, the most powerful labour leader in the United States in his day, and the creator of the modern American trade union, died at 89 in Washington, in 1969. He was passionately concerned with improving the lot of the American miners and was their president for forty years, but one opinion poll found him 'The most unpopular man in the country.' Hatred for him arose largely because of his refusal to honour the no-strike pledges given by the unions after the attack on Pearl Harbor. He called four wartime strikes and was denounced for behaviour little short of treason.

Compton Mackenzie, the Scottish novelist, died in 1972 in Edinburgh, only weeks away from his ninetieth birthday. To the end he retained the 'gaiety and undimmed zest for life of a teenager', as his *Times* obituarist put it. From start to finish his life had been full of bookish activities, beginning with teaching himself to read at the age of 22 months, in 1884, and ending with the publication of the tenth volume of his autobiography, in 1971. In between he wrote 100 novels and many other things besides. His most famous work – because it was beautifully filmed – was *Whisky Galore*. Asked the secret of his health and vigour, when he was in his 80s and was quietly enjoying a drink late one night in his London club,

From Compton Mackenzie's *Whisky Galore*

he gave the memorable reply: 'Never take any exercise.'

Alexander Neill, the Scottish school-reformer and the most distinguished figure in progressive education, died at 89 in 1973. Any child who is today enjoying a less rigid, less ruthless or less monotonous schooling, owes him a debt. At his famous, or notorious, Summerhill School, he allowed the children to govern themselves. Attendance at lessons was optional. There were no rules of authority or staff dominance. He insisted on easy familiarity with the teachers and he believed passionately in letting children develop in their own way, as *they* wished to. Summerhill became so famous that the pupils eventually banned sightseers.

Frederick Thomas, the doyen of British orientalists, who died in 1956, still worked several hours in his garden every day and was at his desk writing and researching until after midnight every night, all through his 80s.

Arturo Toscanini, the Italian orchestral conductor, died at 89 in 1957. He was the chief conductor at the Scala in Milan and later at the New York Philharmonic. As an anti-fascist, he left Europe in 1937 and spent the rest of his professional life with the NBC Symphony Orchestra. He did not return to Italy until his retirement in 1954. One of the greatest conductors of the twentieth century, with driving energy and an amazing musical memory (he never conducted with a score), he enjoyed sixty-eight years of active music-making.

Ethel Walker, the British painter, died at 89 in 1951. She was the most distinguished female artist of her generation.

Frank Lloyd Wright, one of the most original architects of the twentieth century, was criticized for creating only for a rich élite, but his ideas about integrating buildings with their environment contain lessons for all builders in the future. He died in Phoenix, Arizona, in 1959 at the age of 89.

Ninety is the year for enjoying a little finishing canter after the efforts of the race, according to Justice Oliver Wendell Holmes, the American Supreme Court judge, in his ninetieth birthday radio address on March 8, 1931: 'The riders in a race do not stop short when they reach the goal. There is a little finishing canter before coming to a standstill. There is time to hear the kind voice of friends and to say to oneself: "the work is done".' He was, however, slow to take his own advice and did not resign from the Supreme Court until the following January, after which he did enjoy a 'canter' of several years.

According to certain sex studies, 90 is the greatest age for enjoying sexual activity. The Kinsey Reports give 90 as the oldest age at which a *female* subject was sexually active, and a Duke University study found that this was the age of their oldest *male* subject. This figure should not be taken as representing a 'record age' for sex, however. It was simply the oldest age that turned up in moderate samples of the population. As will emerge later (at 94) there are still a few exceptional individuals active a little beyond the four-score-and-ten mark.

Achievement at 90

Margaret Murray, the British archaeologist, was 90 when she became President of the Folklore Society in 1953. She held the position until 1955. She had been interested in folklore for many years and had published books on magic and witchcraft some years earlier. First had come *Witch Cult in Western Europe*, which had caused such strong reactions that, in her next volume on the subject, *The God of the Witches*, she had to include a special request for a cessation of anonymous letters of condemnation. But nothing could stop the indefatigable Dr Murray. She was still publishing books at the age of 100.

Pablo Picasso, the most prolific artist in history, who created over 20,000 works, was still busily drawing and engraving at this age with a seemingly never-ending freshness of line. In Paris there was an unprecedented honour accorded him on his ninetieth birthday, when eight of his works were displayed in the Louvre in the place formerly occupied by da Vinci's *Mona Lisa*.

Life-span of 90

George Bancroft, the 'father of American history', died in Washington in 1891. His reputation was based on his comprehensive ten-volume study of his nation's origins and development. Called *History of the United States*, it took him forty years to complete (from 1834 to 1874), because he was frequently distracted by his political activities.

Clara Barton, the founder of the American Red Cross, died in 1912. She was still actively planning a new project at the time of her death – the establishment of a new Red Cross organization in Mexico, where it was sorely needed.

Daisy Bates, the Irish anthropologist who devoted her life to the cause of the Australian aborigines, died at 90 in Adelaide. She had gone to Australia in 1899 and lived among the aborigines, as a lone European, for thirty-five years. She referred to them as 'the last remnant of palaeolithic man' and they called her their *Kabbarli* or grandmother. She did everything in her power to help them, as they faced the advances and encroachments of the twentieth century, and published an important book on the subject in 1939, *The Passing of the Aborigines*. She refused to live in comfort as she grew older and continued her work in the field until, in her mid-80s, she collapsed and was forced, at last, to spend her final few years away from the tribesmen she loved.

John Boyd-Orr, the Scottish doctor who was so horrified by the malnutrition diseases he found in children from the Glasgow slums that he devoted his whole life to nutritional research, died at 90 in 1971. Described as an 'evangelist of peace through plenty' and as a 'practical farmer whose acres extended around the whole world', he gave himself up to the cause of abolishing starvation and famine. He became the first Director-General of the United Nations Food and Agriculture Organization.

Basil Cameron, the English conductor, died at 90 in 1975. Although he lacked the showy virtuosity of some of his colleagues, his consistency and reliability made him much sought after, with the result that he conducted in many places. Among the orchestras he conducted were the Symphony Orchestras of San Francisco, Seattle, the BBC, London, Budapest, and the Belgian National; also the Royal, Berlin, and London Philharmonics and the Philharmonia.

Giorgio de Chirico, the influential Italian artist, displayed a strange pattern to his life. In a short burst of early activity he produced works that made him a major international figure in the world of art, after which he went into one of the longest declines of any great artist. A forerunner of surrealism and a pioneer of twentieth-century fantastic art, his bizarre, enigmatic dreamscapes of 1911 to 1919 were masterpieces of mysterious atmosphere in which forbidding, deserted townscapes and silent menacing figures, often statues or dummies, await the performance of some indefinable act. Then, when his influence was at its peak, he suddenly pulled away, renounced his early work and spent the next fifty years painting inferior rubbish. He died at 90 in 1978.

Winston Churchill died on January 24, 1965, ten years after his retirement as Prime Minister. He spent his final decade producing his four-volume *History of the English-Speaking Peoples*. An exhibition of sixty-two of his paintings at Burlington House in London attracted 140,000 visitors in 1959. Late in life he developed an interest in the turf and became a racehorse owner. He was heaped with honours, the most unusual of which was being made an honorary citizen of the United States – something which had happened to no one else in history. On his ninetieth birthday there were special celebrations and then, less than two months later, he was dead.

Arthur Evans, the English archaeologist who rose to fame when he discovered the ancient city of Knossos in Crete, died in 1941. In 1899 he had purchased a tract of land that included the site of Knossos and after a year's digging he had unearthed palace ruins covering five and a half acres. He named the Bronze Age civilization that had inhabited the site *Minoan*.

Herbert Hoover, the thirty-first President of the United States, was the unluckiest of all presidents. His period in office should have been one of quiet progress and smooth stability; instead, it was a disaster. He was elected in March 1929 and in October of that year the stock market crashed and the country was plunged into its greatest depression. The result was that he was defeated by Roosevelt at the next election, in 1932, his biggest failure having been his handling of hardship relief. After this he played no important role in public life, although he did visit Hitler in 1938, returning with the advice that the United States should not ally itself with those countries that were opposing fascism. He died in 1964 at the age of 90.

Antony van Leeuwenhoek, the Dutch pioneer of the microscope, was this age when he died in 1723, at Delft. He was the first man ever to observe bacteria and protozoa under a lens. His research on lower forms of life refuted the doctrine of 'spontaneous generation'.

A. E. Matthews, the English actor famous for his elderly character roles in British comedy films, died at 90 in 1960. His *Times* obituary recorded his age as 80, but this was an error based on his own confusion about his age. He often claimed (or pretended) that he was ten years younger than his true age but, as he was touring South Africa as a professional actor in 1889, the older figure must be correct. He was still working in his ninetieth year and when asked how he managed it he replied: 'Easy! I look in the obituary column of *The Times* at breakfast and if my name is not in it, I go off to the studio.' It was said that, at the time, he was the oldest actor acting, anywhere in the world.

Winston Churchill

Florence Nightingale, the 'Lady of the Lamp', who fought endlessly against bungling officialdom to establish nursing as a serious profession with proper training, qualifications and uniforms for nurses, died at 90 in 1910. She was idolized by the troops in the Crimean War and her efforts to improve sanitation, cleanliness and correct feeding of patients started an entirely new epoch in hospital care.

Syngman Rhee, the President of South Korea, died at 90, exiled in Honolulu, in 1965. It was he who requested aid from the United States when his country was invaded from the North by communists in 1950, and this plunged America into a long and costly conflict. He was re-elected in 1956 and 1960, but became increasingly unpopular and had to resign and live in exile from April 1960.

Colonel Sanders, the founder of the Kentucky Fried Chicken food chain, died at this age in Shelbyville, Kentucky, in 1980 - a good advertisement for his products, assuming of course that he ate them himself.

Albert Schweitzer, the French missionary doctor who was awarded the Nobel Peace Prize for his efforts on behalf of 'The Brotherhood of Nations', died at his mission in Lambaréné in Gabon, West Africa. Between 1913 and his death in 1965, he spent nearly all his time there, developing his famous mission hospital from a few simple huts into a large complex, with 350 patients in the hospital itself and another 150 in the leper colony. There were criticisms of his methods from some quarters, but nobody could deny that he was a shining example of self-sacrifice in the service of humanity.

Sophocles, the great Athenian playwright, died in the year 406 BC. He was immensely prolific as a writer and extremely successful in his own time, enjoying a long and creative life. He was the master of classical tragedy, with death a central theme, and was technically the most brilliant of all the ancient Greek dramatists.

Alfred Russell Wallace, the English naturalist who, with Darwin, introduced the concepts of natural selection and evolution, died in 1913. It was on July 1, 1858, that his paper on natural selection was read, along with Darwin's, at a Linnean Society meeting in London, and set the evolutionary ball rolling in the scientific world. His range of interests and the subjects of his many books are amazingly wide: travel, zoology, botany, zoogeography, spiritualism, the supernatural, trade, land rationalization, history, astronomy, politics and vaccination.

Izaak Walton, the English author of *The Compleat Angler*, died at this age during a severe frost, in 1683.

Christopher Wren, the architect of St Paul's Cathedral, London, and the Sheldonian Theatre, Oxford, who was also co-founder and later President of The Royal Society, Professor of Astronomy at Oxford University, and twice a Member of Parliament, died at 90, after a long, busy and successful life, which did not end until 1723.

Ninety-one is the age at which you are a comfort to your neighbours. As Oliver Wendell Holmes explains, a man of this age 'is a picket-guard at the extreme outpost; and the young folks of sixty and seventy feel that the enemy must get by him before he can come near their camp'.

Achievement at 91   Alexandra Baldine-Kosloff, the Prima Ballerina with the Bolshoi Ballet in Moscow, who toured both Europe and the United States, was still actively teaching ballet at this age in the 1980s, in her southern California studio. She taught ninety-minute classes without taking any breaks and insisted on demonstrating even the difficult positions to her students.

Eamon de Valera

Eamon de Valera, the Irish nationalist leader and President of the Republic of Ireland, was still holding the presidency at this advanced age in 1973. He and his wife of sixty-three years then retired to a modest convalescent home near Dublin, where they both died in 1975. During his long life, he had been sentenced to death, reprieved, released and re-imprisoned, but had finally managed to escape. When he became political leader in Ireland, he broke all ties with the British and renamed the Irish Free State *Eire*. He remained neutral during the Second World War and was attacked for calling on the German Ambassador in Dublin to express his condolences on the death of Hitler. He was Prime Minister several times before becoming President in 1959.

Thomas Hobbes, the English philosopher, was still busily writing at 91, just a few months before his death in December 1679. His final years had been active ones. He had written his autobiography when he was 84 and translated Homer's *Odyssey* and *Iliad* when he was 86. He explained his mental liveliness in old age by pointing out that he had always been temperate and had only 'been drunk a hundred times' in over ninety years. He walked a great deal and played tennis even when he was 75. He smoked a pipe of tobacco each day after lunch. In his philosophy he rejected the supernatural and dismissed any religious basis for morality.

Konrad Adenauer

Adolph Zukor was still chairman of Paramount Pictures at 91, a post he accepted in 1936 when he was a mere 63. Prior to that he had been the president of the company for many years and was the most influential man in the film industry.

Life-span of 91   Konrad Adenauer, who had come to power late in life (at 73) died at 91 in 1967. He had always been an anti-Nazi and had been in trouble with Hitler for removing the Nazi flag when the Führer visited his native Cologne. For this he was sacked from his post as Lord Mayor of the city. After the war he founded the Social Democratic Party and led Germany's economic recovery.

Nicholas-François Appert, the French chef who became the 'father of food-canning', died near Paris at this age in 1841.

Enid Bagnold, the British playwright and novelist who wrote *National Velvet* and *The Chalk Garden*, died at this age in 1981.

Albert Baillie, the Dean of Westminster and the Registrar of the Order of the Garter for twenty-seven years, best remembered for his achievement in carrying out the restoration of St George's Chapel, Windsor, died at 91 in 1955.

Ossian Donner, the first Finnish minister to the Court of St James, who was instrumental in establishing Finland as an independent and viable state, died at 91 in 1957.

Alexandre Gustave Eiffel, the French engineer

who designed the 984-foot metal tower in Paris which has made his name famous, died at 91 in 1923, after conducting pioneering studies, late in his life, of aerodynamics.

Jean Henri Fabre, the French entomologist, was one of the great pioneers of accurate field observations of animal behaviour. With immense patience and precision he charted the lives of the insects in the French countryside where he lived. He reported his findings in many publications, the best known of which are *The Life and Love of Insects* and *Social Life in the Insect World,* both of which appeared shortly before the First World War. He died at 91 in 1915.

E.M. Forster, the English author of novels such as *A Room with a View* and *A Passage to India,* collected essays such as *Two Cheers for Democracy,* and the libretto for Benjamin Britten's opera *Billy Budd,* died at 91 in 1970, after which a secret early novel, *Maurice,* was published, revealing his homosexuality.

Sam Goldwyn, the American film tycoon, whose real name was Sam Goldfish, and who had arrived penniless in America as a child, died at 91 in Los Angeles at the beginning of 1974. The secret of his success, he said, was that 'I make my pictures to please myself.' He became so famous for his comments, such as 'A verbal agreement isn't worth the paper it's written on', or 'Anyone seeing a psychiatrist should have his head examined', that such remarks acquired the name of 'Goldwynisms'.

Gustav VI, King of Sweden from 1950 to 1973, and the last Swedish monarch to hold real political power, died at 91, two years after the constitutional reforms that reduced the head of the Swedish royal family to a figurehead role.

Lajos Kossuth, the Hungarian revolutionary leader, who inspired Hungary's struggle for independence from Austria, died at 91 in 1884, in exile. In 1848 he was responsible for abolishing the privileges of the nobles, freeing the peasants and creating a new ministry to run the affairs of the country. In the following year he became virtual dictator of the new republic, but his rule lasted for only a short while – the Austrians returned with Russia's help – and he had to flee abroad.

Somerset Maugham, the bisexual English novelist, playwright and short-story writer, died in the South of France in 1965, suffering from lung congestion. One of the most successful authors ever to write in English, his polished scepticism and dry sense of irony, coupled with an agnostic melancholy, produced writings that were always supremely entertaining, even though they somehow failed as works of literature.

Gilbert Murray, the English classical scholar famous for his impressive translations of the ancient Greek dramatists, died at 91 in 1957.

C.T. Onions, the oldest stipendiary fellow at Oxford University, was the last survivor of the team of lexicographers who worked on the *Oxford English Dictionary.* He joined the *OED* staff in 1895, when he was 22, and died at the age of 91 in 1965.

Zerna Sharp, the American creator of the *Dick and Jane* reading primers used in US classrooms for forty years, died at this age in Frankfort, Indiana, in 1981.

Jean Sibelius

Jean Sibelius, the Finnish composer and the leading musical figure in his country, who was greatly honoured in his own lifetime, died at 91 in 1957. Fascinated by Nordic mythology and landscape, he converted the emotions engendered by them into symphonies and tone-poems of lasting beauty.

De Witt Wallace, the American editor and publisher who founded the *Reader's Digest* magazine in 1922, died at 91 in 1981. In its first year his magazine had 1,500 readers. By 1970 that figure had risen to 20,000,000 in America alone, with many foreign editions as well, making a huge fortune for Wallace – all based on the idea of presenting readers with brief excerpts of writings they did not have the time, or inclination, to read in full.

Somerset Maugham

92

Ninety-two is the age of wistfulness. When he saw a pretty girl walking past, 92-year-old Oliver Wendell Holmes was overheard to say, with a sigh: 'What I wouldn't give to be seventy again!'

American humorist Don Marquis, the creator of *archy and mehitabel*, sees 92 quite differently, as the start of a wild, nothing-to-lose phase of life when one is so old that one can at last do exactly what one wants to do: 'Between the years of ninety-two and a hundred and two, we shall be the ribald, useless, drunken, outcast person we have always wished to be. We shall have a long white beard and long white hair; we shall not walk at all, but recline in a wheel-chair and bellow for alcoholic beverages – in the winter we shall sit before the fire with our feet in a bucket of hot water, a decanter of corn whisky near at hand, and write ribald songs against organized society; strapped to one arm of our chair will be a forty-five calibre revolver, and we shall shoot out the lights when we want to go to sleep, instead of turning them off; when we want air, we shall throw a silver candlestick through the front window and be damned to it; we shall address public meetings (to which we have been invited because of our wisdom) in a vein of jocund malice ... We shall know that the Almost Perfect State is here when the kind of old age each person wants is possible to him. Of course, all of you may not want the kind we want ... some of you may prefer prunes and morality to the bitter end.'

Achievement at 92

Fenner Brockway, the tireless peace campaigner, pacifist, disarmament protagonist, Labour MP and author of many books on left-wing political issues, was still busy in 1980 at the age of 92, when he published *Britain's First Socialists*. This was sixty-seven years after his first book, *Labour and Liberalism*, had appeared.

Life-span of 92

William Angliss, who left England to find his fortune abroad and arrived in Australia at 19 with only a few shillings in his pocket, went on to become a rich pioneer of the frozen meat trade, being one of the first people to attempt shipping meat over long distances by means of refrigeration, as early as the 1880s. He died at this age in 1957.

Alfred Chester Beatty, the mining millionaire, died at 92 in Monaco in 1968. One of the greatest figures in the mining industry, and one of the most successful, he became one of the major art collectors of the twentieth century. He removed his art collection from Britain in 1950 because he felt that the country was becoming dominated by bureaucracy and life there was regulated 'sometimes beyond reason'. When he left he took with him priceless collections of French snuff boxes, early watches, oriental manuscripts, the most valuable stamp collection in the world, thirty-five tons of rare books, and his Impressionist and modern paintings. He did, however, make many charitable bequests, including the establishment of the Royal Cancer Hospital in London.

Beulah Bondi, the American actress whose film career spanned half a century, died in Hollywood at this age in 1981. She specialized in widows, dowagers and grandmothers.

Adeline Bourne, the English suffragette and actress, died at 92 in 1965. At the turn of the century she was acting on the London stage with Mrs Patrick Campbell in plays with a (then) avant-garde feminist theme. Her most famous role was as Oscar Wilde's Salome, in 1911.

James Francis Byrnes, President Truman's Secretary of State from 1945 to 1947, was in charge of American foreign policy at the time that the 'cold war' developed between Russia and the United States. He died at 92 in 1972.

Peter Cooper, the American inventor who built the first American steam locomotive, called *Tom Thumb*, which initiated the great US railroad boom, died at 92 in 1883.

Margaret Deanesly, the British historian who wrote *The History of the Medieval Church*, in 1925, died at 92 in 1977.

John Dewey, the American philosopher and psychologist, died at this age in 1952. He brought a practical approach to philosophy, on the basis that 'the truth is what works'. He wrote and lectured widely on philosophy, ethics, logic, psychology and education – teaching that education is not a preparation for life, it *is* life. School should be a practical demonstration of life, not a formal prologue for it. 'Learning by doing' was one of his influential doctrines of progressive education.

Abraham Flexner, another American educational reformist, also died at this age, in 1959. His 1910 report on medical education in the United States brought about a revolution in the field and raised standards dramatically.

Knut Hamsun, the Norwegian novelist who won the Nobel Prize for his work *Growth of the Soil*, in 1920, was the only internationally famous author who sided with the Nazi tyrants during the Second World War. For his many followers he 'died' on April 9, 1940, when Germany invaded Norway and he joined the Quislings and became a traitor to his own country. After the war he was arrested and interned but was later released because of 'the deterioration in his mental powers'. In his mid-80s he set about writing what proved to be a brilliant report on his experiences during his three years of internment, which made nonsense of the claim that he was suffering from weakened mental powers. He died a second and final time, at 92, in 1952.

Laura Knight, the English painter who specialized in fairgrounds, gypsies, circuses and ballets, usually 'behind the scenes', died at this age in 1970.

Ludwig Koch, the German-born pioneer of recording animal sounds, died at 92 in 1974, at his home near London. The first man ever to record animal noises, he was an anti-Nazi who left Germany in 1936 and settled in England. A sleepless obsessive when recording, he often spent a week on a single subject. Altogether he recorded 200 species of birds and 100 species of other animals.

Walter Matthews, the Dean of St Paul's Cathedral in London from 1934 to 1967, and the author of *Purpose of God*, died at this age in 1973.

John Metcalf, the blind English road-builder known as 'Blind Jack', died at this age in 1810, leaving behind him ninety great-grandchildren.

Robert Moses, the American urban planner, known as 'New York's master-builder', responsible for Lincoln Center, Shea Stadium and the New York Coliseum, died at 92 in 1981.

Philip Noel-Baker, the British peer who won an Olympic Silver Medal in 1920 for the 1,500 metres, and the Nobel Peace Prize in 1959, died at his London home at 92 in 1982. He was a veteran anti-nuclear peace campaigner and a Labour Cabinet Minister who entered the House of Lords in 1977.

Vittorio Orlando, the Palermo-born Italian politician who was Prime Minister of Italy during the latter part of the 1914-18 war, died at 92 in 1952. His bombastic Sicilian behaviour was often ridiculed, but as a man of integrity he was respected. He was one of the Council of Four at Versailles, along with Clemenceau, Lloyd George and Wilson. A fanatical and unrepentant nationalist, his slogan was: 'I Hate Europe'.

Ada Reeve, the English music hall star and one of the gayest (in its old sense) of the Gaiety Girls, spent seventy-six years on the London stage. Her first appearance was in pantomime in 1878 and her last was at the Player's Theatre in 1954, singing songs from her repertory. She died in London in 1966, aged 92.

John Metcalf

Barnes Wallis

Charles Spencelayh, the portrait painter and miniaturist, died at this age in 1958. He mastered the detailed interior scene in very small paintings which were cheerful English pastiches of Dutch interiors. His favourite subjects were Dickensian 'Old Curiosity Shops' crowded with objects.

Cécile Sorel, the French music hall star, like Ada Reeve, her English equivalent, died in September 1966, at this same age. Unlike her English counterpart, however, she began as a serious actress, spending thirty-four years with the Comédie Française, playing in works by Molière, Dumas and Victor Hugo. During this time she already started causing scandals – one night abusing her distinguished audience as 'idiots' in the middle of her performance. A larger-than-life figure, she turned to the music hall when she reached the age of 60 and became the toast of the Casino de Paris. Her social life was spectacular and it was said of her that 'she got to know rather a large number of kings'. Her bed of solid gold and ivory had, at one time or another, supported Marie Antoinette and Madame du Barry. Dictator Mussolini greeted her with 'My dear emissary of Heaven, your visit will mark the highest point of my life.' When Marshal Foch embraced her, it was with such warmth that he left the imprint of his medals on her bare shoulder.

Barnes Wallis, the English inventor, responsible for designing the Airship R-100, which was Britain's most successful airship, died at 92 in 1979. He also invented ten-ton penetration bombs and the Variable Geometry (swing-wing) Aircraft, but he is best remembered as the man who devised the 'ducks and drakes' bouncing-bombs that destroyed the Moehne and Edar Dams in the Second World War, in 1943. This bouncing-bomb invention was immortalized in the feature film *The Dam Busters*. He was a childlike boffin with a vivid imagination that he could apply to solving problems in a most unusual way. In his 90s he was working on a *square* aeroplane that would travel at 5,000 mph.

George Walsh, the silent-screen star who, with his daring athletics, was able to rival Douglas Fairbanks for action scenes, died at 92 in 1981.

Laura Knight

Ninety-three is the greatest age for a sporting triumph. Henry George Miller, an American golfer, was this age when he scored a hole-in-one at the eleventh hole of a golf-course at Anaheim in California. The hole in question was 116 yards long. He had taken up the game in his 50s and went on to prove that it really is an 'old man's sport' by still playing actively at the age of 102.

Another remarkable triumph at 93 was that of orchestral conductor Paul Paray. The seemingly indestructible Frenchman had performed Mendelssohn's Violin Concerto at Yehudi Menuhin's Paris début in 1929 and he conducted it again, fifty years later, with the same soloist, in October 1979, at the opening by Prince Rainier of a new auditorium in Monaco. After the 1979 concert, the 93-year-old Paray flew off the next morning to give *fifteen* concerts in Israel, followed by a tour of Sweden. He died soon after the tour, but nobody could deny that it was a splendid way to go.

Achievement at 93   Vladimir Koppen, the

Russian-born meteorologist, was still busy at 93, trying to complete his huge five-volume work on climatology. He had begun editing it in 1927 and was as active as ever with it just before his death in June 1940. He is best remembered for his systematic method of identifying and mapping the world's climate. He was a strong advocate of Esperanto, spoke it fluently and translated some of his 500 publications into it, in the cause of world peace through a common language.

Leo XIII, who became Pope in 1878, was in such delicate health that there was speculation that his period in office might be brief, especially as he was already nearly 68. But his new role gave him extra staying power and he was still officiating as head of the church at the age of 93, in 1903, after a quarter of a century. He was regarded by many as the most brilliant pope of recent times.

Antonio Stradivari, the world's greatest violin-maker, was still active at the age of 93 with the help of his two sons, Francesco and Omobono. All through his long life he was constantly experimenting and modifying his violins, changing their size, their shape and all their small details. The secret of his special varnish, which is soft in texture and shades from orange to red, has never been discovered. He lived in Cremona in Italy from 1644 to 1737.

Conversion at 93   William Dubois, the

American historian, educator, author, reform leader and philosopher, was 93 when he was converted to communism. He obtained a doctorate from Harvard in 1895 and published his first book, *Souls of Black Folk,* in 1903. He then started crusading for the independence of African colonies and became a leading figure in the National Association for the Advancement of Coloured People. He published many more books for the Negro cause and was editor-in-chief of *The Encyclopaedia of the Negro* from 1933 to 1945. He became increasingly frustrated after years of struggle and eventually went to live in Ghana to work on an *Encyclopaedia Africana,* in 1960. A year later he announced his move to join the Communist Party. Two years later when he was 95, he was awarded Ghanaian citizenship, shortly before his death.

Life-span of 93   Adrian Boult, the English

orchestral conductor, died at this age in 1983. He conducted the Birmingham Municipal Orchestra from 1924 to 1930, the BBC Symphony Orchestra from 1941 to 1949 and the London Philharmonic Orchestra from 1950 to 1957, after which he returned to Birmingham as conductor of the City Orchestra there.

Mary Boyle, the Scottish prehistorian, died at 93 in 1974. For thirty-seven years she had been the collaborator of the Abbé Breuil in his pioneering studies of prehistoric cave art. A wife to him in all but bed, and a co-author, in the true sense, of all his work, she was at his bedside when he died in 1961. She herself wrote the books *In Search of our Ancestors* and *Man Before History.*

Charles Burnell, the English sportsman, was one of the greatest oarsmen of his day, winning many races at the turn of the century and becoming President of the élite Leander Club and the Henley Rowing Club. He held the latter post from 1952 until his death at 93 in 1969.

Lewis Casson, the British actor, director and producer, was, like his wife **Sybil Thorndike,** one of the major figures of the British stage. His first acting as a professional was seen in 1903 and his last – in *Arsenic and Old Lace* – in 1966. His sixty-three years on the stage were almost equalled in length by his marriage to Sybil, which lasted sixty-one years. His acting was slightly overshadowed by that of his wife, who was on stage even longer, from 1903 to 1969. Shaw wrote *St Joan* especially for her, and she was considered the First Lady of British acting for over half a century. Astonishingly, both Lewis Casson and Sybil Thorndike died at the age of 93, he in 1969 and she, seven years later, in 1976.

Charles Chapman, the English illustrator who, for thirty years, drew the Billy Bunter pictures to accompany Frank Richards's famous adventures at Greyfriars School, died at 93 in 1972. Even in his 90s he was still drawing Bunter pictures to amuse his friends and was active until a few weeks before his death. His secret for a long life: cycling, walking and a daily cold bath.

Zachary Cope, the British surgeon who became a world expert on the surgery of the abdomen, died in 1974. The author of many books, in later life he turned to the history of medicine and produced an important biography of Florence Nightingale. He once gave a thirty-minute lecture on the surgery of the duodenal ulcer entirely in verse.

Dudley de Chair, the English admiral who organized the North Sea blockade in the First World War and who was the senior admiral in the Royal Navy at the end of his life, died at 93 in 1958.

Keith Feiling, the British historian who wrote many books during his long professional life, including a biography of Neville Chamberlain and *A History of England*, died at 93 in 1977.

Arthur Gardner, the Oxford hockey blue who became Professor of Medicine at his old university and wrote books on microbes, bacteriology and penicillin, died at this age in 1978.

Duncan Grant, the English painter who was one of the first to be strongly influenced by the French post-Impressionists and who, along with Virginia Woolf, E.M. Forster, Roger Fry and others, became a leading member of the Bloomsbury Group, died at 93 in 1978.

Duncan Grant

Oliver Wendell Holmes, the son of the author of the same name, was one of America's greatest Supreme Court justices. He died in 1935, just two days short of his ninety-fourth birthday. He did not resign from the Supreme Court until January 1932, when he was nearing his ninety-first birthday.

Laurence Housman, the idiosyncratic English author, artist, poet, playwright, essayist and illustrator of fairy-tales, died at 93 in 1959. He had not expected such a long life, since he had made up his mind to die at 67, which he said was his 'lucky number'. Finding himself still vigorously active in his 70s, he wrote a volume of memoirs entitled *The Unexpected Years*, which pin-pointed the many contradictions in his life, making him at one and the same time an idealist and an iconoclast.

Tamara Karsavina, one of the greatest of the Russian ballet dancers, who performed with Nijinsky and had roles specially created for her in such works as *Firebird* and *Petrouchka*, died at 93 in 1978. She married a British diplomat and later became a teacher of Margot Fonteyn.

Oskar Kokoschka, the Austrian painter who moved to London in 1938 to avoid the Nazi regime, died at this age in 1980. He developed a highly personal Expressionist style, vibrant with movement and colour.

John Tenniel, the English cartoonist who is best remembered for his illustrations of Lewis Carroll's *Alice in Wonderland*, in 1865, died within three days of his ninety-fourth birthday, in 1914. For half a century he was the political cartoonist of *Punch*. He was, in fact, *the* political cartoonist of the Victorian era.

Harry Verney, the British politician who wrote a book about Florence Nightingale when he was 90, called *Florence Nightingale at Harley Street*, died at 93 in 1974.

Raoul Walsh, the American film director responsible for more than 100 Hollywood films between 1914 and 1964, when he retired, died at 93 in the winter of 1980–1. He ran away to sea as a boy and later became an assistant to D.W. Griffith. He lost an eye when filming *In Old Arizona*, in 1929, and wore a piratical black eye-patch ever after.

P.G. Wodehouse, the English comic novelist who invented the immortal Bertie Wooster and his 'gentleman's gentleman' Jeeves, died at Long Island at 93 in 1975, six weeks after receiving a much overdue knighthood. This honour due to him as the master of English farce had been delayed because of a naïve blunder he made during the Second World War, when he gave a series of broadcasts from Berlin to the USA. He had been captured by the Germans in 1940 and interned. Released when he was 60, he was forced to stay in Germany and in Berlin he made the fatal radio transmissions. Although they did not favour the Nazis, he was attacked for committing treason and could not return to his beloved England after the war. Instead he became an American citizen and settled in the United States for the final decades of his long life. He was a quiet, gentle, self-effacing man – not at all the picture of a literary giant – who wrote ninety books, twenty film scripts and more than thirty plays and musicals. It is entirely in character that the sole reason for his wartime crisis was his refusal to subject his Pekinese dog 'Wonder' to the brutal British quarantine regulations. It was because of this that he did not return to England from his house in France when the Germans started to over-run western Europe.

Ninety-four is the oldest recorded age for successful paternity, according to the journal of the American Medical Association in 1935. This is, however, an extremely unusual case, most men becoming impotent before they reach their 90s.

For the typical 94-year-old, this is the gerontologist's official age of 'senility', or the 'vegetative period', which lasts until death. Only a few very exceptional people escape the weakening effects of reaching this advanced age.

Achievement at 94

Bertrand Russell, the English philosopher and mathematician, was still active in the field of international peace drives and nuclear disarmament at the age of 94. Ever since the exploding of the first hydrogen bomb, he had devoted his old age to an increasingly shrill attack on the follies of politicians in dragging humanity towards what he called 'universal death'. The climax of his anti-war campaigning came in his mid-90s, when he set up an International War Crimes Tribunal in Stockholm at which a gathering of 'celebrated intellectuals' tried the United States (*in absentia*) for war crimes in Vietnam – and returned a unanimous verdict of *guilty*.

Life-span of 94

Norman Angell, the English political commentator, who was the author of *The Great Illusion*, a controversial book published just before the First World War, in which he put forward the idea that in modern warfare the victor would lose almost as much as the vanquished, died in 1967. Although it was an argument based on cold logic, it was wrongly attacked as a pacifist statement and this caused its author some distress. Wounds were healed, however, when he was awarded the Nobel Peace Prize in 1933. He remained active in the field of political comment until he was in his 90s and was already a nonagenarian when he made his last, two-month-long tour of the United States, lecturing on education for the nuclear age.

Sydney Barnes, the English cricketer, also died at this age in 1967. Up to the time of the First World War, only two bowlers were rated as belonging to the 'supreme class' – Spofforth and Barnes. His career in Test Cricket ended in 1914.

Bernard Baruch, the American financier whose influence as an adviser to men of power was enormous, died at 94 in 1965. Although he rarely held any major office, a measure of his role in the political scene is given by the collection of his letters which he presented to Princeton University in 1964. They included 1,200 letters to him from nine presidents, and no fewer than 700 communications from Churchill.

Bernard Berenson, the American art critic who lived in Italy from 1900 and was the most eminent connoisseur of Renaissance art in the world, died at 94 in 1959.

Robert Cecil, one of the sons of Prime Minister Lord Salisbury, died at 94 in 1958. He is best remembered as one of the founders of the League of Nations, for which work he was awarded the Nobel Peace Prize in 1937.

Edward Craig, the 'Last Victorian of the English Stage', died at 94 in 1966. He was the son of Ellen Terry and spent the first half of his life acting and the second half as a 'professional thwarted genius', referred to by his enemies as the 'spoilt child of artistic Europe'. He had grandiose ideas about theatrical design, but few of them saw the glare of the floodlights. In a broadcast he made at the age of 88, he cheerfully comments: 'Genius and service do not go together. Great Britain therefore showed good sense in not trying to make more use of my genius.'

Archibald Creswell, the outstanding authority on Islamic architecture, died in 1974. A small, neat figure, always dressed in a high starched collar, even in the intense heat of the Middle East, he was much feared for the ferocity of his walking-stick, with which he would thrash anyone found thrashing a donkey, or being cruel to any animal.

Robert Davis, the English inventor who devoted the whole of his life to the development of improved apparatus for breathing under difficult conditions – either high in the air or diving beneath the water – died at 94 in 1965. He also perfected the first successful oxygen breathing apparatus for mining rescue work. He was knighted in 1932 for the valuable work he carried out – to the exclusion of all other interests.

Douglas Fawcett, the mountaineer-philosopher, died at this age in 1960. His two major philosophical works were *The Zermatt Dialogues* and *Oberland Dialogues*, published in the 1920s, in which he put forward a form of idealism called 'Imaginism', in which he presented the imagination as the fundamental reality of the universe. He made an annual ascent of the Matterhorn, but when he was 66 he had a heart attack half-way up. He spent the night in the snow and then, the next morning, completed the climb. But it was his last and, in order to continue enjoying the peaks, he had to learn to fly at 68, so that he could pilot a small plane among the dangerous Alpine currents. When this too was denied him he became a chess champion and was still playing competitively at 94, a few months before his death.

Isobel Field, the stepdaughter of Robert Louis Stevenson, who wrote about him vividly in her autobiography, *This Life I've Loved*, died at 94 in California in 1953.

Karl von Frisch, the Viennese zoologist, whose brilliant researches into the language of honey-bees led to his sharing the Nobel Prize with Konrad Lorenz and Niko Tinbergen, as the three founding fathers of animal ethology, died at the advanced age of 94 in 1981. He lived a simple, almost peasant, life in the open air, spending hours observing the dances of his bees and the way they visited flowers in the countryside. This healthy existence and a quiet, modest, almost shy personality seem to have contributed to his long life, along with that most important ingredient for longevity – success at what he wanted to do. So extraordinary were his results that he exclaimed: 'No competent scientist *ought* to believe these things on first hearing.' However, after several highly competent scientists had repeated his experiments, there was no escaping his amazing discoveries.

Gaston Gallimard, the most influential French publisher of the twentieth century, who published Gide, Valéry, Apollinaire, Camus, Malraux and Sartre, died at 94 in 1975.

Lord Goddard, the Lord Chief Justice of England from 1946 to 1958, who was the embodiment of authority and power on the bench, died at this age in 1971.

Joseph Hooker, the English botanist famous for his *Flora*, helped his father William Hooker to found the Botanical Gardens at Kew and took over from him as director in 1865. He accompanied many expeditions as a field botanist and made detailed records of his observations. He died at 94 in 1911.

Robert Muir, the eminent pathologist, who published his important researches in a book called *Studies on Immunity*, also produced two major textbooks, *Manual of Bacteriology* and *Text-book of Pathology*, during his long career. He died at 94 in 1959.

Jean Prouvost, the French press baron, publisher of *Paris-Soir*, *Marie-Claire* and *Paris-Match* among others, died at this age in 1979.

Marie Rambert, the Polish-born English ballet teacher who founded and directed the Ballet Rambert from 1931 until her death at 94 in 1982, was one of the great figures of the ballet world, from whose school many fine dancers emerged.

Leonard Rogers, the distinguished pioneer of tropical medicine, who made many important advances in the treatment of lethal diseases, managed to survive until the age of 94 despite the fact that he was repeatedly exposed to everything from amoebic dysentery to cholera and leprosy. He was still publishing articles in medical journals until within a few years of his death in 1962.

George Bernard Shaw was a vegetarian, teetotal, non-smoking socialist, but his writings were anything but austere. Their flippancy and irreverent wit irritated many people, but his brilliance as a playwright made it impossible to ignore him. As a boy he had taken little interest in school and came to London as a shy youth of 20. There he took to hanging around the British Museum to improve his education and began writing a series of unsuccessful novels, which publishers refused. Much later, when he began writing for the theatre, he suffered similar failures at first and retreated into the role of a dramatic critic. But then his plays gradually began to attract audiences and by the time he had reached his late 40s he was fully established as a leading playwright. By 1924 he was the most famous playwright in the world and in 1925 he was awarded a Nobel Prize. Plays appeared

George Bernard Shaw

from his pen almost annually and late in his life several, such as *Pygmalion*, *Major Barbara* and *Caesar and Cleopatra*, were filmed, bringing further fame and fortune. But as an old man he was beginning to suffer from the debilitating effects of his strict vegetable diet, and at the age of 82 could only be kept alive by breaking his vegetarian code, much to his distress. With the appropriate injections, however, he was able to enjoy an even longer life and to write several more plays before he died at the age of 94 on November 2, 1950. His goal in life had been to goad people into thinking for themselves, and in that he had certainly succeeded.

Charles Sherrington, the English neurologist and physiologist, who published 200 scientific papers and who won the Nobel Prize in 1932 for his work in studying the human nervous system and for mapping the brain, died at 94 in 1952.

Ben Travers, the English master of bedroom farce, who wrote *A Cuckoo in the Nest*, *Rookery Nook* and *Thark*, among many others, and who dominated the London theatre in his special field in the 1920s and 1930s, died in 1980. He was still writing and staging his plays when in his 90s and, only a few years before his death, was demonstrating on television the daily gymnastic exercises which he claimed kept him fit. Like Shaw, however, it was probably his puckish sense of humour, and his refusal to stop doing the work he loved, that kept him alive and vigorous to such an advanced age.

Henry van der Velde, the Belgian originator of Art Nouveau and a pioneer of the modern movement in applied design, died at 94 in 1957. He revolted against the academic sterility of earlier designers and brought a new freshness to the design of clothes, furniture, glassware, jewellery, ceramics, wallpapers and textiles.

Clough Williams-Ellis, the British architect, who created the village of Portmeirion in North Wales, described as 'one of the most entertaining architectural follies of all time', died at 94 in 1978.

Ninety-five is the oldest age for a virtuoso concert performance. Arthur Rubinstein, the Polish-born American concert pianist, gave his last public concert in April 1982 at the age of 95, eight months before his death. He was a child prodigy who made his début at the end of the nineteenth century and whose public performances spanned nearly ninety years, surely the longest professional career in the history of music.

Life-span of 95

Allan Aynesworth, the English actor, who first appeared on the London stage in *The Red Lamp* in 1887 and ended his career in 1938 in *Victoria Regina*, died at this age in 1959. During his long career he acted in twenty-five London theatres in plays written by over fifty contemporary playwrights. He appeared before five British monarchs: Queen Victoria, King Edward VII, King George V, King George VI and Queen Elizabeth II. He reached his peak of fame as the star of Oscar Wilde's *The Importance of Being Earnest*, in 1895. He kept fit into old age by indulging his country pleasures of fishing, shooting and gardening, and by being 'devoid of ill-humour'.

Archibald Bodkin, the distinguished prosecuting counsel who spent thirty-five years of his life at the Bar before becoming the Director of Public Prosecutions from 1920 to 1930, died at 95 in 1957. He sent thousands of his fellow Britons to prison and not a few to the gallows during his period in office, but in private his personality was described as 'genial and bustling'.

Florence Booth, the Salvation Army leader, was apparently in no hurry to meet her Maker, living to the ripe old age of 95, when, like Archibald Bodkin, she died in 1957. She was married to General Booth, the son of the founder of the Salvation Army, who took over command in 1912. She herself became the leader of the Army's work among women and toiled hard for the cause of female emancipation.

Guy Dain, the British doctor, was chairman of the British Medical Association at the time of the setting up of the National Health Service, during which his quiet, sane voice kept a sensible balance among all the political upheavals. He died at 95 in 1966.

Lewis Halliday, the Adjutant General of the Royal Marines, was the oldest surviving holder of the Victoria Cross. He won it as a member of the guard of the British Legation in Peking in the Boxer Uprising of 1900. He had led an attack in which he was shot through the lung at point-blank range, 'the bullet fracturing and carrying away part of the lung'. Despite his injury he killed three of his assailants and then walked back unaided to the hospital 'so as not to diminish the number of his men engaged in the sortie'. When he was 94 the modest Halliday wrote a letter to the *Royal Marines Journal* stating that this official citation was 'wildly incorrect' – he had in reality only walked back unaided to the defensive wall, after which he accepted help to get to the hos-

Philippe Pétain   Wilfred Rhodes

pital. Despite his severe injury, he did not die until sixty-six years later, in 1966.

George Lanchester, the British pioneer of automobile engineering, who designed and built an experimental car in 1897, died at 95 in 1970. His designs gave birth to the Lanchester armoured car of the First World War, and one of his later cars broke many track records at Brooklands in the 1920s. Daimler eventually took over his Lanchester company.

Mary O'Hara, the American author of *Green Grass of Wyoming, My Friend Flicka* and *Thunderhead*, died at this age in 1980.

Philippe Pétain, Marshal of France, died at this age in 1951. As Commander-in-Chief of the French Armies in the First World War he was a great hero, but he became disgraced in the Second World War. Then, in a political role, he became Prime Minister after France had been overrun by the Nazis and voted for surrender, on the grounds that this would bring softer peace terms and that he could rebuild a new France, even under German domination. He retained as much dignity as possible as a puppet of the Germans, but when the Allies liberated France, he was removed to Germany. On his return to France at the end of the war in 1945 he was tried as a collaborator and a traitor and condemned to death. He was then 89. De Gaulle – perhaps remembering the Marshal's heroic deeds in the First World War – reduced the sentence to life imprisonment, and he was transferred to the Island of Yeu, fifteen miles off the French coast, and detained there until his death six years later.

Wilfred Rhodes, the famous English cricketer who was the oldest person to play Test Cricket, died at 95 in 1973. He played for England against the West Indies on April 12, 1930, when he was 52 years and 165 days old. Next to W. G. Grace he was considered to be England's greatest all-round cricketer. After he retired in 1930 his eyesight went and he was soon completely blind, but continued to visit Test Matches to enjoy them – and to follow them closely – by sound only.

Fred Russell, the English ventriloquist who was

performing from 1886 to 1952, a professional career lasting sixty-six years, from the heyday of music hall to the advent of television, died at 95 in 1957.

Leopold Stokowski, the flamboyant orchestral conductor, died in his sleep at 95 in 1977. Born in London, the son of an immigrant Polish cabinet maker and his Irish wife, he soon proved himself to be a precocious musical talent, learning to play the violin, piano and organ before he was 13. When he was still in his 20s he had become the conductor of the Cincinnati Symphony Orchestra. From there he moved to Philadelphia where, over thirty years, he created a symphony orchestra that became world famous. In the late 1930s his fame increased when he performed on film as the conductor in Walt Disney's *Fantasia*. Although he was criticized by the musical purists for mixing classical music and cartoons, the film proved to be a 'classic' in its own right and was repeatedly re-released around the world. In his private life, Stokowski gave as much of his time and his considerable energy to seducing women as he did to his music. He had affairs with a wide variety of women, from society snobs to actresses and chambermaids. His penchant for the female students at the Curtis Institute of Music in Philadelphia gave it the nickname of 'Coitus Institute'. He referred to his women as 'angels of mercy who rejuvenate'. In 1937 he seduced Greta Garbo who reported that she felt electricity going through her 'from head to toe'. When he was 63 he married for the third time – on this occasion to the 23-year-old heiress Gloria Vanderbilt. She divorced him when he was 80, after which he returned to his round of casual affairs until his death at 95.

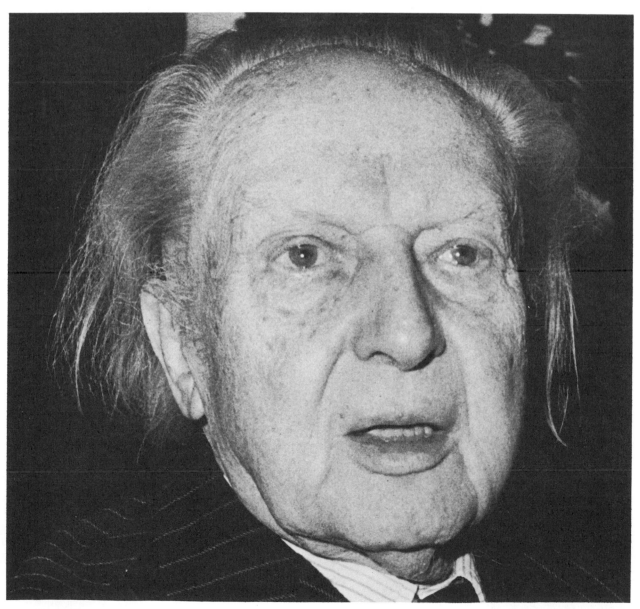

Leopold Stokowski

Ninety-six is the age at which Grandma Moses held her first exhibition of paintings in London, at the Matthieson Gallery, following her enormous success in the United States. She was still actively painting at this age and continued to do so for several more years.

Life-span of 96 **Havergal Brian**, the English composer, was a tragic example of a creative artist who despite being almost totally ignored all his life, nevertheless persisted in his work, year in and year out. He completed no fewer than thirty-two symphonies, sometimes for orchestras so vast that nobody could afford to stage them. His persistence in composing hugely ambitious, unplayed works until his death at 96 in 1972 can only be described as heroic. **Pablo Casals,** the Spanish master of the cello, known particularly for his interpretations of the works of Bach, died in 1973. When the fascists won the Spanish Civil War he exiled himself and lived in the French Pyrenees and in Puerto Rico. When he was 80 he married Marta Montanez, a young Puerto Rican student of his, who cared for the great man in his later years. For those who believe that smoking leads inevitably to an early death, it has to be mentioned that Casals, at 96, was still puffing away on his 'apparently inextinguishable pipe'.

Pablo Casals

Havergal Brian

Ninian Comper, the Scottish architect who specialized in religious buildings and liked to design them 'from the altar outwards', died at 96 in 1960. His best-known works are St Mary, Wellingborough, the Crypt Chapel of St Mary Magdalene, Paddington, and the Warrior's Chapel in Westminster Abbey.

Kathleen Courtney, who was active in Women's Suffrage until 1914, then worked for the League of Nations, and later became chairman of the United Nations Association, died in 1974. A formidable chairman of any meeting, stern but courteous, she cut like a knife through the butter of confused discussion.

Friedrich Forster, the German educationalist and political writer whose books were publicly burnt by Hitler in 1933, died at 96 in 1966, in a Swiss sanatorium near Zurich. He had to flee the Nazis in the 1930s and went to the USA, but all his long life he laboured to attack German militarism. When 95 he was praised by the German Roman Catholic Church as 'the Patriarch of the Good Germany'.

Mary Grey, the Welsh-born actress, who enjoyed a long and distinguished career on the stage, both in London and New York, died at 96 in 1974. Her performances covered a large part of the twentieth century and her highest achievement was as Madame Ranevsky, to John Gielgud's Trofimov, in Chekhov's *The Cherry Orchard*, in 1925.

Fletcher Harvey, the American physicist whose work resulted in the development of stereophonic

Philippa Strachey, *right*

sound, died at this age in 1981.

Arthur Mann, the English journalist who was editor of the *Yorkshire Post* for twenty years and led a fight in his paper against appeasement before the Second World War, died in 1972, a few days after his ninety-sixth birthday. He was a fearless campaigner who, when under fire from local politicians for attacking Prime Minister Chamberlain, fought vigorously for editorial independence. He is one of several notable exceptions to the rule that hard-pressed newspaper editors burn themselves out quickly and die of stress-related diseases at an early age.

Charles Ponsonby, the English politician and Parliamentary Private Secretary to Sir Anthony Eden during the Second World War, died at 96 in 1976.

Stanley Reed, the editor of *The Times of India* from 1907 to 1923, who later became a British MP, died at this age in 1969.

J.W. Robertson Scott, the founder of the *Countryman*, died at 96 in 1962. In 1927 he decided to create a quarterly magazine devoted to rural life. Perversely, he was a teetotal, non-smoking, meat-shunning man who was anti-hunting, shooting, fishing and racing – a puritan quite unlike his typical readership.

Ernest Shepard, the famous English illustrator, whose sketches for *Winnie-the-Pooh* in 1926 and *Wind in the Willows* in 1931 brought him world-wide recognition, died at 96 in 1976. His work as an artist ranged from his first painting exhibited at the Royal Academy in 1901, to his illustrations for *The Pooh Party Book* in 1971 – a professionally active life extending for seven decades.

Philippa Strachey, the English suffragette who was a pioneer in the struggle for equal rights for women, was a sister of Lytton Strachey, the biographer. In 1907 she became secretary of the London Society for Women's Suffrage, being the born administrator in the group. Unlike many of her pioneer 'sisters', she was able to live to see the full fruits of the work of the early suffragettes. She died at 96 in 1968.

An illustration by Ernest Shepard for *Wind in the Willows*

Ninety-seven is the age for giving up alcohol, according to Mrs M. Cruttenden of Southend. She became teetotal at this age, in case hard drinking damaged her health and shortened her life . . . but she changed her mind when she reached 100.

Achievement at 97
Winifred Rushforth, the Scottish doctor and psychoanalyst, was still active at this age, conducting dream therapy groups in Edinburgh. She went to India as a medical missionary in 1909 where Gandhi taught her to spin. Later, Carl Jung became her friend and her major influence. She returned to Scotland where she founded the Davidson Institute. After half a century of work there, she had plans, at 97, for starting up a Dream Group Movement 'to help everyone tap the hidden wisdom of their unconscious mind'. Faced with an injury from a fall, she commented: 'Not yet, God, not yet. I've an awful lot left to do.'

Life-span of 97
Roger Baldwin, one of the founders of the American Civil Liberties Union, died at this age in 1981, at Ridgewood, New Jersey.

Mary Gilmore, the best-known female poet in Australia, died in 1962 at the age of 97. Her most famous book was her first one, called *Married*. She was an active figure in the Labour movement and for over twenty years edited a woman's page in the Sydney *Worker*. She was created a Dame in 1937.

Heinrich Jordan, the Hanover-born director of the Zoological Museum at Tring, became a naturalized British subject in 1911. He was an entomologist of outstanding abilities and was made President of the Royal Entomological Society in 1929. His work foreshadowed the 'new systematics' by a quarter of a century. He was a kindly man with a puckish sense of humour, and died at 97 in 1959.

Harry Knox, the British soldier who was made a general during the First World War, became ADC to the King in 1925, Lieutenant of the Tower of London in 1933, and Adjutant-General to the Forces in 1935, was sacked by the Secretary of State for War (Hore-Belisha) in 1937. General Knox was an arch-traditionalist who hated any new ideas for the military, which brought him into conflict with progressive minds, such as Hore-Belisha's, and the conflict between the two men became so intense that there was no alternative but to dismiss him. As a sop he was made Governor of the Royal Hospital in Chelsea in 1938.

La Belle Otero, one of the last surviving courtesans of the Gay Nineties, died in Nice in 1965 at the age of 97. She once owned a fabulous hoard of jewels that had been showered on her by royalty and the aristocracy in her courtesan days, but her fondness for gambling at the casinos had gradually used it up. Gambling had become her one great passion in later life – a replacement for the excitement of sex. In her lifetime she made and lost $25,000,000. She was Spanish in origin, the illiterate daughter of a prostitute. At 11 she was so violently raped by the local shoemaker, who broke her pelvis in the process, that she was never able to bear children. She drifted into prostitution herself, but became a sensational dancer as well and a star of the music halls. When she appeared in New York in 1890, the critics raved about her and she soon became the most sought-after courtesan in Europe, charging as much as $10,000 for a night's work in the bedroom. Among her reputed lovers were King Edward VII, King Leopold II of Belgium, the Kaiser, Tsar Nicholas of Russia, Prince Albert of Monaco, Prince Nicholas of Montenegro, King Alfonso of Spain, the Shah of Persia, the Khedive of Cairo, and France's Aristide Briand, who later became a Nobel Laureate.

Edward Pease, one of the founders of the Fabian Society, died at this age in 1955. It was in his rooms that the Fabian Society was born and he contributed a great deal to the development of Fabian socialism. He published *A History of the Fabian Society* in 1916.

John Rockefeller, the American founder of Standard Oil, died at Ormond Beach, Florida, in 1937. He and his son, John D. Rockefeller, gave away $930,000,000 in benefactions. By 1892 he was the world's richest man, worth almost $1,000,000,000, having built his first oil refinery twenty-nine years earlier, in 1863. After 1897 he devoted himself entirely to philanthropy.

John Rockefeller

Bertrand Russell, the English philosopher and mathematician, whose *Principia Mathematica*, published just before the First World War, was a major scholarly work dealing with the relationship between mathematics and logic, died at 97 in 1970. In 1945 he had published a popular work called *The History of Western Philosophy*, and he was awarded the Nobel Prize for Literature in 1950, but it was for his attacks on the accepted conventions of society that he became best known to the public. All his life he had opposed rigid thinking in the realms of sex, education, religion, women's rights, politics and especially warfare. He advocated free love and open marriages at a time

Bertrand Russell

when such suggestions were considered scandalous. And he practised what he preached, having many affairs throughout his long life. He said he had 'abnormally strong sexual urges'. Religion he rejected as useless, when he was 18. Warfare appalled him and his pacifism in the First World War landed him in prison in 1918 as a 'security risk'. Traditional education he abhorred, insisting on more progressive methods. And nuclear weapons became his great hatred in his final years of campaigning. All his life he struggled to free himself from the austere puritanism of the Presbyterian upbringing imposed upon him by his grandmother, following the death of both his parents when he was still an infant. In many respects he succeeded and he certainly enjoyed his long life, his dying sentiment being: 'I do so hate to leave this world.'

Mary Sargant-Florence, the English mural painter, disproved Michelangelo's dictum that 'oil painting is for women and fresco for men'. She became an expert at the difficult art of *buon' fresco*, or true fresco, in which the paint is applied to the plaster background while it is still wet. She died at 97 in 1954.

Joe Smith, the American comedian, who was one of the partners in the 'Smith and Dale' vaudeville comedy team for seventy-three years, died at Englewood, New Jersey, in 1981.

Carl Vinson, from Georgia, was the American politician who held the record for the longest service in the House of Representatives – a total of fifty years. He died at 97 in 1981.

La Belle Otero

Ninety-eight is a year for slowing down, according to American farmer Wally Lattimer. At this age he has decided to get up later in the morning, at 6 am instead of before dawn, and to work a shorter day, twelve hours instead of fourteen. Having outstayed his wives, he now lives alone on his forty-acre Kansas farm, with its eleven-acre garden. He claims that his secret is his calm temperament: 'I don't worry and I don't get mad.'

Life-span of 98

Blanche Clough, the principal of Newnham College, Cambridge, in the 1920s, and a pioneer of women's emancipation, died at this age in 1960.

John Garner, Vice-President of the United States from 1933 to 1941, died in 1967, two weeks before his ninety-ninth birthday. He was known by the nickname of 'Cactus Jack' and in his old age gained the reputation of being something of a political sage. Despite considerable wealth, he always lived a simple life, spending most of his time quietly on his Texas ranch.

Isocrates, the Athenian author and teacher, who specialized in writing orations for others, and whose eloquence attracted a steady stream of pupils from far and wide, lived from 436 BC to 338 BC. He was 98 when, filled with grief and despair because Greece had been defeated in battle and lost its independence, he starved himself to death. The event was referred to by John Milton in the lines: '. . . as that dishonest victory/at Chaeronea, fatal to liberty,/Killed with report that old man eloquent.'

John Laing, the British civil engineer who was President of the building company John Laing and Son and also President of the London Bible Society, died at 98 in 1978.

John Garner

Scott Lidgett, usually considered to be the greatest Methodist since John Wesley, died in 1953. He was President of the United Methodist Church and wrote a number of widely read religious books. His volume *Salvation* was published when he was 97, in 1952. A man of childlike simplicity, his memory and his conversation remained active and lively to the end.

Joseph Paul-Boncour, who was Prime Minister of France briefly in 1932, died at 98 in 1972. Short, erect and with a mane of white hair, using fiery gestures whenever he spoke, he was the dream figure of every French political cartoonist. An internationalist, he was a major figure at the League of Nations and again at the United Nations – always attempting to encourage good sense at the international level.

Eden Phillpotts, the English novelist and playwright, died at Exeter in 1960, aged 98. He wrote *The Farmer's Wife*, which was staged in wartime London, in 1917, and revived many times afterwards. His first authorship dates from 1888 and he was still active in 1944, when he wrote *The Changeling*, at the age of 82. He was a Devonshire writer – honest, simple and unassertive, whose long life reflects the untormented philosophy of the man.

Fred Streeter, the English broadcaster and master-gardener, died at 98 in 1975. The son of a shepherd, he started broadcasting in the 1930s and became the nation's gardening expert, answering thousands of letters a month until the time he died, with over forty years spent as *the* expert.

Lionel Tertis, the viola virtuoso, who refused to allow his instrument to play 'second fiddle' to the violin, died in 1975. He and his friend Pablo Casals

Fred Streeter

were born on exactly the same day, but Tertis lived sixteen months longer. He designed a special 'Tertis Model Viola', of which sixty were built, and also a new type of cello.

Madeleine Vionnet, the French couturière who exerted a major influence on fashion in the 1920s and 1930s, and invented the revolutionary 'Bias cut' in 1926, died at 98 in 1975. As a result of her 'Bias cut', the principle of 'form-fit by hang' was established and the first slip-over-the-head dresses were on the market in a very short space of time. From 1919 to 1939 she reigned over French couture, from a vast mansion in the Avenue Montaigne.

Maxime Weygand, the French general, who was Chief of the General Staff in 1940 and Commander-in-Chief in all theatres, died in 1965. He advised Pétain to demand an armistice from the Germans, but in 1942 he was arrested by the Nazis and imprisoned in Germany. Released in 1945 and returned to France, he was an embittered man. It was a sad end for the soldier who had been Foch's right-hand man during the First World War, but he retired to his beautiful Château of Coatamour near Morlaix and devoted his time to writing his memoirs.

John Laing

Ninety-nine is an age for self-indulgence, taking it easy before the triumph of scoring a century. One 99-year-old, when asked the secret of his longevity, replied with great fervour: 'Never deny yourself anything.' The man in question was attending the funeral of his younger brother who had just died at the mere age of 97. The older man was complaining that his young brother had refused to heed his warning 'that theological research is not compatible with longevity ... God does not mean us to pry into these matters.' To a 99-year-old, even 97 appears young. After the funeral, the old man went to Trinity College, where his brother had been a fellow, and solemnly drank his customary half-bottle of port.

Fred Dean, a retired theatre commissionaire, was looking forward to his own form of self-indulgence at his special birthday party. Convinced that it was his hundredth, he had put an invitation in his window asking everyone to help him celebrate it. When his neighbours checked to make sure that the usual message from the Queen was on its way, they discovered that the old man was in fact about to be 99, not 100 as he imagined. In order not to disappoint him, they went ahead with the celebrations and told him nothing. MP Cyril Smith aided the deception by sending the old man a letter of congratulations from Westminster, in the hope that he would accept it as the usual, formal message. If Fred lasts out to his real 100th birthday, he will get another one, and only then learn the truth.

Achievement at 99

Dora Booth, a Major in the Salvation Army and a granddaughter of General Booth, the founder of the Army, faced what might have been the daunting prospect of taking part in a television talk show at the age of 99, in December 1982. She was accompanied by her two (slightly) younger sisters. The programme host, Russell Harty, asked her the question: 'If your grandfather had been alive would he have come with you?' and was promptly put in his place by the answer 'No. He would have been doing something much more important.' Despite her age, Major Dora was fully alert and her performance was described by a leading critic as 'the most endearing event of the week'. Her only problem was that she kept forgetting her host's famous name.

Life-span of 99

George Coopland, the medieval historian, was 99 years and 269 days old – only ninety-six days short of his hundredth birthday – when he died in 1975. He had been Professor of Medieval History in Cairo in the late 1920s and at Liverpool in the late 1930s and early 1940s. He wrote many books, from 1914 to 1969, including *Serfdom and Feudalism*, in 1928, and *The Tree of Battles*, in 1949. The only clue to his unusual longevity is that he listed his 'recreation' in his *Who's Who* entry as 'country life'.

Daniel Mannix, the Roman Catholic Archbishop of Melbourne, lived almost as long: 99 years and 246 days. He died in 1963, still tall, lean and erect. His longevity secret was that he walked the three miles between his home and his cathedral every day, even in his 90s. He became Archbishop of Melbourne in 1917 and was soon politically active as Australia's chief spokesman for *Sinn Fein*, having spent the first half of his long life in Ireland. He opposed Australian conscription for the First World War, arguing that it was a 'trade war' – a conflict between rival capitalists. In 1920, when the war was over, and at the height of the 'Irish Troubles', he went to the USA, where huge audiences across the States had gathered to see him. Extraordinary scenes occurred as he left on the liner *Baltic* from New York. There were, according to the *New York Times*, 'hisses, cheers, fist fights and the flash of revolvers'. Bonfires were lit on the Irish coast to greet him, but two destroyers were sent to intercept the *Baltic* and he was refused permission to land in Ireland. One of the destroyers took him to Penzance and the following year, after visiting Rome, he went home to Melbourne.

Martin Routh, the English academic, who was President of Magdalen College, Oxford, from April 1791 until December 1854, was, amazingly, still in office when he died, at the age of 99. Since he had been a college fellow for sixteen years before he became President, he must have been an Oxford don for an astounding total of seventy-nine years – probably a 'world teaching record'.

Titian, the Venetian master who has been called the 'founder of modern painting', is reputed to have lived to the age of 99, although some authorities have thrown doubt on this. Traditionally, he was born in Venice in 1477 and died there in 1576. He was the greatest of all the Venetian painters and his liberated

Titian

Fred Dean inviting everyone to help celebrate his 99th birthday, believing it to be his 100th

use of colour, and broad brush-strokes, was revolutionary for his day. The most prolific of all the great artists of earlier times, he has about 1,000 paintings attributed to him and, even allowing for the feeling that perhaps as many as 250 of these are spurious, his output was staggering. He was immensely popular and successful in his own lifetime, so that he could afford to make a grand gesture. None was grander than the one he made when Henry III, King of France, visited his studio. Titian gave him, as a present, all the pictures of which he had enquired the price. In technique, one unusual aspect of his work is that he used his fingers as well as his brushes – something more commonly associated with the avant-garde artists of the twentieth century. Titian was the most brilliant manipulator of paint in relation to colour, texture, surface, harmony, tone and luminosity, of any of the great European masters.

One hundred is the year of the Queen's Message. Everyone who reaches this age in Britain receives an official congratulation from the Queen on their hundredth birthday. This symbolic moment of triumph over age is something that nonagenarians look forward to during the years running up to the great day. In some cases it may even give them something to live for – and there certainly seem to be fewer deaths at 99 than one might expect, as if, having got so close, the 99-year-olds are determined to hang on that little bit longer.

In a recent year, the number of Queen's Telegrams sent out was counted and the total came to 399. The chances of putting one hundred candles on a birthday cake are, for British males, 12,500 to one; for British females, 2,500 to one.

(The demise of the English telegram will not stop the ritual of the 'Queen's message of good wishes'. It will continue to be sent, under the name of 'telemessage'. Such are the wonders of postal bureaucracy.)

Achievement at 100
Estelle Winwood can consider it an achievement simply to have reached the age of 100, as it had repeatedly been predicted that she would come to a 'decadent, no-good, early end'. She was the English actress who scandalized New York in 1919 by wearing lipstick off-stage as well as on. She was originally trained at the Liverpool Repertory, and appeared in forty Broadway plays and many films. In her later years she lived in California where, even at 100, she was smoking sixty cigarettes a day, drinking sherry, dining out most nights and playing bridge three times a week. Married four times, she said that even at 100 she was still 'waiting for something wonderful to happen'. She claimed that she had never had a woman friend, with the exception of Tallulah Bankhead, commenting: 'I don't like women. I think they're silly.'

Wilf Nixon at the Fulham Football Club lunch

Eubie Blake

Misfortune at 100
Wilf Nixon, the English footballer, suffered an unfortunate accident during the festivities on his 100th birthday. The former Fulham goalkeeper was taken ill before the Saturday match at Fulham on October 23, 1982, when a piece of meat became lodged in his throat at his celebration lunch. Instead of watching the match as a special guest of honour, he had to be rushed to hospital, and although the problem was easily dealt with, he missed his great occasion.

Life-span of 100
Eubie Blake, the American ragtime pianist and pioneer of boogie-woogie, died in New York on February 13, 1983, aged 100. Best remembered as the writer of the song 'I'm Just Wild About Harry', he had celebrated his 100th birthday the week before, with the words: 'If I'd known I was gonna live this long, I'd have taken better care of myself.'

Harry Brittain, the founder of the Commonwealth Press Union, the Pilgrims Club and the Tariff Reform League, among other things, was reputed to have sat on more committees during his long life than any other man. Trained as a lawyer, he became an MP and journalist, always seeking new challenges. A colourful character, who wore a carnation so consistently that he had one named after him, Sir Harry eventually died at the age of 100 in July 1974.

George Higginson, the oldest general in British military history, was still vigorous and alert on his 100th birthday, when he attended an inspection by the Duke of Connaught of the first battalion of the Grenadier Guards at Windsor. Although he had officially retired at the age of 67, in 1893, he continued to advise Queen Victoria on military matters for some years to come and he did not start work on his

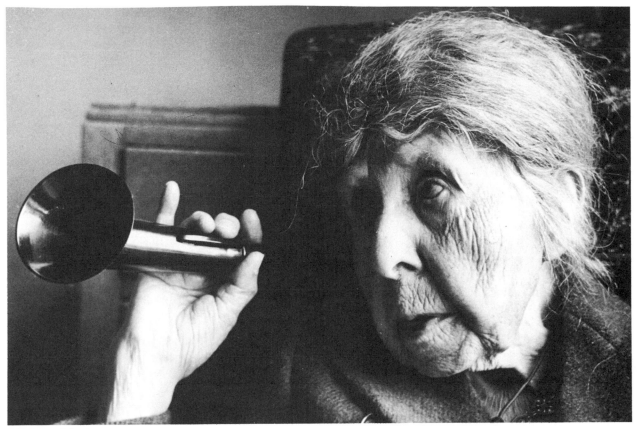

Margaret Murray

military memoirs until he was approaching 90. *Seventy-one Years of a Guardsman's Life* was published in 1916 to mark his ninetieth birthday. He came from a rich and distinguished military family, fought right through the Crimean War, including the sieges of Balaclava, Sevastopol and Inkerman and, despite the high mortality in the battles, the killer diseases such as cholera; and severe food shortages, he arrived home in good health. Even at 100, *The Times* reported that he appeared like a man of only 70. Modesty and zest were two features of his life that he claimed helped his life-span, in addition to exercises with weighted Indian clubs (in his 90s) and long walks. He was a non-smoker, but enjoyed meat and drank in moderation. He was born on June 21, 1826, and died on February 1, 1927.

Margaret Murray, the British archaeologist, died at this age in November 1963. No other person can have sailed around the Cape in 1866 and then published a book about her life's adventures in the year 1963. Born in Calcutta, Dr Murray travelled back and forth between India and Germany several times before she was 25 (in 1888). At the age of 17 she was the only white woman working as a ward sister in a Calcutta hospital, but then she turned to archaeology and became a student of Egyptology under Sir Flinders Petrie, in London in 1894. By 1904 she had joined him for excavations in the field and continued with her field-work, in various countries, until her retirement in 1932. Even then, she kept on digging and making field trips, well into her 70s. She wrote many books, including *The Splendour that was Egypt*, when she was 86, and *The Genesis of Religion* and *My First Hundred Years* – both published when she was already 100 years old.

Henry Ridley, the English botanist, was still describing new species of plants in publications when well into his 90s. A short, stout, vivacious man and an inveterate story-teller, he was the person responsible for the establishment of the rubber plantation industry in Malaya. He was Director of Botany in Malaya for twenty-three years and had the rubber plants sent out to the colony from Kew Gardens. He died about seven weeks before his 101st birthday, in 1956.

James Swinburne, a pioneer in the development of plastics, died in 1958. He was beaten by one day in the rush to secure patents on the earliest form of commercially important plastic – by Dr Baekeland – but later became the chairman of Bakelite Ltd. He made two sidetracks in the form of books entitled *Population and the Social Problem*, where he urged that only by studying the 'principle of population' could one solve the problem of poverty, and then, at the age of 92, *The Mechanism of the Watch*.

Ellaline Terriss, the widow of the famous actor Sir Seymour Hicks, died at 100 in 1971. She had become a Gaiety Girl in 1895 and later toured the world. She was said to have performed the best ever Alice in Lewis Carroll's *Alice in Wonderland*.

Over one hundred is the age of the ultimate survivors, people who have defied the ageing process to an unusual degree. Every year in England about twenty men and 100 women die at these extreme ages, but few of them are still very active. There are some amazing, isolated exceptions to this rule, but most 100+ individuals suffer from greatly weakened faculties and are often confined to their beds.

> Few animals can outlive human beings. The major exception is the tortoise family. The greatest age for a tortoise accepted by the *Guinness Book of Records* is 152+ years for a male Marion's Tortoise. It was born some time before 1766, when it was shipped from the Seychelles to Mauritius by the Chevalier de Fresne. It was given by him to the Port Louis army garrison, where it lived out its long life, until 1918 when it was accidentally killed.

Life-span of 101 Grandma Moses, the famous American 'Naïve' painter, died at the age of 101 in December 1961. She had been painting for thirty-four years. Her pictures were first discovered by a New York art dealer in 1938, when she was 78, and the following year one of them was shown at New York's Museum of Modern Art in an exhibition titled *Contemporary Unknown American Painters*. In 1940 she had her own show in New York and in 1949, at the age of 89, she found herself famous, with awards being showered upon her, including one presented to her personally by President Truman. After this, she held more than 200 exhibitions, across the world, and she was still actively painting at the age of 100. Her immense success stimulated enormous interest in 'Sunday Painters' and virtually started a new genre, although she is overshadowed in historical importance, of course, by the great Douanier Rousseau.

Gwladys Norton-Griffiths, who died at 101 in 1974, started out as a singer, was given a prize by Brahms, and was approved of by a young music critic called George Bernard Shaw. Later in life, she became a political wife and a famous London hostess. She was a wife, mother and grandmother of MPs. During the Second World War she moved back into London at the height of the Blitz, to help at the Westminster Hospital, when most other people who were free to do so were leaving the city. During her life she travelled to almost every corner of the globe and was still making long trips up to the age of 93. At 95 she bought her first typewriter, because she could no longer hold a pen, and she was still typing letters on it when she was 100.

William Stone, who died at 101 in 1958, was one of the last members of a dying breed – the rich, carefree, bachelor 'man-about-town'. He boasted that he had not missed a London first night at the theatre in forty years. He travelled extensively, indulging his favourite hobby of butterfly-hunting, and became a knowledgeable naturalist. At the age of 100, he was still

very active and was then the senior member of the Athenaeum, the Oxford and Cambridge Club, the United Universities' Club, the Reform Club, the Garrick Club, the Hurlingham Club, the Bachelors' Club, the Linnean Society, the Chemical Society, the Royal Geographical Society and the Zoological Society. When he died, a lifestyle which had already become antique died with him.

Life-span of 102 Theodore Taylor, the Yorkshire mill-owner who was a pioneer of industrial profit-sharing, died at 102 in 1952. In 1896 he had introduced the idea that his 2,000 employees should share in the profits from his company's woollen manufacturing, and it had proved immensely successful. He believed that it was the only way to eliminate industrial strife, and continued to preach his concept for many years, against considerable opposition. Even at the age of 99 he was still speaking in public about it, and travelled to London to press his case. He was also instrumental in suppressing the opium trade between India and China and was the leader of the anti-opium movement in Britain. His recipe for longevity was keeping both the body and the mind busy, even in advanced old age. At the age of 100, he was working a seven- to eight-hour day at his factory, twice a week. He was also a non-smoker and teetotaller.

Adolph Zukor

Life-span of 103 Edith Beck of Hampshire gave up smoking on her 103rd birthday, because she thought it was time to look after her health. Surprisingly, there appears to be no consistent rule about smoking and longevity.

James Chapman, the market gardener, died at this age in March 1975. Gardening, like walking, *does* appear to have a strong link with extreme longevity.

Adolph Zukor, the American film producer who died at 103 in 1976, was the longest-lived of all Hollywood's leading figures. Diminutive but dynamic, he was a one-man history of the movies. Starting with the Penny Arcade in 1903 and rising rapidly to become President of Paramount Pictures, he was the promoter of the very first full-length feature film, which was *Queen Elizabeth*, starring Sarah Bernhardt. Zukor was still listed as chairman of the Board Emeritus at Paramount when he was 103 years old. In 1949 he was given a special Academy Award for his services to the cinema.

Life-span of 104 Rickard Christophers, who laid the foundations of the world-wide malaria-eradication programme, through his lifelong pioneering studies, died at 104 in 1978. From 1910 to 1922 he was in charge of the Central Malaria Bureau in India, after which he became Professor of Malarial Studies at London University and took charge of the Experimental Malaria Unit at the London School of Hygiene and Tropical Medicine. He was also the author of important books on malaria.

Life-span of 110 Frederick Butterfield, who was a dispensing chemist in the West End of London, serving Queen Victoria's household, was born a year before Lincoln's assassination and lived to become the oldest male in Britain. He died at 110 in a Harrogate nursing home in 1974. His secret for a long life seemed to be 'regular habits'. He rose at 6 am every morning of his life and after washing and dressing always read three newspapers. He voted Conservative in every election from 1890 to 1974 inclusive, making him the longest-voting Tory in history.

Toka Miyata, the oldest woman in Japan, died at the age of 110 in March 1982. She did not need to enter an old people's home for medical aid until she was 103. Even then she was never sick, merely frail, and always enjoyed her trips in a wheelchair around the nursing home in Matsuhashi.

Life-span of 111 Katherine Plunket, an Irish estate-owner, was able to recall, in her last years, that more than a century earlier she had been dandled on the knee of Sir Walter Scott. She had perfect health until the age of 102, when she became bedridden with bronchitis. Up to that point she had always overseen the running of her ancestral estate in County Louth. She recovered quickly from her illness and insisted on having the newspapers read to her daily. Her secrets for a long life: never dwell on the past – live in the present and contemplate the future; keep an unruffled, carefree attitude to life; eat well – she was no vegetarian puritan, and was particularly fond of game, always feasting on turkey, plum pudding and champagne on her many birthdays.

Life-span of 116 Hadj Mohammed El Mokri, who was Grand Vizier of Morocco for nearly fifty years, died in 1957 at a reputed age of 116 years. This age is accepted in his *Times* obituary, but not in the *Guinness Book of Records* which will only admit that he was 'more than 112'. He was a member of the regency council when the Sultan ben Arafa withdrew to Tangier in September 1955. He must therefore have been actively in office at the astonishing age of 114. But when independence came to Morocco in 1956 his office was abolished and he died the following year. He had acted for many years as the moderate mediator between the Sultan and the Resident-General. He once led a delegation to the court of Napoleon III and he took another one to the French Government in 1955 to discuss independence for Tangier. Since he travelled there by aircraft, he must, at 114, have been the oldest air passenger in history. In personality he was said to be wise, subtle, pious, dignified, witty and delicate.

Life-span of 118 Shigechiyo Izumi of Japan is recognized as the oldest man who has ever lived. All the longer claims – and there are many – are now regarded as spurious, or at least doubtful. He was born at Asan on the Japanese island of Tokunoshima and he was recorded as a six-year-old child in 1871, when Japan's first census was taken. A great devotee of television in his later years, he gives his secret of long life as 'not worrying'. He celebrated his 118th birthday on June 29, 1983, and was still alive at the time of writing.

Shigechiyo Izumi, the oldest man in the world, on the eve of his 118th birthday

Conclusion

Two contradictory messages emerge from the pages of this book. The first is that there are typical features of each age. As members of the human species we are *likely* to behave in certain predictable ways as we pass, year by year, from birth to death. The opening section of each 'age-page' in the book gives a sketch of what is most usual at each age from 0 to 100 +. Being biological organisms, we are bound to share many basic ageing rates and ageing processes. At this level we are all very similar.

The second message contradicts this. The later part of each 'age-page' reveals the wonderful variability of the human population. Although there are typical ways of behaving at every age, there are always exceptions to the rule – individuals who break the pattern, being either unusually early or unusually late in their progress from childhood to old age. There are the amazingly mature children, who become virtuosos at tender ages and there are childlike adults who refuse to age, either physically or mentally, at the usual rate. The significance of these two messages is clear enough: age rules exist, but they are there to be broken. Exceptional individuals will always ignore them and will write symphonies at 9 and elope at 90.

What makes an exceptional individual is also clear. For children who develop very early, it seems that, in addition to a good brain, there must be a doting, fanatical parent, ready to devote many hours to their training. For adults who remain vivaciously childlike even in old age, there has to be a sustained enthusiasm and optimism for some aspect of life. Retirement and 'sitting in the sunset' is a surefire killer. People who want a long life with an alert old age should never contemplate retirement. If they are forcibly retired they should immediately immerse themselves in some new and absorbing activity.

Comments such as these inevitably lead to the temptation to end this book with a set of suggestions about improved life-span. After reading the thousands of biographies needed to compile these pages, a number of special qualities have emerged that seem to be associated with the enjoyment of a longer, healthier life. Some are obvious enough, others are not. Here, in conclusion, are what appear to be the major factors that favour longevity in the human species.

If you happen to be lucky enough to come from a long line of octogenarians (or better), then you have a greater chance of a long life yourself. If your ancestors and close relatives have a poor health record, with early deaths from such weaknesses as heart trouble, diabetes, cancer or strokes, then you yourself will be less favoured in the longevity stakes.

Some people are naturally more mobile and physically active, while others are lazier and more sedentary. The high-mobility group are at a considerable advantage, providing their activities are not the result of stressed agitation. Individuals whose daily lifestyle involves them in regular muscular effort live longer than those who spend all day sitting down or flopping out. Among exceptional individuals who lived to a *very* old age, it is remarkable how many of them were vigorous walkers. There is a special reason why such activities as walking and gardening prolong life so spectacularly. They are both 'non-intensive' forms of all-over bodily movement. The more earnest exercisers, such as joggers and runners, display a conscious or unconscious anxiety about their health. If they take exercise too seriously it will work against them, unless, of course, they are young professional athletes or sportsmen in the peak of condition. Such young professionals are not driven on by a nagging anxiety about their health, but by the competitive demands of their sport. But older individuals who take up some intensive form of gymnastic or athletic activity are usually people who fear declining health and feel driven to protect themselves. Sadly, it seems their whole attitude is wrong, if the longevity champions are anything to go by. It is crucially important that physical exercise – as we grow past the young-sportsman stage – should be *ex*tensive rather than *in*tensive and, above all, that it should be *fun*. Walking or gardening that is done for the pure pleasure of the activity is worth a hundred times more than some boring exercise done out of a sense of 'health-duty'.

It follows from the last point that a calm temperament favours longevity. Those who are sharply aggressive, emotionally explosive or naggingly anxious are at a grave disadvantage. White-hot anger and nail-biting anxiety are the undertaker's friends. The easy-going, relaxed, emotionally stable person will nearly always outlive his easily enraged or worry-ridden neighbour. Those who find that their way of life drives them mad with fury, frustration or fear should seek other employment.

It is important to make a distinction between being calmly relaxed and being passively lazy. Relaxation does not imply lack of enthusiasm, nor does it contradict the idea of passionate interest in a person's favourite preoccupations. Indeed, zest for living and an eagerness to pursue one's chosen subjects are vital elements in ensuring a long life. It goes without saying that *retirement* from such activities is to be avoided at all costs. If a job or occupation has to be given up, then it must be replaced immediately with some new and equally absorbing hobby or task that each day offers an exciting new challenge. Pessimism and cynicism are negative elements; optimism stretches lives.

One of the greatest weaknesses to avoid is living in the past. Thinking about 'the good old days', complaining about how the world is deteriorating, and criticizing the younger generation, are sure signs of an early funeral. Those individuals who live in the present and contemplate the future with interest have a much better chance of a long life than those who wallow in nostalgia.

Being successful is a great life-stretcher. A sufficient degree of success can even override typical life-shorteners such as obesity and a fondness for drink. But it is important that, in gaining the success, the individuals should not over-stress themselves or place themselves in a prolonged state of tension. And success must always be measured in personal terms. A hill-shepherd may feel just as successful, in his own way, as a world-famous Nobel Laureate. Success must not be confused with fame.

Long-lived individuals seem to be more concerned with what they *do* than who they *are*. They live outside themselves rather than dwelling egocentrically on their own personalities. Self-centred vanity is foreign to their thinking. They may be proud of their work, their inventions or their creations, but not of themselves.

In personal habits, the long-lived are generally rather moderate and middle-of-the-road. Extremes of diet are not common. Man is an omnivorous species and a mixed diet seems to favour longevity. Long-lived vegetarians appear to survive despite their diet rather than because of it. Simple foods generally win over fancy cuisine, but confirmed gastronomes may well argue that quality of life is more important than sheer quantity of years. Puritanical arguments about smoking and drinking have little to support them. Many very long-lived individuals enjoy both nicotine and alcohol – in moderation. If one listened to the propaganda about the link between smoking and lung cancer, it would be hard to conceive of anyone who enjoyed cigarettes surviving to a ripe old age, yet this does happen. The explanation has to do with an individual's susceptibility to cancer at the outset. A cancer-prone individual will obviously be at greater risk if he or she smokes. But a cancer-resistant person can apparently blow smoke-rings for a hundred years without any cancer damage to the lungs. The catch is that we are not yet expert at categorizing people in this respect, early in life. So smoking remains a longevity game-of-chance. In fairness, however, it must be pointed out that, for the cancer-resistant individual, smoking may act as a helpful calming device and, as already mentioned, keeping calm is a life-stretcher.

Most long-lived people have a well developed sense of self-discipline. At first sight, this might seem to conflict with the suggestion that a calm, easy-going person enjoys a longer life, but there need be no contradiction here. Self-discipline does not imply a harsh, military-style, self-imposed masochism. All it means is that the individual orders his or her life and imposes a pattern on the simple events of the day. The man who lives long because he walks a mile a day does so because he does it *every* day, as part of an organized existence. The man who lies around for six days and then feels so guilty that he walks seven miles on the next day, lacks the simple level of self-discipline that favours an extended life-span.

Over and over again, during the writing of this book, it emerged that long life goes with 'a twinkle in the eye'. A sense of humour, an almost mischievous impishness, and a feeling that life is fun, are strong weapons against the ageing process. The sour-faced puritan and the solemnly earnest bore soon begin to lose ground, leaving their more amused and amusing contemporaries to enjoy the last laugh.

Finally, nothing is to be gained by a head-in-the-sand avoidance of the facts of life and death. The healthiest solution is to accept that one's span on earth is limited and then to live every day, *in the present*, to the full. Nostalgia and procrastination have always been, and will always be, the two greatest enemies of a fruitful, enjoyable existence on this planet.

Index

The references are to the ages under which each person is listed.

Picture Credits

The author and publishers would like to thank the following sources for permission to reproduce the photographs in this book:

0   Baby, John Topham Picture Library
1   Child, John Topham Picture Library; Headline, John Frost Newspaper Collection
2   Princess Anne, John Topham Picture Library; Edgar Allan Poe, BBC Hulton Picture Library
3   Elizabeth Taylor, Peter Falk, John Topham Picture Library; Wolfgang Amadeus Mozart, Mansell Collection
4   Louis XIV, Mansell Collection; Jean-Paul Sartre, Gustav Mahler, BBC Hulton Picture Library; Malcolm X, John Topham Picture Library
5   Dalai Lama, John Topham Picture Library; Lillian Gish, Kobal Collection/Cinegate Films; Natalie Wood, Kobal Collection/TCF
6   Shirley Temple, Kobal Collection/Paramount; K'ang Hsi, Mansell Collection; Peter Pan, John Topham Picture Library
7   Fred Astaire, Kobal Collection/RKO; Winston Churchill, BBC Hulton Picture Library; The Clowns, Kobal Collection/SGC Titanus
8   Charlie Chaplin, Kobal Collection/United Artists; James III, Mansell Collection; Johnny Sheffield, Kobal Collection/MGM
9   Annie Oakley, BBC Hulton Picture Library; Ellen Terry, National Portrait Gallery; Genghis Khan, Mansell Collection
10 Tatum O'Neal, Kobal Collection/Paramount; Elizabeth Taylor, Kobal Collection/MGM; Ingres, John Topham Picture Library

11 Ludwig van Beethoven, George Frederick Handel, Mansell Collection
12 Sabu, London Films; Brooke Shields, Kobal Collection/Paramount
14 Picasso drawing, © Picasso, Spadem; Henry VII, National Portrait Gallery
15 Sue Lyon, Kobal Collection/MGM
16 Lady Jane Grey, National Portrait Gallery
17 Judy Garland, Kobal Collection/MGM
18 Queen Victoria, Mansell Collection; Tutankhamen, Camera Press
19 Jane Austen, National Portrait Gallery; Yvonne Goolagong, BBC Hulton Picture Library; Erté design, Mary Evans Picture Library/© Erté, Spadem
20 Jane Russell, Kobal Collection/United Artists; Rossetti's Prosperine, The Tate Gallery; Lauren Bacall, Kobal Collection/Warner Brothers
21 Brigitte Bardot, Rex Features; Eddie Cochran, Yves Saint-Laurent, John Topham Picture Library
22 Alfred the Great, Mansell Collection; Oswald Mosley, Nureyev with Fonteyn, John Topham Picture Library
23 Jesse Owens, BBC Hulton Picture Library; Saint Francis, Mansell Collection; Orson Welles, Kobal Collection/RKO
24 Audrey Hepburn, Kobal Collection/Paramount; David Garrick, Mansell Collection
26 Nero, Mansell Collection; Samuel Pepys, National Portrait Gallery; Jean Harlow, Kobal Collection/MGM

27 Yuri Gagarin, John Topham Picture Library; Rupert Brooke, National Portrait Gallery; Robert Burns, Mansell Collection; Janis Joplin, Rex Features
28 Billie Jean King, Sporting Pictures, UK
29 Mrs Beeton, Anne Boleyn, National Portrait Gallery
30 Marlon Brando; Kobal Collection/Columbia
31 Isambard Kingdom Brunel, National Portrait Gallery; Rudolph Valentino, Kobal Collection/Paramount
32 Amelia Bloomer, Orville Wright, Mansell Collection; Sean Connery, Kobal Collection/United Artists
33 Martin Luther, Mansell Collection
34 Clint Eastwood, Kobal Collection/United Artists; Pullman Dining Car, Mansell Collection
35 Blondin, BBC Hulton Picture Library
36 Adolf Hitler, Kobal Collection/Tobis; Andy Warhol, John Topham Picture Library; Henry Purcell, Mansell Collection; Marilyn Monroe, Kobal Collection/United Artists
37 Richard Burton, Kobal Collection/20th Century-Fox; Nell Gwynne, Mansell Collection
38 Amelia Earhart, Mansell Collection
39 William I, Cleopatra, Mansell Collection
41 Brendan Behan, BBC Hulton Picture Library
42 Elvis Presley, Kobal Collection/CBS; Edward VIII, Mansell Collection
43 Roman Polanski, John Topham Picture Library
44 Winnie-the-Pooh, John Topham Picture Library/Line illustration by E.H. Shepard, copyright under the Berne Convention, In the U.S.A. Copyright 1926 E.P. Dutton & Co. Inc. Copyright renewal © 1954 by A. A. Milne; Friedrich Nietszche, Mansell Collection
45 Rasputin, Mansell Collection
46 Oscar Wilde, National Portrait Gallery
47 Edith Piaf, Kobal Collection
48 Harold Wilson, John Topham Picture Library
49 Puccini, Mary Evans Picture Library; David Niven, Kobal Collection/United Artists
51 Gordon of Khartoum, Mansell Collection; Dirk Bogarde, Kobal Collection/Warner Brothers
52 Colette, Mary Evans Picture Library
53 Princess Grace, Kobal Collection/Warner Brothers; René Descartes, Mansell Collection
54 Bill Haley, Kobal Collection
55 Henry VIII, Mansell Collection
57 George Frederick Handel; Mansell Collection; Humphrey Bogart, Kobal Collection/Warner Brothers
58 John Milton, Charles Dickens, Thomas More, John Knox, Mansell Collection
59 Clark Gable, Kobal Collection/United Artists
60 Geoffrey Chaucer, Mansell Collection
61 *Hindenburg*, Mansell Collection
62 Cary Grant, Kobal Collection/MGM; Aristotle, Mansell Collection
63 Nikita Khrushchev, Rex Features; Charles Laughton, Kobal Collection/MGM; Benjamin Disraeli, Mansell Collection
64 Laurence Olivier, Edith Evans, Kobal Collection/Rank
65 Laurel and Hardy, Kobal Collection/MGM
66 John Betjeman, Camera Press; *The Garden of Earthly Delights* by Hieronymus Bosch, Cooper Bridgeman Art Library/Prado, Madrid; Thomas Cranmer, Mansell Collection; Edgar Rice Burroughs, Kobal Collection
67 W.G. Grace, George IV, Mansell Collection
68 Merle Oberon and Laurence Olivier, Kobal Collection/Sam Goldwyn; Joshua Reynolds, National Portrait Gallery; Stanley Spencer, Private Collection
69 Ronald Reagan, Kobal Collection/Universal; Buster Keaton, Kobal Collection/MGM
70 Francis Chichester, John Topham Picture Library; Copernicus, Mansell Collection
71 Louis Armstrong, John Topham Picture Library; Maurice Utrillo, Camera Press
73 Bela Lugosi, Kobal Collection/Universal; W.B. Yeats, National Portrait Gallery
74 George Frederick Handel, Guiseppe Garibaldi, Mansell Collection; Stan Laurel, Kobal Collection/MGM; Nehru, John Topham Picture Library
75 Archimedes, Alexander Graham Bell, Mansell Collection; Duke Ellington, Camera Press; Chico and Harpo Marx, Kobal Collection/Paramount
76 Cecil Beaton, John Topham Picture Library; Albert Einstein, JPA Publicity
78 Mahatma Gandhi, Mansell Collection
79 Charles de Gaulle, William Harvey, Mansell Collection; Louis Mountbatten, Rex Features; Edward G. Robinson, Kobal Collection/Warner Brothers
80 'Top-Shot', Kobal Collection/Warner Brothers
81 George III, Queen Victoria, Mansell Collection; The Kennedy Clan, Rex Features
82 William Gladstone, Mansell Collection; Francisco Franco, Rex Features
83 Henry Ford, The Ford Motor Company
84 *The Snail* by Matisse, The Tate Gallery; Thomas Edison's laboratory, Mansell Collection; Clement Attlee, John Topham Picture Library
85 Queen Mary, Mansell Collection
86 Édouard Daladier, Paul von Hindenburg, Mansell Collection; Jomo Kenyatta, John Topham Picture Library
87 Chiang Kai-shek, Thomas Hardy, Mansell Collection; Groucho Marx, Kobal Collection/Paramount
88 Mae West, Kobal Collection/20th Century-Fox
89 Arthur Rubinstein, Courtesy of Eva Rubinstein; *Whisky Galore*, Kobal Collection/Ealing Studios
90 Winston Churchill, John Topham Picture Library
91 Eamon de Valera, Jean Sibelius, Mansell Collection; Konrad Adenauer, Somerset Maugham, John Topham Picture Library
92 Laura Knight, Barnes Wallis, John Topham Picture Library; John Metcalf, Mansell Collection
93 Duncan Grant, John Topham Picture Library
94 George Bernard Shaw, John Topham Picture Library
95 Philippe Pétain, Wilfred Rhodes, Mansell Collection; Leopold Stokowski, John Topham Picture Library
96 Havergal Brian, Popperfoto; Pablo Casals, Mansell Collection; Wind in the Willows, Line illustration by E.H. Shepard, copyright under the Berne Convention, In the U.S.A. Copyright 1933 Charles Scribner's Sons, Renewal copyright © 1961 E.H. Shepard; Philippa Strachey, Mary Evans Picture Library
97 John Rockefeller, La Belle Otero, Mansell Collection; Bertrand Russell, John Topham Picture Library
98 John Garner, *Illustrated London News*; John Laing, John Topham Picture Library; Fred Streeter, BBC
99 Titian, Mansell Collection; Fred Dean, *Manchester Evening News*
100 Wilf Nixon, Gary Pearce; Eubie Blake, Popperfoto; Margaret Murray, John Topham Picture Library
100+ Adolph Zukor, Kobal Collection; Shigechiyo Izumi, John Topham Picture Library